Lecture Notes in Computer Science   13858

The series Lecture Notes in Computer Science (LNCS), including its subseries Lecture Notes in Artificial Intelligence (LNAI) and Lecture Notes in Bioinformatics (LNBI), has established itself as a medium for the publication of new developments in computer science and information technology research, teaching, and education.

LNCS enjoys close cooperation with the computer science R & D community, the series counts many renowned academics among its volume editors and paper authors, and collaborates with prestigious societies. Its mission is to serve this international community by providing an invaluable service, mainly focused on the publication of conference and workshop proceedings and postproceedings. LNCS commenced publication in 1973.

Ivan Georgiev · Maria Datcheva ·
Krassimir Georgiev · Geno Nikolov
Editors

Numerical Methods and Applications

10th International Conference, NMA 2022
Borovets, Bulgaria, August 22–26, 2022
Proceedings

 Springer

Editors
Ivan Georgiev ⓘ
Institute of Mathematics and Informatics
Bulgarian Academy of Sciences
Sofia, Bulgaria

Maria Datcheva ⓘ
Institute of Mechanics
Bulgarian Academy of Sciences
Sofia, Bulgaria

Krassimir Georgiev
Institute of Information and Communication
Technologies
Bulgarian Academy of Sciences
Sofia, Bulgaria

Geno Nikolov ⓘ
Sofia University "St. Kliment Ohridski"
Sofia, Bulgaria

ISSN 0302-9743   ISSN 1611-3349 (electronic)
Lecture Notes in Computer Science
ISBN 978-3-031-32411-6   ISBN 978-3-031-32412-3 (eBook)
https://doi.org/10.1007/978-3-031-32412-3

This Springer imprint is published by the registered company Springer Nature Switzerland AG
The registered company address is: Gewerbestrasse 11, 6330 Cham, Switzerland

Preface

The tenth edition of the International Conference on Numerical Methods and Applications (NM&A'22), held every four years in Bulgaria since 1984, took place in the beautiful resort of Borovets in the period 22–26 August 2022. The nice weather during the event and the fresh air of the Rila Mountains contributed tangibly to the creative atmosphere of the conference, giving the participants the opportunity to present their latest achievements, exchange ideas, continue existing or start new fruitful scientific collaborations. During NM&A'22, there were discussed a wide range of issues related to both recent theoretical advances in numerical methods and their applications to the mathematical modelling of a broad variety of processes and phenomena.

There were four special sessions featured in the scientific program:

Numerical Search and Optimization
Organizers: Stefka Fidanova (Bulgarian Academy of Sciences), Gabriel Luque (University of Malaga, Spain).
Large-Scale Models: Numerical Methods, Parallel Computations and Applications
Organizers: Krassimir Georgiev (Bulgarian Academy of Sciences), Zahari Zlatev (Aarhus University, Denmark), Maya Neytcheva (Uppsala University, Sweden).
Advanced Material Modelling and Simulations
Organizers: Maria Datcheva, Roumen Iankov, Svetoslav Nikolov (Bulgarian Academy of Sciences).
Computational Biology
Organizers: Nevena Ilieva, Sylvi-Maria Gurova (Bulgarian Academy of Sciences), Roumen Anguelov (University of Pretoria, South Africa).

Moreover, five invited lectures were received with great interest, namely:

Optimal Learning of Solutions of PDEs (Peter Binev, University of South Carolina, USA);
Richardson Extrapolation and its Variants (István Faragó, Eötvös Loránd University, Hungary);
Why do we sleep? - Multi-physics Problems Related to Brain Clearance (Kent-Andre Mardal, University of Oslo, Norway);
Non-overlapping Domain Decomposition Methods of Optimal Computational Complexity (Svetozar Margenov, Bulgarian Academy of Sciences);
A Posteriori Error Estimates in Finite Element Method by Preconditioning (Ludmil Zikatanov, Penn State University, USA).

Each of the papers in this volume has passed a thorough blind review procedure. We thank all the authors who contributed to the volume as well as the reviewers for their precious help in selecting the volume's chapters.

NM&A'22 was organized by the Institute of Mathematics and Informatics of the Bulgarian Academy of Sciences in cooperation with the Faculty of Mathematics and

Informatics of Sofia University "St. Kliment Ohridski" and the Institute of Information and Communication Technologies of the Bulgarian Academy of Sciences. The success of NM&A'22 would not have been possible without the joint efforts and hard work of many colleagues from various institutions and organizations. We are grateful to all members of the Organizing and Scientific Committees. We are also thankful to the local staff for the excellent service.

Conference website: http://www.math.bas.bg/~nummeth/nma22/.

Ivan Georgiev
Maria Datcheva
Krassimir Georgiev
Geno Nikolov

The original version of the book was revised: An error in the affiliation of Maria Datcheva and Krassimir Georgiev was corrected. The correction to the book is available at https://doi.org/10.1007/978-3-031-32412-3_31

Organization

Program Committee

Chairpersons

Maria Datcheva	Institute of Mechanics, BAS, Bulgaria
Ivan Georgiev	Institute of Mathematics and Informatics, BAS, Bulgaria
Krassimir Georgiev	Institute of Information and Communication Technologies, BAS, Bulgaria
Geno Nikolov	Sofia University "St. Kliment Ohridski", Bulgaria

Members

Andrey Andreev	Technical University of Gabrovo, Bulgaria
Emanouil Atanassov	Institute of Information and Communication Technologies, BAS, Bulgaria
Peter Binev	University of South Carolina, USA
Jan Buša	Technical University of Kosice, Slovakia
Remigijus Ciegis	Vilnius University, Lithuania
Pasqua D'Ambra	Institute for Applied Computing, CNR, Italy
Ivan Dimov	Institute of Information and Communication Technologies, BAS, Bulgaria
Stefka Dimova	Sofia University St. Kliment Ohridski, Bulgaria
István Faragó	Eötvös Loránd University, Hungary
Stefka Fidanova	Institute of Information and Communication Technologies, BAS, Bulgaria
Irina Georgieva	Institute of Mathematics and Informatics, BAS, Bulgaria
Clemens Hofreither	Johann Radon Institute for Computational and Applied Mathematics, AAS, Austria
Snezhana Gocheva-Ilieva	University of Plovdiv Paisii Hilendarski, Bulgaria
Jean-Luc Guermond	Texas A&M University, USA
Raphaèle Herbin	Aix-Marseille University, France
Oleg Iliev	Fraunhofer ITWM, Germany
Nevena Ilieva	Institute of Information and Communication Technologies, BAS, Bulgaria
Bosko Jovanovic	University of Belgrade, Serbia

Tzanio Kolev	Lawrence Livermore National Laboratory, USA
Natalia Kolkovska	Institute of Mathematics and Informatics, BAS, Bulgaria
Johannes Kraus	University of Duisburg-Essen, Germany
Raytcho Lazarov	Texas A&M University, USA
Ivan Lirkov	Institute of Information and Communication Technologies, BAS, Bulgaria
Kent-Andre Mardal	University of Oslo, Norway
Svetozar Margenov	Institute of Information and Communication Technologies, BAS, Bulgaria
Pencho Marinov	Institute of Information and Communication Technologies, BAS, Bulgaria
Svetoslav Markov	Institute of Mathematics and Informatics, BAS, Bulgaria
Piotr Matus	National Academy of Sciences of Belarus, Belarus
Petar Minev	University of Alberta, Canada
Mihail Nedjalkov	Vienna University of Technology, Austria
Kalin Penev	Solent University, UK
Bojan Popov	Texas A&M University, USA
Stefan Radev	Institute of Mechanics, BAS, Bulgaria
Pedro Ribeiro	University of Porto, Portugal
Wil Schilders	Eindhoven University of Technology, The Netherlands
Siegfried Selberherr	Vienna University of Technology, Austria
Slavcho Slavchev	Institute of Mechanics, BAS, Bulgaria
Mikhail Todorov	Technical University of Sofia, Bulgaria
Vidar Thomée	Chalmers University of Technology, Sweden
Peter Vabishchevich	North-Eastern Federal University, Russia
Yuri Vassilevski	Marchuk Institute of Numerical Mathematics, RAS, Russia
Ivan Yotov	University of Pittsburgh, USA
Ludmil Zikatanov	Pennsylvania State University, USA

Organizing Committee

Ivan Georgiev (Chairperson)	Institute of Mathematics and Informatics, BAS, Bulgaria
Ivan Bazhlekov	Institute of Mathematics and Informatics, BAS, Bulgaria
Ivan Hristov	Sofia University St. Kliment Ohridski, Bulgaria
Maria Datcheva	Institute of Mechanics, BAS, Bulgaria

Elena Lilkova   Institute of Information and Communication
                                 Technologies, BAS, Bulgaria
Milena Dimova   University of National and World Economy,
                                 Bulgaria
Veselina Vucheva   Institute of Mathematics and Informatics, BAS,
                                 Bulgaria

Contents

SUSHI for a Bingham Flow Type Problem

Wassim Aboussi[1,2], Fayssal Benkhaldoun[1] (ID), and Abdallah Bradji[1,3(✉)] (ID)

[1] LAGA, Sorbonne Paris Nord University, Villetaneuse, Paris, France
{aboussi,fayssal}@math.univ-paris13.fr, abdallah.bradji@gmail.com,
abdallah.bradji@etu.univ-amu.fr
[2] LAMA, Faculty of Sciences, Sidi Mohamed Ben Abdellah University, Fez, Morocco
[3] Department of Mathematics, Faculty of Sciences,
Badji Mokhtar-Annaba University, Annaba, Algeria
abdallah.bradji@univ-annaba.dz
https://www.math.univ-paris13.fr/ fayssal/,
https://www.i2m.univ-amu.fr/perso/abdallah.bradji/

Abstract. We establish a nonlinear finite volume scheme for a Bingham Flow Type Problem. The equation is a nonlinear parabolic one and it is a simplified version of the Bingham visco-plastic flow model [8, Pages 38–40]. The space discretization is performed using SUSHI (Scheme using Stabilization and Hybrid Interfaces) developed in [9] whereas the time discretization is uniform. We first prove a discrete *a priori* estimate. We then, prove the existence of at least one discrete solution using the Brouwer fixed point theorem. The uniqueness of the approximate solution is shown as well. We subsequently prove error estimates of order one in time and space in $L^\infty(L^2)$ and $L^2(H^1)$–discrete norms.

Keywords: Bingham Flow Type Problem · SUSHI · A priori estimate · Error estimates

MSC2010: 65M08 · 65M12 · 65M15 · 35L20

1 Problem to Be Solved and the Aim of This Note

Let us consider the following simplified Bingham flow in cylinders introduced in [8, (15), Page 38]:

$$\rho u_t(\boldsymbol{x},t) - \mu \Delta u(\boldsymbol{x},t) - g\nabla \cdot \left(\frac{\nabla u}{|\nabla u|}\right)(\boldsymbol{x},t) = C, \quad (\boldsymbol{x},t) \in \Omega \times (0,T), \quad (1)$$

with initial and Dirichlet boundary conditions:

$$u(\boldsymbol{x},0) = u^0(\boldsymbol{x}), \quad \boldsymbol{x} \in \Omega \quad \text{and} \quad u(\boldsymbol{x},t) = 0, \quad (\boldsymbol{x},t) \in \partial\Omega \times (0,T). \quad (2)$$

Supported by MCS team (LAGA Laboratory) of the "Université Sorbonne- Paris Nord".

As we can see that there is no sense to the previous equation when $\nabla u = 0$. A regularized problem, see [8, (16), Page 39], can be given by, for $(\boldsymbol{x}, t) \in \Omega \times (0, T)$

$$\rho \partial_t u_\varepsilon(\boldsymbol{x}, t) - \mu \Delta u_\varepsilon(\boldsymbol{x}, t) - g \nabla \cdot \left(\frac{\nabla u_\varepsilon}{\sqrt{\varepsilon^2 + |\nabla u_\varepsilon|^2}} \right)(\boldsymbol{x}, t) = C, \qquad (3)$$

where

$$u_\varepsilon(\boldsymbol{x}, 0) = u^0(\boldsymbol{x}), \quad \boldsymbol{x} \in \Omega \quad \text{and} \quad u_\varepsilon(\boldsymbol{x}, t) = 0, \quad (\boldsymbol{x}, t) \in \partial\Omega \times (0, T). \qquad (4)$$

The parameters μ and g denote respectively the viscosity and plasticity yield, see [8, Page 37].

It is proved in [8, Theorem 1, Page 39] that the problem (1)–(2) can be written in the following variational inequation formulation: $u(t) \in H_0^1(\Omega)$, for a.e. $t \in (0, 1)$, for all $v \in H_0^1(\Omega)$:

$$\rho \int_\Omega \partial_t u(\boldsymbol{x}, t)(v(\boldsymbol{x}) - u(\boldsymbol{x}, t)) d\boldsymbol{x} + \mu \int_\Omega \nabla u(\boldsymbol{x}, t)(\nabla v(\boldsymbol{x}) - \nabla u(\boldsymbol{x}, t)) d\boldsymbol{x}$$

$$+ g(j(v) - j(u)) \geq C \int_\Omega (v(\boldsymbol{x}) - u(\boldsymbol{x}, t)) d\boldsymbol{x}, \qquad (5)$$

with

$$u(0) = u^0 \qquad (6)$$

and

$$j(v) = \int_\Omega |\nabla v|(\boldsymbol{x}) d\boldsymbol{x}. \qquad (7)$$

In addition to this, the following error estimate, between the solution u_ε of (3)–(4) and the solution u of (5)–(7), holds

$$\|u_\varepsilon(t) - u(t)\|_{L^2(\Omega)} \leq \sqrt{\frac{gm(\Omega)}{\mu\lambda_0} \left(1 - \exp\left(-\frac{2\mu\lambda_0}{\rho} t \right) \right)^{\frac{1}{2}}} \sqrt{\varepsilon}, \qquad (8)$$

with λ_0 the smallest eigenvalue of the operator $-\Delta$.
Estimate (8) implies that

$$\lim_{\varepsilon \to 0} u_\varepsilon = u \quad \text{in} \quad L^\infty(0, T; L^2(\Omega)).$$

To study the model (3)–(4) "rigorously", let us write it under the following general version of quasilinear parabolic equation (see [3, page 202] and [7, page 431])

$$u_t(\boldsymbol{x}, t) - \alpha \Delta u(\boldsymbol{x}, t) - \nabla \cdot \mathcal{F}(\nabla u)(\boldsymbol{x}, t) = \mathcal{G}(\boldsymbol{x}, t), \quad (\boldsymbol{x}, t) \in \Omega \times (0, T), \qquad (9)$$

where Ω is an open bounded polyhedral subset in $\mathbb{R}^d$, with $d \in \mathbb{N}^* = \mathbb{N} \setminus \{0\}$, $T > 0$, $\alpha > 0$, $\mathcal{F}$ is a function defined on $\mathbb{R}^d$ into $\mathbb{R}^d$, and $\mathcal{G}$ is a real function defined on $\Omega \times (0, T)$.

Initial condition is given by, for a given function u^0 defined on Ω

$$u(0) = u^0. \qquad (10)$$

Homogeneous Dirichlet boundary conditions are given by

$$u(\boldsymbol{x},t) = 0, \qquad (\boldsymbol{x},t) \in \partial\Omega \times (0,T). \tag{11}$$

The function $\mathcal{F}$ is assumed to satisfy:

- $\mathcal{F} \in \mathcal{C}^1(\mathbb{R}^d)$. We set $\mathcal{F} = (F_1, \ldots, F_d)$, we assume that, for some $M > 0$,

$$F_1, \ldots, F_d \text{ and their first derivatives are bounded by } M. \tag{12}$$

- For all $\theta_1, \theta_2 \in \mathbb{R}^d$

$$(\mathcal{F}(\theta_1) - \mathcal{F}(\theta_2), \theta_1 - \theta_2)_{L^2(\Omega)^d} \geq 0. \tag{13}$$

It is worthy to mention that the regularized equation (3) is a particular case in (9) by taking $\mathcal{F}(\theta) = \dfrac{g\theta}{\rho\sqrt{\varepsilon^2 + |\theta|^2}}$ which satisfies the hypotheses (12)–(13). It is also useful to notice that a similar hypothesis to (13) is widely used in the literature of Leray–Lions type operator $-\nabla \cdot \mathbf{a}(\boldsymbol{x}, u, \nabla u)$, see for instance [5, (2.85c), Pages 52–53].

2 Space and Time Discretizations and Some Preliminaries

Definition 1. (Space discretization, cf. [9]). Let Ω be a polyhedral open bounded subset of $\mathbb{R}^d$, where $d \in \mathbb{N} \setminus \{0\}$, and $\partial\Omega = \overline{\Omega} \setminus \Omega$ its boundary. A discretization of Ω, denoted by $\mathcal{D}$, is defined as the triplet $\mathcal{D} = (\mathcal{M}, \mathcal{E}, \mathcal{P})$, where:

1. $\mathcal{M}$ is a finite family of non empty connected open disjoint subsets of Ω (the "control volumes") such that $\overline{\Omega} = \cup_{K \in \mathcal{M}} \overline{K}$. For any $K \in \mathcal{M}$, let $\partial K = \overline{K} \setminus K$ be the boundary of K; let $\mathrm{m}\,(K) > 0$ denote the measure of K and h_K denote the diameter of K.
2. $\mathcal{E}$ is a finite family of disjoint subsets of $\overline{\Omega}$ (the "edges" of the mesh), such that, for all $\sigma \in \mathcal{E}$, σ is a non empty open subset of a hyperplane of $\mathbb{R}^d$, whose $(d-1)$–dimensional measure is strictly positive. We also assume that, for all $K \in \mathcal{M}$, there exists a subset $\mathcal{E}_K$ of $\mathcal{E}$ such that $\partial K = \cup_{\sigma \in \mathcal{E}_K} \overline{\sigma}$. For any $\sigma \in \mathcal{E}$, we denote by $\mathcal{M}_\sigma = \{K, \sigma \in \mathcal{E}_K\}$. We then assume that, for any $\sigma \in \mathcal{E}$, either $\mathcal{M}_\sigma$ has exactly one element and then $\sigma \subset \partial\Omega$ (the set of these interfaces, called boundary interfaces, denoted by $\mathcal{E}_{\text{ext}}$) or $\mathcal{M}_\sigma$ has exactly two elements (the set of these interfaces, called interior interfaces, denoted by $\mathcal{E}_{\text{int}}$). For all $\sigma \in \mathcal{E}$, we denote by $\boldsymbol{x}_\sigma$ the barycentre of σ. For all $K \in \mathcal{M}$ and $\sigma \in \mathcal{E}$, we denote by $\mathbf{n}_{K,\sigma}$ the unit vector normal to σ outward to K.
3. $\mathcal{P}$ is a family of points of Ω indexed by $\mathcal{M}$, denoted by $\mathcal{P} = (\boldsymbol{x}_K)_{K \in \mathcal{M}}$, such that for all $K \in \mathcal{M}$, $\boldsymbol{x}_K \in K$ and K is assumed to be $\boldsymbol{x}_K$–star-shaped, which means that for all $\boldsymbol{x} \in K$, the property $[\boldsymbol{x}_K, \boldsymbol{x}] \subset K$ holds. Denoting by $d_{K,\sigma}$ the Euclidean distance between $\boldsymbol{x}_K$ and the hyperplane including σ, one assumes that $d_{K,\sigma} > 0$. We then denote by $\mathcal{D}_{K,\sigma}$ the cone with vertex $\boldsymbol{x}_K$ and basis σ.

The time discretization is performed with a uniform mesh with constant step $k = \dfrac{T}{N+1}$ with $N \in \mathbb{N} \setminus \{0\}$. We denote the mesh points by $t_n = nk$, for $n \in [\![0, N+1]\!]$. We denote by ∂^1 the discrete first time derivative given by

$$\partial^1 v^{j+1} = \frac{v^{j+1} - v^j}{k}.$$

Throughout this paper, the letter C stands for a positive constant independent of the parameters of discretizations.

We use the finite volume space considered in [6, Definition 5.1, Page 2037], that is the space $\mathcal{H}_\mathcal{D} \subset L^2(\Omega)$ of functions which are constant on each control volume K of $\mathcal{M}$. We associate any $\sigma \in \mathcal{E}_{\text{int}}$ with a family of real numbers $\left(\beta_\sigma^K\right)_{K \in \mathcal{M}}$ (this family contains in general at most $d+1$ nonzero elements) such that

$$1 = \sum_{K \in \mathcal{M}} \beta_\sigma^K \quad \text{and} \quad x_\sigma = \sum_{K \in \mathcal{M}} \beta_\sigma^K x_K. \tag{14}$$

Then, for any $u \in \mathcal{H}_\mathcal{D}$, we set $u_\sigma = \displaystyle\sum_{K \in \mathcal{M}} \beta_\sigma^K u_K$, for all $\sigma \in \mathcal{E}_{\text{int}}$ and $u_\sigma = 0$, for all $\sigma \in \mathcal{E}_{\text{ext}}$.

In order to analyze the convergence, we need to consider the size of the discretization $\mathcal{D}$ defined by $h_\mathcal{D} = \sup\{\text{diam}(K), K \in \mathcal{M}\}$ and the regularity of the mesh given by (see [9, (4.1)–(4.2), Page 1025])

$$\theta_\mathcal{D} = \max\left(\max_{\sigma \in \mathcal{E}_{\text{int}}, K, L \in \mathcal{M}} \frac{d_{K,\sigma}}{d_{L,\sigma}}, \max_{K \in \mathcal{M}, \sigma \in \mathcal{E}_K} \frac{h_K}{d_{K,\sigma}}, \max_{K \in \mathcal{M}, \sigma \in \mathcal{E}_K \cap \mathcal{E}_{\text{int}}} \frac{\sum_{L \in \mathcal{M}} |\beta_\sigma^L| \, |x_\sigma - x_L|^2}{h_K^2}\right). \tag{15}$$

The scheme we shall present is based on the discrete gradient of [9]. For $u \in \mathcal{H}_\mathcal{D}$, we define, for all $K \in \mathcal{M}$, for a.e. $x \in \mathcal{D}_{K,\sigma}$

$$\nabla_\mathcal{D} u(x) = \nabla_K u + \left(\frac{\sqrt{d}}{d_{K,\sigma}} \left(u_\sigma - u_K - \nabla_K u \cdot (x_\sigma - x_K)\right)\right) \mathbf{n}_{K,\sigma}, \tag{16}$$

where

$$\nabla_K u = \frac{1}{\text{m}(K)} \sum_{\sigma \in \mathcal{E}_K} \text{m}(\sigma) \left(u_\sigma - u_K\right) \mathbf{n}_{K,\sigma}.$$

We define the bilinear form $\langle \cdot, \cdot \rangle_F$ defined on $\mathcal{H}_\mathcal{D} \times \mathcal{H}_\mathcal{D}$ by

$$\langle u, v \rangle_F = \int_\Omega \nabla_\mathcal{D} u(x) \cdot \nabla_\mathcal{D} v(x) dx.$$

3 Formulation of a Finite Volume Scheme for the Problem (9)–(11)

We now define a nonlinear finite volume scheme for (9)–(11). The unknowns of this scheme are denoted by $\{u_\mathcal{D}^n; n \in [\![0, N+1]\!]\}$ and they are expected to approximate the values:

$$\{u(x_K, t_n); n \in [\![0, N+1]\!], K \in \mathcal{M}\}.$$

The finite volume scheme we present needs the regularity assumptions $u^0 \in \mathcal{C}^2(\overline{\Omega})$ and $\mathcal{G} \in \mathcal{C}\left([0,T]; \mathcal{C}(\overline{\Omega})\right)$. These assumptions can be weakened. The principles of the finite volume scheme are based on the approximation of (9), (10), and (11) and they can be sketched as follows:

1. **Approximation of initial condition** (10). The discretization of initial condition (10) can be performed as: Find $u_{\mathcal{D}}^0 \in \mathcal{H}_{\mathcal{D}}$ such that for all $v \in \mathcal{H}_{\mathcal{D}}$

$$\langle u_{\mathcal{D}}^0, v \rangle_F = -\left(\Delta u^0, v\right)_{L^2(\Omega)}. \tag{17}$$

2. **Approximation of** (9). For any $n \in [\![0, N]\!]$, find $u_{\mathcal{D}}^{n+1} \in \mathcal{H}_{\mathcal{D}}$ such that, for all $v \in \mathcal{H}_{\mathcal{D}}$

$$\left(\partial^1 u_{\mathcal{D}}^{n+1}, v\right)_{L^2(\Omega)} + \alpha \langle u_{\mathcal{D}}^{n+1}, v \rangle_F + \left(\mathcal{F}(\nabla_{\mathcal{D}} u_{\mathcal{D}}^{n+1}), \nabla_{\mathcal{D}} v\right)_{L^2(\Omega)^d} = \left(\mathcal{G}(t_{n+1}), v\right)_{L^2(\Omega)}. \tag{18}$$

4   Convergence Analysis of Scheme (17)–(18)

In this section, we provide a discrete *priori estimate*, prove the existence and uniqueness for the scheme (17)–(18), and we provide new error estimates in $L^\infty(L^2)$ and $L^2(H^1)$–discrete norms under assumption that the problem (9) – (11) has a "smooth" solution.

Theorem 1. (New error estimates for (17)–(18)). *Let Ω be a polyhedral open bounded subset of $\mathbb{R}^d$, where $d \in \mathbb{N} \setminus \{0\}$. Assume that the function $\mathcal{F}$ involved in equation (9) satisfies the hypotheses (12)–(13) and that the problem (9)–(11) admits a solution u satisfying $u \in \mathcal{C}^2([0,T]; \mathcal{C}^2(\overline{\Omega}))$. Let $k = \dfrac{1}{N+1}$, with a given $N \in \mathbb{N} \setminus \{0\}$, be the constant step and $t_n = nk$, for $n \in [\![0, N+1]\!]$, be the temporal mesh points. Let $\mathcal{D} = (\mathcal{M}, \mathcal{E}, \mathcal{P})$ be a discretization in the sense of Definition 1. Assume that $\theta_{\mathcal{D}}$ satisfies $\theta \geq \theta_{\mathcal{D}}$. Let $\nabla_{\mathcal{D}}$ be the discrete gradient given by (16). Then, there exists a unique solution $(u_{\mathcal{D}}^n)_{n=0}^{N+1} \in \mathcal{H}_{\mathcal{D}}^{N+2}$ to scheme (17)–(18) and the following error estimates hold:*

- $L^\infty(L^2)$*-error estimate:*

$$\max_{n=0}^{n=N+1} \|u(t_n) - u_{\mathcal{D}}^n\|_{L^2(\Omega)} \leq C(k + h_{\mathcal{D}}). \tag{19}$$

- $L^2(H^1)$*-error estimate:*

$$\left(\sum_{n=0}^{n=N+1} k \|\nabla u(t_n) - \nabla_{\mathcal{D}} u_{\mathcal{D}}^n\|_{L^2(\Omega)^d}^2\right)^{\frac{1}{2}} \leq C(k + h_{\mathcal{D}}). \tag{20}$$

The following theorem provides a new discrete *a priori estimate* for the problem (17)–(18):

Theorem 2. (New discrete a priori estimate). *Assume the same hypotheses of Theorem 1, except the one concerned with the regularity of the exact solution u which is not needed to obtain a priori estimate for the problem (17)–(18). Then, any solution $(u_{\mathcal{D}}^n)_{n=0}^{N+1} \in \mathcal{H}_{\mathcal{D}}^{N+2}$ for scheme (17)–(18) is satisfying the following a priori estimates:*

$$\max_{n=0}^{N+1} \|u_{\mathcal{D}}^n\|_{L^2(\Omega)} \leq \frac{\sqrt{T}\left(M + C_{\mathrm{p}}\|\mathcal{G}\|_{\mathcal{C}([0,T];\mathcal{C}(\overline{\Omega}))}\right)}{\sqrt{\alpha}} + \|u_{\mathcal{D}}^0\|_{L^2(\Omega)} \tag{21}$$

and

$$\left(\sum_{n=0}^{N+1} k\alpha \|\nabla_{\mathcal{D}} u_{\mathcal{D}}^n\|_{L^2(\Omega)^d}^2\right)^{\frac{1}{2}} \leq \frac{\sqrt{T}\left(M + C_{\mathrm{p}}\|\mathcal{G}\|_{\mathcal{C}([0,T];\mathcal{C}(\overline{\Omega}))}\right)}{\sqrt{\alpha}}$$
$$+ \|u_{\mathcal{D}}^0\|_{L^2(\Omega)}, \tag{22}$$

where C_{p} is the constant which appears in the discrete Poincaré inequality.

Proof. Taking $v = u_{\mathcal{D}}^{n+1}$ in (18) yields

$$\left(\partial^1 u_{\mathcal{D}}^{n+1}, u_{\mathcal{D}}^{n+1}\right)_{L^2(\Omega)} + \alpha \|\nabla_{\mathcal{D}} u_{\mathcal{D}}^{n+1}\|_{L^2(\Omega)^d}^2 + \left(\mathcal{F}(\nabla_{\mathcal{D}} u_{\mathcal{D}}^{n+1}), \nabla_{\mathcal{D}} u_{\mathcal{D}}^{n+1}\right)_{L^2(\Omega)^d}$$
$$= \left(\mathcal{G}(t_{n+1}), u_{\mathcal{D}}^{n+1}\right)_{L^2(\Omega)}. \tag{23}$$

Using hypothesis (12), equality (23) implies that

$$\left(\partial^1 u_{\mathcal{D}}^{n+1}, u_{\mathcal{D}}^{n+1}\right)_{L^2(\Omega)} + \alpha \|\nabla_{\mathcal{D}} u_{\mathcal{D}}^{n+1}\|_{L^2(\Omega)^d}^2$$
$$\leq M \|\nabla_{\mathcal{D}} u_{\mathcal{D}}^{n+1}\|_{L^2(\Omega)^d} + \|\mathcal{G}\|_{\mathcal{C}([0,T];\mathcal{C}(\overline{\Omega}))} \|u_{\mathcal{D}}^{n+1}\|_{L^2(\Omega)}. \tag{24}$$

Gathering this with the identity $a(a-b) = \frac{1}{2}\left((a-b)^2 + a^2 - b^2\right)$ and using the discrete Poincaré inequality [9, Lemma 5.4, Page 1038] yield

$$\|u_{\mathcal{D}}^{n+1}\|_{L^2(\Omega)}^2 - \|u_{\mathcal{D}}^n\|_{L^2(\Omega)}^2 + 2k\alpha \|\nabla_{\mathcal{D}} u_{\mathcal{D}}^{n+1}\|_{L^2(\Omega)^d}^2$$
$$\leq 2kM \|\nabla_{\mathcal{D}} u_{\mathcal{D}}^{n+1}\|_{L^2(\Omega)^d} + 2k\|\mathcal{G}\|_{\mathcal{C}([0,T];\mathcal{C}(\overline{\Omega}))} \|u_{\mathcal{D}}^{n+1}\|_{L^2(\Omega)}$$
$$\leq 2k\left(M + C_{\mathrm{p}}\|\mathcal{G}\|_{\mathcal{C}([0,T];\mathcal{C}(\overline{\Omega}))}\right) \|\nabla_{\mathcal{D}} u_{\mathcal{D}}^{n+1}\|_{L^2(\Omega)^d}. \tag{25}$$

Using inequality $ab \leq a^2/2 + b^2/2$, inequality (25) implies that

$$\|u_{\mathcal{D}}^{n+1}\|_{L^2(\Omega)}^2 - \|u_{\mathcal{D}}^n\|_{L^2(\Omega)}^2 + k\alpha \|\nabla_{\mathcal{D}} u_{\mathcal{D}}^{n+1}\|_{L^2(\Omega)^d}^2$$
$$\leq \frac{k\left(M + C_{\mathrm{p}}\|\mathcal{G}\|_{\mathcal{C}([0,T];\mathcal{C}(\overline{\Omega}))}\right)^2}{\alpha}. \tag{26}$$

Summing the previous inequality over $n \in [\![0, J-1]\!]$, with $J \in [\![1, N+1]\!]$, yields

$$\|u_{\mathcal{D}}^J\|_{L^2(\Omega)}^2 - \|u_{\mathcal{D}}^0\|_{L^2(\Omega)}^2 + \sum_{n=0}^{J-1} k\alpha \|\nabla_{\mathcal{D}} u_{\mathcal{D}}^{n+1}\|_{L^2(\Omega)^d}^2$$
$$\leq \frac{T\left(M + C_{\mathrm{p}}\|\mathcal{G}\|_{\mathcal{C}([0,T];\mathcal{C}(\overline{\Omega}))}\right)^2}{\alpha}. \tag{27}$$

This implies that

$$\|u_{\mathcal{D}}^J\|_{L^2(\Omega)}^2 + \sum_{n=0}^{J-1} k\alpha \|\nabla_{\mathcal{D}} u_{\mathcal{D}}^{n+1}\|_{L^2(\Omega)^d}^2$$
$$\leq \frac{T\left(M + C_{\mathrm{p}}\|\mathcal{G}\|_{\mathcal{C}([0,T];\mathcal{C}(\overline{\Omega}))}\right)^2}{\alpha} + \|u_{\mathcal{D}}^0\|_{L^2(\Omega)}^2. \tag{28}$$

This with the inequality $\sqrt{a+b} \leq \sqrt{a} + \sqrt{b}$ imply the desired estimates (21) and (22). This completes the proof of Theorem 2. □

Proof of Theorem 1

1. Existence for scheme (17)–(18). First we remark that (17) has a unique solution $u_{\mathcal{D}}^0 \in \mathcal{H}_{\mathcal{D}}$ thanks to the fact that the matrix involved in (17) is square and also to the fact that $\|\nabla \cdot \|_{L^2(\Omega)^d}$ is a norm on $\mathcal{H}_{\mathcal{D}}$.

Let us now prove the existence of $u_{\mathcal{D}}^1$. To this end, we consider the unique solution $U_{\mathcal{D}} \in \mathcal{H}_{\mathcal{D}}$, for a given $W_{\mathcal{D}} \in \mathcal{H}_{\mathcal{D}}$, for all $v \in \mathcal{H}_{\mathcal{D}}$

$$\left(\frac{1}{k}(U_{\mathcal{D}} - u_{\mathcal{D}}^0), v\right)_{L^2(\Omega)} + \alpha \langle U_{\mathcal{D}}, v \rangle_F$$
$$= -(\mathcal{F}(\nabla_{\mathcal{D}} W_{\mathcal{D}}), \nabla_{\mathcal{D}} v)_{L^2(\Omega)^d} + (\mathcal{G}(t_{n+1}), v)_{L^2(\Omega)}. \tag{29}$$

One sets $U_{\mathcal{D}} = \mathbb{F}^1(W_{\mathcal{D}})$, so that $\mathbb{F}^1$ is an application from the space $\mathcal{H}_{\mathcal{D}}$ into $\mathcal{H}_{\mathcal{D}}$. Due to the presence of $\mathcal{F}$, the application $\mathbb{F}^1$ is not linear. We remark that $U_{\mathcal{D}}$ is solution for (18) when $n = 0$ if and only if $U_{\mathcal{D}}$ is a fixed point to $\mathbb{F}^1$. Taking $v = U_{\mathcal{D}}$ in (29) yields

$$\left(\frac{1}{k}(U_{\mathcal{D}} - u_{\mathcal{D}}^0), U_{\mathcal{D}}\right)_{L^2(\Omega)} + \alpha \|\nabla_{\mathcal{D}} U_{\mathcal{D}}\|_{L^2(\Omega)^d}^2$$
$$= -(\mathcal{F}(\nabla_{\mathcal{D}} W_{\mathcal{D}}), \nabla_{\mathcal{D}} U_{\mathcal{D}})_{L^2(\Omega)^d} + (\mathcal{G}(t_{n+1}), U_{\mathcal{D}})_{L^2(\Omega)}. \tag{30}$$

Using inequality $a(a - b) = \frac{1}{2}\left((a - b)^2 + a^2 - b^2\right)$, (30) implies that

$$\|U_{\mathcal{D}}\|_{L^2(\Omega)}^2 - \|u_{\mathcal{D}}^0\|_{L^2(\Omega)}^2 + 2k\alpha\|\nabla_{\mathcal{D}} U_{\mathcal{D}}\|_{L^2(\Omega)^d}^2$$
$$\leq -2k\left(\mathcal{F}(\nabla_{\mathcal{D}} W_{\mathcal{D}}), \nabla_{\mathcal{D}} U_{\mathcal{D}}\right)_{L^2(\Omega)^d} + 2k\left(\mathcal{G}(t_{n+1}), U_{\mathcal{D}}\right)_{L^2(\Omega)}. \tag{31}$$

Using the discrete Poincaré inequality together with assumption (12) and inequality $ab \leq a^2/2 + b^2/2$ (as in (25)–(26)), (31) yields

$$\|U_{\mathcal{D}}\|_{L^2(\Omega)}^2 \leq \delta^1, \tag{32}$$

where, for each $n \in [\![1, N+1]\!]$, we set

$$\delta^n = \|u_{\mathcal{D}}^0\|_{L^2(\Omega)}^2 + \frac{nk\left(M + C_{\mathrm{p}}\|\mathcal{G}\|_{\mathcal{C}([0,T];\mathcal{C}(\overline{\Omega}))}\right)^2}{\alpha} \tag{33}$$

Consequently, $\mathbb{F}^1(\mathbf{B}(0, \sqrt{\delta^1})) \subset \mathbf{B}(0, \sqrt{\delta^1})$, where $\mathbf{B}(0, \sqrt{\delta^1})$ is the closed ball in $\mathcal{H}_\mathcal{D}$ of center 0 and radius $\sqrt{\delta^1}$ with respect to the norm $\|\cdot\|_{L^2(\Omega)}$. In addition to this, the application $\mathbb{F}^1$ can be proved to be continuous; indeed, let $U_\mathcal{D} = \mathbb{F}^1(W_\mathcal{D})$ and $\overline{U}_\mathcal{D} = \mathbb{F}^1(\overline{W}_\mathcal{D})$. This gives

$$\left(U_\mathcal{D} - \overline{U}_\mathcal{D}, v\right)_{L^2(\Omega)} + \alpha k \langle U_\mathcal{D} - \overline{U}_\mathcal{D}, v \rangle_F$$
$$= -k \left(\mathcal{F}(\nabla_\mathcal{D} W_\mathcal{D}) - \mathcal{F}(\nabla_\mathcal{D}\overline{W}_\mathcal{D}), \nabla_\mathcal{D} v\right)_{L^2(\Omega)^d}. \tag{34}$$

Taking $v = U_\mathcal{D} - \overline{U}_\mathcal{D}$ in (34) and using assumption (12) yield

$$\|U_\mathcal{D} - \overline{U}_\mathcal{D}\|^2_{L^2(\Omega)} + \alpha k \|\nabla_\mathcal{D}(U_\mathcal{D} - \overline{U}_\mathcal{D})\|^2_{L^2(\Omega)^d}$$
$$\leq k M \|\nabla_\mathcal{D}(W_\mathcal{D} - \overline{W}_\mathcal{D})\|_{L^2(\Omega)^d} \|\nabla_\mathcal{D}(U_\mathcal{D} - \overline{U}_\mathcal{D})\|_{L^2(\Omega)^d}. \tag{35}$$

This gives

$$\alpha \|\nabla_\mathcal{D}(U_\mathcal{D} - \overline{U}_\mathcal{D})\|_{L^2(\Omega)^d} \leq M \|\nabla_\mathcal{D}(W_\mathcal{D} - \overline{W}_\mathcal{D})\|_{L^2(\Omega)^d}. \tag{36}$$

Therefore $\mathbb{F}^1$ is continuous on $\mathcal{H}_\mathcal{D}$ with respect to the norm $\|\nabla\cdot\|_{L^2(\Omega)^d}$ which implies that $\mathbb{F}^1$ is continuous on $\mathcal{H}_\mathcal{D}$ with respect to the norm $\|\cdot\|_{L^2(\Omega)}$ (since $\mathcal{H}_\mathcal{D}$ is a finite dimensional space and therefore all the norms are equivalent). The application $\mathbb{F}^1$ is then satisfying the hypotheses of fixed point theorem (see for instance [5, Theorem D. 2, Page 469] and references therein). This implies the existence of a solution $u_\mathcal{D}^1 \in \mathbf{B}(0, \sqrt{\delta^1})$ to scheme (18) when $n = 0$. The uniqueness of $u_\mathcal{D}^1$ can by justified by assumption that there exist two solutions $u_\mathcal{D}^1$ and $\overline{u}_\mathcal{D}^1$ to scheme (18) when $n = 0$. This implies that, for all $v \in \mathcal{H}_\mathcal{D}$

$$\left(u_\mathcal{D}^1 - \overline{u}_\mathcal{D}^1, v\right)_{L^2(\Omega)} + \alpha k \langle u_\mathcal{D}^1 - \overline{u}_\mathcal{D}^1, v \rangle_F$$
$$+ k \left(\mathcal{F}(\nabla_\mathcal{D} u_\mathcal{D}^1) - \mathcal{F}(\nabla_\mathcal{D}\overline{u}_\mathcal{D}^1), \nabla_\mathcal{D} v\right)_{L^2(\Omega)^d} = 0. \tag{37}$$

Taking $v = u_\mathcal{D}^1 - \overline{u}_\mathcal{D}^1$ in (37) yields

$$\|u_\mathcal{D}^1 - \overline{u}_\mathcal{D}^1\|^2_{L^2(\Omega)} + \alpha k \|\nabla_\mathcal{D}(u_\mathcal{D}^1 - \overline{u}_\mathcal{D}^1)\|^2_{L^2(\Omega)^d}$$
$$+ k \left(\mathcal{F}(\nabla_\mathcal{D} u_\mathcal{D}^1) - \mathcal{F}(\nabla_\mathcal{D}\overline{u}_\mathcal{D}^1), \nabla_\mathcal{D} u_\mathcal{D}^1 - \nabla_\mathcal{D}\overline{u}_\mathcal{D}^1\right)_{L^2(\Omega)^d} = 0. \tag{38}$$

Using assumption (13), (38) implies that $\|u_\mathcal{D}^1 - \overline{u}_\mathcal{D}^1\|_{L^2(\Omega)} = \|\nabla_\mathcal{D}(u_\mathcal{D}^1 - \overline{u}_\mathcal{D}^1)\| = 0$. This gives $u_\mathcal{D}^1 = \overline{u}_\mathcal{D}^1$. This means that there exists a unique solution $u_\mathcal{D}^1$ to scheme (18) when $n = 0$ and this solution is satisfying the following bound

$$\|u_\mathcal{D}^1\|^2_{L^2(\Omega)} \leq \delta^1, \tag{39}$$

Reasoning now by induction: we assume that there exists a unique solution $u_\mathcal{D}^m$ to scheme (18) when $n = m-1$ which satisfies the bound $\|u_\mathcal{D}^m\|^2_{L^2(\Omega)} \leq \delta^m$ and prove that there exists a unique solution $u_\mathcal{D}^{m+1}$ to (18) when $n = m$ which satisfies the bound $\|u_\mathcal{D}^{m+1}\|^2_{L^2(\Omega)} \leq \delta^{m+1}$. For this reason, we follow the same above approach followed to prove the existence and uniqueness of $u_\mathcal{D}^1$ to scheme (18) when $n = 0$

which satisfies the bound (39). To this end, we consider the unique solution $U_{\mathcal{D}} \in \mathcal{H}_{\mathcal{D}}$, for a given $W_{\mathcal{D}} \in \mathcal{H}_{\mathcal{D}}$, for all $v \in \mathcal{H}_{\mathcal{D}}$

$$\left(\frac{1}{k}(U_{\mathcal{D}} - u_{\mathcal{D}}^m), v\right)_{L^2(\Omega)} + \alpha \langle U_{\mathcal{D}}, v \rangle_F$$
$$= -(\mathcal{F}(\nabla_{\mathcal{D}} W_{\mathcal{D}}), \nabla_{\mathcal{D}} v)_{L^2(\Omega)^d} + (\mathcal{G}(t_{n+1}), v)_{L^2(\Omega)}. \tag{40}$$

One sets $U_{\mathcal{D}} = \mathbb{F}^{m+1}(W_{\mathcal{D}})$, so that $\mathbb{F}^{m+1}$ is an application from the space $\mathcal{H}_{\mathcal{D}}$ into $\mathcal{H}_{\mathcal{D}}$. We remark that $U_{\mathcal{D}}$ is solution for (18) when $n = m$ if and only if $U_{\mathcal{D}}$ is a fixed point to $\mathbb{F}^{m+1}$. Taking $v = U_{\mathcal{D}}$ in (40) and using inequality $a(a - b) = \frac{1}{2}\left((a - b)^2 + a^2 - b^2\right)$ to get

$$\|U_{\mathcal{D}}\|_{L^2(\Omega)}^2 - \|u_{\mathcal{D}}^m\|_{L^2(\Omega)}^2 + 2k\alpha \|\nabla_{\mathcal{D}} U_{\mathcal{D}}\|_{L^2(\Omega)^d}^2$$
$$\leq -2k\left(\mathcal{F}(\nabla_{\mathcal{D}} W_{\mathcal{D}}), \nabla_{\mathcal{D}} U_{\mathcal{D}}\right)_{L^2(\Omega)^d} + 2k\left(\mathcal{G}(t_{n+1}), U_{\mathcal{D}}\right)_{L^2(\Omega)}. \tag{41}$$

Using the discrete Poincaré inequality together with assumption (12) and inequality $ab \leq a^2/2 + b^2/2$, (41) implies that

$$\|U_{\mathcal{D}}\|_{L^2(\Omega)}^2 - \|u_{\mathcal{D}}^m\|_{L^2(\Omega)}^2 \leq \frac{k\left(M + C_{\mathrm{p}}\|\mathcal{G}\|_{\mathcal{C}([0,T];\mathcal{C}(\overline{\Omega}))}\right)^2}{\alpha}. \tag{42}$$

This gives, using the assumption $\|u_{\mathcal{D}}^m\|_{L^2(\Omega)}^2 \leq \delta^m$, $\|U_{\mathcal{D}}\|_{L^2(\Omega)}^2 \leq \delta^{m+1}$.

Consequently, $\mathbb{F}^{m+1}(\mathbf{B}(0, \sqrt{\delta^{m+1}})) \subset \mathbf{B}(0, \sqrt{\delta^{m+1}})$. The application $\mathbb{F}^{m+1}$ can be justified to be continuous as done above for $\mathbb{F}^1$. Therefore the application $\mathbb{F}^{m+1}$ is satisfying the hypotheses of fixed point theorem and consequently there exists a solution $u_{\mathcal{D}}^{m+1} \in \mathbf{B}(0, \sqrt{\delta^{m+1}})$ to scheme (18) when $n = m$. The uniqueness of $u_{\mathcal{D}}^{m+1}$ can be justified as done above for $u_{\mathcal{D}}^1$.

2. Proof of estimates (19)–(20). To prove the error estimates (19)–(20), we compare the scheme (17)–(18) with the following auxiliary scheme: For any $n \in [\![0, N + 1]\!]$, find $\bar{u}_{\mathcal{D}}^n \in \mathcal{H}_{\mathcal{D}}$ such that

$$\langle \bar{u}_{\mathcal{D}}^n, v \rangle_F = (-\Delta u(t_n), v)_{L^2(\Omega)}, \quad \forall v \in \mathcal{H}_{\mathcal{D}}. \tag{43}$$

2.1. Comparison between the solutions of (43) and the solutions of the continuous problem (9)–(11). The following convergence results can be either deduced from [9] or using the fact $\partial^1 \bar{u}_{\mathcal{D}}^n$ is satisfying the same scheme (43) with right hand side $-\Delta \partial^1 u(t_n)$ instead of $-\Delta u(t_n)$:

– Discrete $L^\infty(L^2)$ and $L^\infty(H^1)$–error estimates. For all $n \in [\![0, N + 1]\!]$

$$\|u(t_n) - \bar{u}_{\mathcal{D}}^n\|_{L^2(\Omega)} + \|\nabla u(t_n) - \nabla_{\mathcal{D}} \bar{u}_{\mathcal{D}}^n\|_{(L^2(\Omega))^d} \leq Ch_{\mathcal{D}}. \tag{44}$$

– $\mathcal{W}^{1,\infty}(L^2)$–error estimate:

$$\max_{n=1}^{N+1} \|\partial^1(u(t_n) - \bar{u}_{\mathcal{D}}^n)\|_{L^2(\Omega)} \leq Ch_{\mathcal{D}}. \tag{45}$$

2.2. Comparison between the schemes (17)–(18) and (43). Let us define the error $\eta_{\mathcal{D}}^n = u_{\mathcal{D}}^n - \bar{u}_{\mathcal{D}}^n$. Comparing (43) with the first scheme in (17) and using the fact that $u(0) = u^0$ (see (10)) imply that $\eta_{\mathcal{D}}^0 = 0$. Writing scheme (43) in the level $n + 1$, multiplying the result by α, and subtracting the resulting equation from (18) to get, for all $n \in [\![0, N]\!]$ and for all $v \in \mathcal{H}_{\mathcal{D}}$

$$\left(\partial^1 u_{\mathcal{D}}^{n+1}, v\right)_{L^2(\Omega)} + \alpha \langle \eta_{\mathcal{D}}^{n+1}, v \rangle_F + \left(\mathcal{F}(\nabla_{\mathcal{D}} u_{\mathcal{D}}^{n+1}), \nabla_{\mathcal{D}} v\right)_{L^2(\Omega)^d}$$
$$= \left(\mathcal{G}(t_{n+1}) + \alpha \Delta u(t_{n+1}), v\right)_{L^2(\Omega)}. \tag{46}$$

Subtracting $\left(\partial^1 \bar{u}_{\mathcal{D}}^{n+1}, v\right)_{L^2(\Omega)}$ from both sides of (46) yields

$$\left(\partial^1 \eta_{\mathcal{D}}^{n+1}, v\right)_{L^2(\Omega)} + \alpha \langle \eta_{\mathcal{D}}^{n+1}, v \rangle_F + \left(\mathcal{F}(\nabla_{\mathcal{D}} u_{\mathcal{D}}^{n+1}), \nabla_{\mathcal{D}} v\right)_{L^2(\Omega)^d}$$
$$= \left(\mathcal{G}(t_{n+1}) + \alpha \Delta u(t_{n+1}) - \partial^1 \bar{u}_{\mathcal{D}}^{n+1}, v\right)_{L^2(\Omega)}. \tag{47}$$

From equation (9), we deduce that

$$\alpha \Delta u + \mathcal{G} = u_t - \nabla \cdot \mathcal{F}(\nabla u). \tag{48}$$

Substituting $\alpha \Delta u + \mathcal{G}$ by its value of (48) in (47) yields

$$\left(\partial^1 \eta_{\mathcal{D}}^{n+1}, v\right)_{L^2(\Omega)} + \alpha \langle \eta_{\mathcal{D}}^{n+1}, v \rangle_F = \mathbb{T}_1 + \mathbb{T}_2, \tag{49}$$

where

$$\mathbb{T}_1 = -\left(\mathcal{F}(\nabla_{\mathcal{D}} u_{\mathcal{D}}^{n+1}), \nabla_{\mathcal{D}} v\right)_{L^2(\Omega)^d} - (\nabla \cdot \mathcal{F}(\nabla u)(t_{n+1}), v)_{L^2(\Omega)} \tag{50}$$

and

$$\mathbb{T}_2 = \left(u_t(t_{n+1}) - \partial^1 \bar{u}_{\mathcal{D}}^{n+1}, v\right)_{L^2(\Omega)}. \tag{51}$$

The following "dual consistency", see for instance [2, Page 1300] and references therein, holds

$$\left| (\nabla \cdot \mathcal{F}(\nabla u)(t_{n+1}), v)_{L^2(\Omega)} + (\mathcal{F}(\nabla u)(t_{n+1}), \nabla_{\mathcal{D}} v)_{L^2(\Omega)} \right|$$
$$\leq C h_{\mathcal{D}} \|\nabla_{\mathcal{D}} v\|_{L^2(\Omega)^d}. \tag{52}$$

On the other hand, using the triangle and the Cauchy-Schwarz inequalities together with (45) and the fact that $\max_{n=1}^{N+1} |\partial^1 u(t_n) - u_t(t_n)|$ is order k yield

$$|\mathbb{T}_2| \leq C(h_{\mathcal{D}} + k)\|v\|_{L^2(\Omega)}. \tag{53}$$

Gathering now (49)–(53) leads to

$$\left(\partial^1 \eta_{\mathcal{D}}^{n+1}, v\right)_{L^2(\Omega)} + \alpha \langle \eta_{\mathcal{D}}^{n+1}, v \rangle_F$$
$$+ \left(\mathcal{F}(\nabla_{\mathcal{D}} u_{\mathcal{D}}^{n+1}) - \mathcal{F}(\nabla u)(t_{n+1}), \nabla_{\mathcal{D}} v\right)_{L^2(\Omega)^d}$$
$$\leq C(h_{\mathcal{D}} + k)\|v\|_{L^2(\Omega)} + C h_{\mathcal{D}} \|\nabla_{\mathcal{D}} v\|_{L^2(\Omega)^d}. \tag{54}$$

This with error estimate (44) and assumption (12) imply that

$$\left(\partial^1 \eta_{\mathcal{D}}^{n+1}, v\right)_{L^2(\Omega)} + \alpha \langle \eta_{\mathcal{D}}^{n+1}, v \rangle_F$$
$$\leq C(h_{\mathcal{D}} + k)\|v\|_{L^2(\Omega)} + Ch_{\mathcal{D}}\|\nabla_{\mathcal{D}} v\|_{L^2(\Omega)^d}. \tag{55}$$

Taking $v = \eta_{\mathcal{D}}^{n+1}$ in (55) and using inequality $a(a-b) \geq \frac{1}{2}\left(a^2 - b^2\right)$ and assumption (13) give (recall that $\eta_{\mathcal{D}}^n = u_{\mathcal{D}}^n - \bar{u}_{\mathcal{D}}^n$)

$$\|\eta_{\mathcal{D}}^{n+1}\|_{L^2(\Omega)}^2 - \|\eta_{\mathcal{D}}^n\|_{L^2(\Omega)}^2 + 2k\alpha\|\nabla_{\mathcal{D}}\eta_{\mathcal{D}}^{n+1}\|_{L^2(\Omega)^d}^2$$
$$\leq Ck(h_{\mathcal{D}} + k)\|\eta_{\mathcal{D}}^{n+1}\|_{L^2(\Omega)} + Ckh_{\mathcal{D}}\|\nabla_{\mathcal{D}}\eta_{\mathcal{D}}^{n+1}\|_{L^2(\Omega)^d}. \tag{56}$$

Using the discrete Poincaré inequality together and inequality $ab \leq a^2/2 + b^2/2$, (56) implies that

$$\|\eta_{\mathcal{D}}^{n+1}\|_{L^2(\Omega)}^2 - \|\eta_{\mathcal{D}}^n\|_{L^2(\Omega)}^2 + k\alpha\|\nabla_{\mathcal{D}}\eta_{\mathcal{D}}^{n+1}\|_{L^2(\Omega)^d}^2 \leq Ck(h_{\mathcal{D}} + k)^2. \tag{57}$$

Summing the previous inequality over $n \in [\![0, J-1]\!]$, with $J \in [\![1, N+1]\!]$, and using the facts that $\eta^0 = 0$ and $\sum_{n=0}^{N+1} k \leq T$ yield

$$\|\eta_{\mathcal{D}}^J\|_{L^2(\Omega)}^2 + \sum_{n=0}^{J-1} k\alpha\|\nabla_{\mathcal{D}}\eta_{\mathcal{D}}^{n+1}\|_{L^2(\Omega)^d}^2 \leq C(h_{\mathcal{D}} + k)^2. \tag{58}$$

This implies that

$$\max_{n=0}^{N+1} \|\eta_{\mathcal{D}}^n\|_{L^2(\Omega)} \leq C(h_{\mathcal{D}} + k) \text{ and } \left(\sum_{n=0}^{J-1} k\|\nabla_{\mathcal{D}}\eta_{\mathcal{D}}^{n+1}\|_{L^2(\Omega)^d}^2\right)^{\frac{1}{2}} \leq C(h_{\mathcal{D}} + k). \tag{59}$$

Using now error estimates (44), the triangle inequality, and estimates (59) yield the desired estimates (19)–(20). This completes the proof of Theorem 1. □

Remark 1. [**A uniqueness result**] Error estimate (19) implies the uniqueness of any possible solution $u \in \mathcal{C}^2([0,T]; \mathcal{C}^2(\overline{\Omega}))$ for problem (9)–(11) due to the uniqueness of the limit.

The existence of a solution for problem (9)–(11), in some weak sense, can be proved maybe via *a priori* estimates similar to (21)–(22) of Theorem 2 together with some tools of compactness, see [5] with perhaps some slight modifications for the scheme (17)–(18). In fact, the scheme (17) needs the strong regularity $u^0 \in H^2(\Omega)$ which can be changed to a scheme (or simply a convenient interpolation for the initial data u^0) which needs less regularity. The r.h.s of scheme (18) also needs the continuity of $\mathcal{G}$ in time which can be changed for instance to $\left(\frac{1}{k}\int_{t_n}^{t_{n+1}} \mathcal{G}(t)dt, v\right)_{L^2(\Omega)}$. In this way, we will only need the regularity $L2$ in both time and space, see some related details in [5].

The subject of the existence of a weak solution to problem (9)–(11) is an interesting task to work on in the near future.

5   Conclusion and Perspectives

We considered a nonlinear cell centered finite volume scheme (the unknowns are located on the centers of the control volumes at each time level) for a simple Bingham Flow Type equation which was considered in [8]. This equation is a nonlinear parabolic. The space discretization is performed using SUSHI (Scheme using Stabilization and Hybrid Interfaces) developed in [9] whereas the time discretization is uniform. We first proved a new discrete bound for any possible approximate solution. We then showed rigorously the existence and uniqueness of the approximate solution. We have proved optimal error estimates, under assumption that the exact solution is smooth, in the discrete norms of $L^{\infty}(L^2)$ and $L^2(H^1)$.

There are several interesting perspectives for the present work. Among them we plan to work in the near future on the following related tasks:

- Extension to general SUSH in which the unknowns are located on the centers and on the edges of the centers of the control volumes.
- Consider a general nonlinear parabolic equation which encompasses the simple Bingham Flow Type equation we considered in this work.
- Consider a system of Bingham Flow model instead of only single equation.
- Prove an error estimate in $L^{\infty}(H^1)$.
- Consider a linearized scheme version for the one considered here.
- Prove the convergence towards the solution of a weak formulation.
- Use Crank-Nicolson to improve the order in time.
- Extension to a thixotropic Bingham model and to other non-Newtonian models as Herschel–Bulkley and Casson.

References

1. Baranger, J., Machmoum, A.: Existence of approximate solutions and error bounds for viscoelastic fluid flow: characteristics method. Comput. Methods Appl. Mech. Engrg. **148**(1–2), 39–52 (1997)
2. Bradji, A.: An analysis for the convergence order of gradient schemes for semilinear parabolic equations. Comput. Math. Appl. **72**(5), 1287–1304 (2016)
3. Brezis, H.: Analyse fonctionnelle. (French) [[Functional analysis]] Théorie et applications. [Theory and applications] Collection Mathématiques Appliquées pour la Maîtrise. [Collection of Applied Mathematics for the Master's Degree] Masson, Paris (1983)
4. Droniou, J., Goldys, B., Le, K.-Ngan: Design and convergence analysis of numerical methods for stochastic evolution equations with Leray-Lions operator. IMA J. Numer. Anal. **42**(2), 1143–1179 (2022)
5. Droniou, J., Eymard, R., Gallouët, T., Guichard, C., Herbin, R.: The Gradient Discretisation Method. Mathématiques et Applications, 82, Springer Nature Switzerland AG, Switzerland (2018)
6. Eymard, R., Gallouët, T., Herbin, R., Linke, A.: Finite volume schemes for the biharmonic problem on general meshes. Math. Comput. **81**(280), 2019–2048 (2012)
7. Evans, L.C.: Partial Differential Equations. Graduate Studies in Mathematics, American Mathematical Society, 19 (1998)

8. Dean, E.J., Glowinski, R., Guidoboni, G.: Review on the numerical simulation of Bingham visco-plastic flow: Old and new results. J. Nonnewton. Fluid Mech. **142**(1–3), 36–62 (2007)
9. Eymard, R., Gallouët, T., Herbin, R.: Discretization of heterogeneous and anisotropic diffusion problems on general nonconforming meshes. IMA J. Numer. Anal. **30**(4), 1009–1043 (2010)
10. Eymard, R., Gallouët, T., Herbin, R.: A cell-centred finite-volume approximation for anisotropic diffusion operators on unstructured meshes in any space dimension. IMA J. Numer. Anal. **26**, 326–353 (2006)
11. Aberqi, A., Aboussi, W., Benkhaldoun, F., Bennouna, J., Bradji, A.: Homogeneous incompressible Bingham viscoplastic as a limit of bi-viscosity fluids. J Elliptic Parabol Equ (2023). https://doi.org/10.1007/s41808-023-00221-z

Two-Phase Algorithm for Solving Vehicle Routing Problem with Time Windows

Adis Alihodzic[1(✉)], Eva Tuba[2], and Milan Tuba[3]

[1] University of Sarajevo, Sarajevo, BiH, Bosnia and Herzegovina
adis.alihodzic@pmf.unsa.ba
[2] Singidunum University, Belgrade, Serbia
etuba@acm.org
[3] State University of Novi Pazar, Novi Pazar, Serbia
mtuba@acm.org

Abstract. It is well-known that determining the optimal number of vehicles and searching for the shortest distances produced with them presents an NP-hard problem. Many distribution problems in real-world applications can be expressed as vehicle routing problems with time windows (VRPTW), where the main objective is to minimize the fleet size and assign a sequence of customers to each truck of the fleet minimizing the total distance travelled, such that all customers are being served and the total demand served by each truck does not exceed its capacity. To address this problem, the efficient two-phase approach is applied. The proposed method in the first phase divides the fleet of customers into smaller clusters by using the prominent K-Means algorithm. In the second phase, the famous state-of-the-art CPLEX solver is applied to find the shortest routes between customers and the smallest number of vehicles that travel them. The experimentation has been done with the 56 instances composed of the 100 customers Solomon's benchmark. The outcomes demonstrate that our two-phase approach can deliver in real-time sub-optimal solutions very close to the best ones, which for almost all instances of all classes are superior compared to other heuristics.

Keywords: K-Means Algorithm · Vehicle Routing Problem with Time Windows (VRPTW) · Combinatorial Optimization Problems (COPs) · Traveling Salesman Problem (TSP) · CPLEX Solver

1 Introduction

In a time of increasing fuel prices, transportation is an essential factor that directly affects the quality of life in today's fashionable society. In logistics, inefficient vehicle routes imply more energy consumption, higher distribution cost, and environmental pollution. For example, the distribution costs at groceries delivery can lead to an increase in the product price up to 70% [7]. In combinatorial optimization, the problem of seeking an efficient vehicle route for distributing goods from depot to customers is usually called a Vehicle Routing Problem

© The Author(s), under exclusive license to Springer Nature Switzerland AG 2023
I. Georgiev et al. (Eds.): NMA 2022, LNCS 13858, pp. 14–25, 2023.
https://doi.org/10.1007/978-3-031-32412-3_2

(VRP) [15]. The vehicle routing problem (VRP) is one of the most critical problems in distribution logistics [21]. It is well-known that it belongs to the class of NP-hard problems and is a generalization of the famous Traveling Salesman Problem (TSP). Indeed, even locating a feasible solution to the VRPTW with a fixed fleet size is itself an NP-complete problem [18]. Theoretical research and practical applications of this problem began in 1959 with the Truck Dispatching Problem (TDP), from the need to rationalize fuel supply to gas stations, which was first posed by Dantzig and Ramser [6]. A few years later, the development of specific algorithms began for the vehicle routing problem, such as algorithms based on dynamic programming with time and space relaxation [3]. The VRP represents the basis of many issues that differ in terms of the number and type of constraints and has a considerable economic significance. Besides the basic of well-known conventional vehicle routing problem (VRP), many variations can be found in the literature [11]. This paper focuses on one of these variants: the Vehicle Routing Problem with Time Windows (VRPTW). VRPTW can be expressed as the problem of creating the least-cost routes for a fleet of identical vehicles from one depot to a set of geographically dispersed points. The routes must be made so that each customer is visited only once by precisely one vehicle within a given time interval, all routes start and end at the depot, and the overall demands of all points on one particular path must not exceed the capacity of the vehicle. In addition, it is assumed that each vehicle covers just one route. The VRPTW has multiple objectives: to minimize the number of vehicles required and the entire travel time and distance incurred by the fleet. Some of the most practical applications of the VRPTW include bank deliveries [12], postal deliveries [13], national franchise restaurant services [17], school bus routing [1] and so on. The VRPTW is still the subject of intensive research for both exact methods, and approximate algorithms [10]. Several heuristics based on different principles have appeared, including saving, geographical distance, and so on [4]. In the past couple of decades, many compelling methods for addressing the VRPTW can be found in the literature [9]. In addition to existing techniques, the development of novel methods which will properly face routing problems is a hot topic for the current research community. In line with this, we report the efficient two-phase approach for solving a multiple-vehicle routing problem with time windows. The proposed method in the first phase divides the fleet of customers into smaller clusters by using the prominent K-means algorithm. After that, in the second phase of the algorithm, the CPLEX solver is applied to each cluster to generate a solution, i.e. the shortest routes between customers and the smallest number of vehicles that travels them. The constructed solution's quality is enhanced by replacing the customers located on the edges of the obtained clusters, which are very short distances from each other. The experimentation has been done with the 56 instances composed of the 100 customers Solomon's benchmark, which can be found at website http://web.cba.neu.edu/~msolomon/ problems.htm. The outcomes demonstrate that our hybrid approach can deliver in real-time sub-optimal solutions very close to their optimal ones, which are superior to many heuristics in literature [8]. In this way, we will prove that our mixed method, which uses K-means and CPLEX solver, is a promising strategy to tackle the notable VRPTW. In subsequent sections, we foremost introduce ourselves to the mathematical definition of the VRPTW, while Sect. 3 presents our proposed method for solving the VRPTW. In Sect. 4, simulation results are reported. Finally, our endings are concerned in the last part of the paper.

2   Mathematical Problem Formulation for the VRPTW

The essential components of each VRP are warehouses (depots), vehicles, and users (locations) to be served (visited). In this section, we define a model of the VRPTW in which the primary goal is to minimize the number of vehicles, while the secondary aim is to reduce the total distance or time, whereby all customers are visited precisely once, and at the end of the route, each vehicle must return to the warehouse from which it started. This problem is most often defined through the network $G = (V, A)$ in which the vertices of set V are clients, and the arcs of set A present travel costs between elements of set V [2]. The set V contains n customers marked with numbers $1, 2, \cdots, n$, and two nodes 0 and $n + 1$, representing the depot. Set of arcs A is the set of all possible connections between the nodes of set $C = V \setminus \{0, n+1\}$. All feasible vehicle routes correspond to paths in G, which begin from node 0 and end at node $n+1$. Each pair $(i, j) \in A$, $i \neq j$ is assigned non-negative travel cost c_{ij} and time t_{ij} which are symmetrical, i.e. $c_{ij} = c_{ji}$ and $t_{ij} = t_{ji}$. Also, a demand request d_i is defined at each location of i-th customer, where zero demands are defined for nodes 0 and $n + 1$, that is, $d_0 = d_{n+1} = 0$. Each vehicle is assigned a fixed capacity Q. The set of all vehicles is denoted by K and their size by $|K|$. A time window $[a_i, b_i]$ is associated with each customer i that composes a VRPTW instance. A vehicle arriving before the earliest possible departure E from the depot must wait, and a customer arriving later than the latest possible arrival L at the depot will not be served. This type of service time window is called a hard time window. The windows for both depots are the same, i.e. $[a_0, b_0] = [a_{n+1}, b_{n+1}] = [E, L]$. Hence, a vehicle should not leave the depot before E and must return by L. In this way, a route would not be feasible if a vehicle tries to serve any customer after the upper limit L of the range $[E, L]$. On the other hand, a route would be feasible if the vehicle reaches a customer before the lower limit E. In this last special situation, the customer cannot be served before this limit, so the vehicle should wait until E to initiate the delivery. Apart from this temporal window, the problem can also consider the service time s_i of the i-th customer. For the nodes 0 and $n + 1$, service times are equal zero, i.e. $s_0 = s_{n+1} = 0$. The parameter s_i presents the vehicle's service time, which should be spent at the i-th customer to perform the delivery properly. This should be considered if the vehicle arrives on time for the next customer. So the mathematical model of VRPTW is presented as follows:

$$\min \sum_{k \in K} \sum_{i \in V} \sum_{j \in V} c_{ij} x_{ijk}$$

subject to:

$$x_{iik} = 0, \qquad \forall i \in V, \forall k \in K \qquad (1)$$

$$\sum_{k \in K} \sum_{j \in V} x_{ijk} = 1, \qquad \forall i \in C, i \neq j \tag{2}$$

$$\sum_{j \in V} x_{0jk} = 1, \qquad \forall k \in K \tag{3}$$

$$\sum_{i \in V} x_{ihk} - \sum_{j \in V} x_{hjk} = 0, \qquad \forall k \in K, \ \forall h \in C \tag{4}$$

$$\sum_{i \in V} x_{in+1k} = 1, \qquad \forall k \in K \tag{5}$$

$$x_{ijk}(w_{ik} + s_i + t_{ij} - w_{jk}) \leq 0, \qquad \forall k \in K, \ \forall i \in V, \ \forall j \in V \tag{6}$$

$$a_i \sum_{j \in V} x_{ijk} \leq w_{ik} \leq b_i \sum_{j \in V} x_{ijk}, \qquad \forall k \in K, \ \forall i \in C \tag{7}$$

$$E \leq w_{ik} \leq L, \qquad \forall k \in K, \ \forall i \in \{0, n+1\} \tag{8}$$

$$\sum_{i \in C} d_i \sum_{j \in V} x_{ijk} \leq Q, \qquad \forall k \in K \tag{9}$$

$$\sum_{j \in C} x_{0jk} \leq |K|, \qquad \forall k \in K \tag{10}$$

$$x_{ijk} \in \{0,1\}, \qquad \forall i \in V, \ \forall j \in V, \ \forall k \in K \tag{11}$$

A given model include flow variables x_{ijk}, and time variables w_{ik}, where a flow variable x_{ijk} is equal to one if arc (i,j) is used by vehicle k, and zero otherwise. The time variable w_{ik} defines the start of service at a customer i when serviced by vehicle k. Constraint (1) claims that all diagonal elements are equals to zero, while constraint (2) says that every customer will be visited by exactly one vehicle. Next, constraints (3) and (5) guarantee that each vehicle will start from depot 0 and end at depot $n+1$, respectively. Condition (4) defines a flow of the path to be gone by vehicle k. Additionally, constraints (6)–(8) ensure schedule feasibility concerning time window constraints. Precisely, any vehicle k which travelling from customer i to customer j must not arrive later than $w_{ik} + s_i + t_{ij}$ at customer j. Also, for a given k, constraints (7) force $w_{ik} = 0$ whenever customer i is not visited by vehicle k. Constraint (9) is a capacity constraint where each vehicle can be loaded to its capacity. Constraint (10) characterises the usage of the lowest number of vehicles. The last condition (11) sets binary constraints on the flow variables. It is essential to highlight that constraint (6) usually results in non-convex optimization due to quadratic terms. Thanks to the binary restrictions (11), condition (6) can be linearized as

$$w_{ik} + s_i + t_{ij} - w_{jk} \leq (1 - x_{ijk})M_{ij}, \qquad \forall k \in K, \ \forall i \in V, \ \forall j \in V \tag{12}$$

where M_{ij} are large constants which can be changed by $\max\{b_i + s_i + t_{ij} - a_j, 0\}$.

Table 1. The client's demand and time window as well distances between them.

Client's request and time window				The distances between clients									
ID	Demand	Service time	Time window	d_{ij}	0	1	2	3	4	5	6	7	8
0	0	0	[0, 50]	0	0	40	60	75	90	200	100	160	80
1	2	1	[1, 4]	1	40	0	65	40	100	50	75	110	100
2	1.5	2	[4, 6]	2	60	65	0	75	100	100	75	75	75
3	4.5	1	[1, 2]	3	75	40	75	0	100	50	90	90	150
4	3	3	[4, 7]	4	90	100	100	100	0	100	75	75	100
5	1.5	2	[3, 5.5]	5	200	50	100	50	100	0	70	90	75
6	4	2.5	[2, 5]	6	100	75	75	90	75	70	0	70	100
7	2.5	3	[5, 8]	7	160	110	75	90	75	90	70	0	100
8	3	0.8	[1.5, 4]	8	80	100	75	150	100	75	100	100	0

3   Our Proposed Approach for Solving the VRPTW

It is known that the VRPTW is NP-hard because the time complexity of the search space in the case of visiting n customers is proportional to $O(n!)$. Specifically, in the case of 100 customers, to find the smallest number of shortest feasible routes (those which meet the constraints of vehicle capacity and time window), it is necessary to calculate 100! permutations and pick the one that produces the smallest number of shortest routes. Those routes are being used by vehicles to serve all customers exactly once. In particular, let be given $n = 8$ customers, and $k = 3$ vehicles as in the test example in the paper [14], where data about constraints, customers and distances between them are given in Table 1. The sign 0 in the mentioned table indicates a virtual customer or depot (warehouse). From the set of all permutations (8! =40320), the permutation 4-6-1-3-7-2-5-8 gives the best feasible solution, which is not unique since there are other permutations with the same solution, such as permutation 7-2-5-8-4-6-1-3. From the permutation 4-6-1-3-7-2-5-8, two routes **0-4-6-1-3-0** and **0-7-2-5-8- 0** are being generated, so the total length of those routes is equal 845. Even on such a small example, it can be seen that the search space grows exponentially due to the complexity of the VRPTW, so for many customers, an exact solution cannot be got in real-time. Therefore, it is required to reduce the search space by clustering clients into a certain number of clusters using known clustering methods, representing the first step of our approach.

3.1   Utilizing the K-Means Algorithm for Customer Clustering

The first step of our approach is the application of the K-Means algorithm as a clustering method for grouping consumers into clusters due to the Euclidean distance between them so that each cluster contains those customers who are the closest in terms of geographic distance [22]. Since any partitioning of the original set concerns the quality of the solution, in this paper, the number of clusters is fixed at three because the size of clusters greater than three gives a lower quality solution. In contrast, the number of clusters less than three does not sufficiently

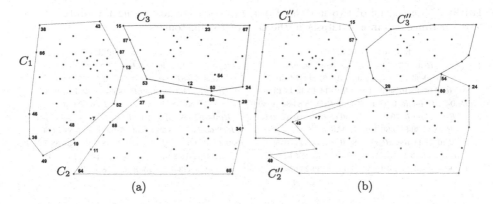

Fig. 1. (a) The design of clusters C_1, C_2 and C_3 by applying the K-means algorithm to the instance R101 composed of 100 customers. (b) The modification of clusters C_1, C_2 and C_3 have been created by moving neighbouring close clients in order to gain a better solution.

reduce the search space, so the solution cannot be obtained in a reasonable time. Therefore, the first step of our approach is to cluster the set of customers into three clusters C_1, C_2 and C_3. One example of clustering the instance R101 of Solomon's dataset was illustrated in Fig. 1 a). By dividing the R101 instance into three parts, three clusters have been generated such that the first cluster C_1 has 42 customers, the second one C_2 contains 33 consumers, and the last one C_3 counts 25 clients. Applying the second step of our approach (this step will be described below) realised by the CPLEX Solver implies that nine vehicles are necessary to visit the customers of cluster C_1 for 3.93 s, while the overall length of the determined routes is 663.107. All nine routes of cluster C_1 on which the mentioned vehicles move has shown in Table 2. In order to visit all customers of cluster C_2, it was necessary to employ seven vehicles because seven routes with a total length of 617.405 were found in a time of 2.59 s. For the last cluster C_3, all its customers were served by five vehicles, i.e. five routes were generated with a total length of 417.676 for 0.84 s. Table 2 shows the calculated routes for clusters C_1, C_2 and C_3 in its second column. Therefore, it was necessary to employ 21 vehicles to serve all customers of the instance R101 by using routes from the second column of Table 2 whose total length is 1698.188. Those routes were generated in real-time by our approach. From Fig. 1 a), division of the instance R101 by using the simple version of the K-Means algorithm produces three convex hulls such as $CH_1 = \{49, 19, 52, 13, 87, 43, 38, 86, 46, 36\}$, $CH_2 = \{64, 65, 34, 29, 68, 28, 27, 88, 11\}$ and $CH_3 = \{12, 80, 24, 67, 23, 15, 57, 53\}$. Various algorithms such as Fuzzy C-Means clustering, K-means++, K-median and others will probably generate diverse clusters and convex hulls, which ultimately refer to creating routes with a smaller total distance and fewer vehicles needed to serve all clients. The quality of produced solutions can be improved if it is noted that customers of all clusters are candidates for mutual exchange if and

Table 2. The usage of two models of our approach for generating the routes of the R101 instance Solomon's dataset in the case of 100 customers.

C	K-Means + CPLEX Solver	Improved K-Means + CPLEX Solver
1	R_1:0-82-7-8-46-17-0;R_2:0-59-99-94-96-0	R_1:0-36-47-19-8-46-17-0;R_2:0-95-98-16-86-91-100-0
	R_3:0-36-47-19-49-48-0;R_4:0-5-61-85-37-93-0	R_3:0-45-82-18-84-60-89-0;R_4:0-59-99-94-96-0
	R_5:0-95-98-16-86-91-100-0;R_6:0-14-44-38-43-13-0	R_5:0-14-44-38-43-13-0;R_6:0-92-42-15-87-57-97-0
	R_7:0-52-18-6-0;R_8:0-92-42-87-97-0;R_9:0-45-83-84-60-89-0	R_7:0-5-83-61-85-37-93-0; R_8:0-52-6-0
2	R_1:0-27-69-76-79-3-68-0; R_2:0-63-64-90-20-32-70-0	R_1:0-62-11-90-20-32-70-0;R_2:0-33-29-78-34-35-77-0
	R_3:0-33-30-51-9-66-1-0; R_4:0-28-29-78-34-35-77-0	R_3:0-65-71-81-50-68-0;R_4:0-27-69-76-79-3-54-24-80-0
	R_5:0-62-11-10-0; R_6:0-31-88-0; R_7:0-65-71-81-50-0	R_5:0-30-51-9-66-1-0;R_6:0-63-64-49-48-0;R_7:0-31-88-7-10-0
3	R_1:0-2-21-73-41-56-74-58-0;R_2:0-72-75-22-54-24-80-0	R_1:0-72-75-22-74-58-0; R_2:0-2-21-73-41-56-4-0
	R_3:0-12-40-53-0;R_4:0-15-57-26-0;R_5:0-39-23-67-55-4-25-0	R_3:0-39-23-67-55-25-0; R_4:0-28-12-40-53-26-0

only if they are located very close to the boundary of their convex hulls. This fact help to reduce the number of vehicles, generating a smaller total length of the routes. Namely, let customers of clusters C_1 and C_2 be first considered. Then candidates of cluster C_1 which are closest to convex hull CH_1 and at the shortest distance of the convex hull CH_2 are 7, 19, 48, 49 and 52. By moving the customers 7, 48 and 49 from cluster C_1 into cluster C_2, clusters C_1 and C_2 are modified into new clusters $C_1' = C_1 \setminus \{7, 48, 49\}$, and $C_2' = C_2 \cup \{7, 48, 49\}$, respectively. Next, by considering clusters C_1' and C_3, clients from cluster C_3 that can move to cluster C_1' are 2, 15, 53, 57, and 58 because they are very close to convex hull CH_3 and at the same time they are located at the shortest distance from convex hull CH_1. By moving candidates 15 and 57 from cluster C_3 to cluster C_1, these clusters are modified, and two new clusters are created such as $C_1'' = C_1' \cup \{15, 57\}$ and $C_3'' = C_3 \setminus \{15, 57\}$. It remains to modify the clusters C_2' and C_3'. The candidates of cluster C_2' that can be transferred to cluster C_3' are 19, 28 and 68, while candidates of cluster C_3' that can be moved to cluster C_2' are 12, 24, 26, 53, 54 and 80. By moving customer $28 \in C_2'$ to C_3', and customers 24, 54 and 80 from C_3' to C_2', new clusters are obtained such as $C_2'' = \{24, 54, 80\} \cup C_2' \setminus \{28\}$, and $C_3'' = C_3' \setminus \{24, 54, 80\} \cup \{28\}$. In this way, three new clusters C_1'', C_2'' and C_3'' were generated, whose geometry in the form of simple polygons stands in Fig. 1 b). The produced geometry returns a smaller number of vehicles and better routes by applying the CPLEX Solver in the second step of our approach. Namely, by exploiting mentioned solver to new clusters C_1'', C_2'' and C_3'', the number of vehicles for cluster C_1'' becomes 8, and the total length of the routes is equal to 638.849, which is certainly a better result in comparison with the one related to the cluster C_1. For clusters C_2'' and C_3'', the number of vehicles is 7 and 4, respectively, while the total length of all routes for these two clusters is 698.425 and 313.525, respectively. Finally, the total number of vehicles to serve the clusters C_1'', C_2'' and C_3'' is 19. At the same time, the total length is equal to 1650.799, representing an improvement compared to the previously obtained suboptimal solution. The routes of the new clusters C_1'', C_2'' and C_3'' determined with the improved version of our approach there are in the third column of Table 2. Thus, the proposed modification of the original clusters C_1, C_2 and C_3 produced by using the basic K-Means algorithm, as a result, generates a new type of clustering shown in Fig. 1 b).

3.2   Seeking Exact Feasible Solutions by the CPLEX Solver

The second phase of our proposed approach was based on parallel servicing of the customers included in the clusters produced in the first part of our approach. Servicing customers in the context of obtaining the shortest total distance and the least number of vehicles in IBM ILOG CPLEX Optimization Solver [5] is presented in the form of a binary integer programming problem, as we can see in Fig. 2. From the code shown in Fig. 2, we can note that depending on the benchmark problem being processed following parameters are set: customers C, their number nC and demands $d[C]$, vehicles K, their number nK and capacity q, customers and depots V, service times $s[V]$, as well as time window constraints $a[V]$ and $b[V]$. An arbitrary element $t[i][j]$ of the matrix $t[V][V]$ is calculated as $cost[i][j] + s[i]$ for $i \neq j$, otherwise is zero. The object X is a decision binary 3D vector whose components are being determined by the CPLEX solver. Specifically, for the example of 8 customers whose data is shown in Table 1, two vehicles were necessary to serve all customers, which is written in the form of object X in this way $X[1][0][4] = 1$, $X[1][4][6] = 1$, $X[1][6][1] = 1$, $X[1][1][3] = 1$, $X[1][3][9] = 1$, $X[2][0][7] = 1$, $X[2][7][2] = 1$, $X[2][2][5] = 1$, $X[2][5][8] = 1$, $X[2][8][9] = 1$, where 0 and 9 represent depots. The CPLEX Solver, based on techniques such as branch-and-bound and branch-and-cut, can determine an optimal solution in terms of the smallest number of the shortest routes in real-time, as we will see in the experimental analysis.

```
int nC=...;//size of customers      float a[V]=...;//earliest arrival time of vehicle
int nK=...;//size of vehicles       float b[V]=...;//latest arrival time of vehicle
float q=...;//cap.of vehicles       float s[V]=...;//service time
range C=1..nC;//set of customers    float t[V][V];//time equals cost + service time
range K=1..nK;//set of vehicles     dvar float W[K][V];//time variables for waiting time
float d[C]=...;//demands of cust.    dvar boolean x[K][V][V];//binary flow variables
range V=0..nC+1;//cust. and depot.  float cost[V][V];//travel cost or distance or time

minimize sum(k in K,i in V, j in V) cost[i][j]*x[k][i][j];
subject to{
  forall(k in K, i in V) C1: x[k][i][i]==0;
  forall(i in C) C2: sum(k in K, j in V) x[k][i][j]==1;
  forall(k in K) C3: sum(j in V) x[k][0][j]==1;
  forall(h in C, k in K) C4: sum(i in V) x[k][i][h]==sum(j in V) x[k][h][j];
  forall(k in K) C5:sum(i in V) x[k][i][nC+1] == 1;
  forall(i in V, j in V, k in K) C6: W[k][i]+s[i]+t[i][j]-10000*(1-x[k][i][j])<=W[k][j];
  forall(i in V, k in K) C7C8: a[i]<=W[k][i]<=b[i];
  forall(k in K) C9:sum( i in C, j in V) d[i]*x[k][i][j]<=q;
  C10: sum (k in K, j in V) x[k][0][j] <= nK;
}
```

Fig. 2. The OPL Code For Solving VRPTW.

Table 3. The application of our approach for fetching the optimal distribution of customers by clusters of the class R, as well seeking the shortest distance between them and picking the optimal number of vehicles visiting all customers.

Inst.	Optimal solutions		K-Means No. Cust.			Improved K-Means and CPLEX									Total	
						CLUSTER 1			CLUSTER 2			CLUSTER 3				
	N_v	T_d	C_1	C_2	C_3	N_c	N_v	T_d	N_c	N_v	T_d	N_c	N_v	T_d	N_v	T_d
R101	20	1637.7	32	40	28	41	8	638.85	38	7	698.43	21	4	313.53	19	1650.80
R102	18	1466.6	32	40	28	29	5	506.02	43	7	596.64	28	5	385.02	17	1487.68
R103	14	1208.7	32	40	28	34	4	431.32	42	6	544.16	24	4	240.93	14	1216.40
R104	11	971.5	32	40	28	41	4	437.26	36	3	306.99	23	3	248.18	10	992.42
R105	15	1355.3	32	40	28	27	4	434.27	45	6	576.83	28	4	382.43	14	1393.52
R106	13	1234.6	32	40	28	32	4	452.42	48	6	581.47	20	2	229.58	12	1263.48
R107	11	1064.6	32	40	28	32	4	459.11	40	4	399.75	28	3	231.54	11	1090.39
R108	NA	NA	32	40	28	31	3	328.34	42	4	361.41	27	3	265.09	10	954.84
R109	13	1146.9	32	40	28	38	5	478.34	32	4	387.16	30	3	315.29	12	1180.78
R110	12	1068	32	40	28	42	5	515.98	38	4	353.03	20	2	230.89	11	1099.90
R111	12	1048.7	32	40	28	27	3	300.96	43	4	378.89	30	4	382.21	11	1062.07
R112	NA	NA	32	40	28	32	3	325.88	48	5	437.49	20	2	207.85	10	971.23

N_v: number of vehicles, T_d: total distance, N_c: number of customers by a cluster.

4 Experimental Analysis

In this part, experimental analysis was accomplished to evaluate the effectiveness of our approach using 56 instances composed of the 100 customers in Solomon's benchmark as the test cases [19]. Our approach has been implemented in C# programming language, wherein in the first phase, we create the disjoint clusters of a smaller number of customers by using the K-Means algorithm, and in the second phase, we use IBM ILOG CPLEX Optimizer 22.1.0 to address the VRPTW on these clusters, which was expressed in the form of binary integer programming (BIP). This optimizer is being called from C# in conjunction with Concert Technology. All experiments were conducted on a PC with Intel(R) Core(TM) i7-3770K K 3.5 GHz processor with 32GB of RAM running under the Windows 11 x64 operating system. Solomon's benchmark contains six datasets with information about geographical data, vehicle capacity and tightness, customers' demand and positioning of the time windows. Datasets R include randomly generated customers, uniformly distributed over a square. Datasets C contain clustered customers, and the time windows are generated by the known vehicle routes for these customers. Datasets RC have mixed customers of datasets C and R. In the remaining of this section, the experimental results are presented. From the statistical data shown in Table 3, we can observe that our hybrid approach based on improved K-Means and CPLEX Solver generates very acceptable sub-optimal solutions for instances R101 to R112 of class R. Although CPLEX Solver produces exact solutions, due to the search space reduction performed by the K-Means, the solutions obtained by our proposed approach are approximate, and they are comparable to the ones generated by the best heuristics http://web.cba.neu.edu/~msolomon/heuristi.htm. The main benefit of our method is

Table 4. The calculated outcomes for the instances of the classes R, C and RC in the case of 100 customers.

Inst.	Opt. Sol.		Our Sol.		Inst.	Opt. Sol.		Our Sol.		Inst.	Opt. Sol.		Our Sol.	
	N_v	T_d	N_v	T_d		N_v	T_d	N_v	T_d		N_v	T_d	N_v	T_d
R201	8	1143.2	4	1263.3	C105	10	827.3	10	828.9	RC103	11	1258.0	11	1282.5
R202	NA	NA	4	1085.0	C106	10	827.3	10	828.9	RC104	NA	NA	10	1139.5
R203	NA	NA	3	942.7	C107	10	827.3	10	828.9	RC105	15	1513.7	14	1552.0
R204	NA	NA	3	755.7	C108	10	827.3	10	828.9	RC106	NA	NA	12	1376.3
R205	NA	NA	3	1049.3	C109	10	827.3	10	828.9	RC107	12	1207.8	11	1244.6
R206	NA	NA	3	927.7	C201	3	589.1	3	591.6	RC108	11	1114.2	11	1135.9
R207	NA	NA	3	812.8	C202	3	589.1	3	591.6	RC201	9	1261.8	4	1417.5
R208	NA	NA	2	727.7	C203	3	588.7	3	591.2	RC202	8	1092.3	4	1170.7
R209	NA	NA	3	933.6	C204	3	588.1	3	590.6	RC203	NA	NA	3	1068.9
R210	NA	NA	3	978.5	C205	3	586.4	3	588.9	RC204	NA	NA	3	798.5
R211	NA	NA	3	794.7	C206	3	586.0	3	588.5	RC205	7	1154.0	4	1375.6
C101	10	827.3	10	828.9	C207	3	585.8	3	588.3	RC206	NA	NA	3	1146.3
C102	10	827.3	10	828.9	C208	3	585.8	3	588.3	RC207	NA	NA	3	1073.4
C103	10	826.3	10	828.1	RC101	15	1619.8	15	1651.0	RC208	NA	NA	3	840.3
C104	10	822.9	10	824.8	RC102	14	1457.4	13	1508.3					

its clarity and the generation of sub-optimal results in real-time, unlike complicated heuristics whose execution time is significantly longer. Also, compared to other approximate methods in literature [8], our method yields better outcomes for almost all instances of all classes, both in terms of a smaller number of routes or vehicles and in the context of the total distance. Namely, according to the numerical results shown in Table 3 and Table 4 which optimize Solomon's 100-customer instances, as well outcomes published in the paper [8], our hybrid method generates better results for 17 instances of class R compared to the following algorithms S-PSO-VRPTW [8], HGA [20], OCGA [16] and ACO TS [23]. For the remaining six instances, such as R103, R104, R105, R108, R109 and R202, better results have been achieved by the algorithms such as HGA, ACO-TS, S-PSO-VRPTW, ACO-TS, ACO-TS and S-PSO- VRPTW, respectively. Regarding class C, our method calculated better or equal outcomes for all instances compared to the cited algorithms. The exception is instance C208, where the ACO-TS induces the same number of vehicles as our method but a slightly shorter overall distance. Finally, in the case of the RC class, the OCGA and S-PSO-VRPTW algorithms produce better outcomes than our method for instances RC108 and RC205 and the ACO-TS algorithm for instances RC101 and RC103. On the other hand, our method produces far superior outcomes for the remaining 12 instances out of 16 in total. Regarding the fetched numerical results by our approach, it is essential to highlight that for all instances of classes R, C and RC, the list of feasible routes whose sub-optimal results (N_v, T_d) exist in Table 3 and Table reftab4 can being downloaded from the link https://sites. google.com/view/vrptwaalihodzic. These routes do not violate the capacity and time window constraints. Based on what has been stated so far, it obeys that the presented results verify the quality of our proposed approach for solving the VRPTW in the case of almost all instances mentioned classes, and they are competitive with the best ones in the literature.

5   Conclusion

This article presents a hybrid approach based on the composition between the K-Means algorithm and the CPLEX Solver optimizer. In the first phase of our method, K-Means divides the customers into three clusters, while in the second phase, CPLEX Solver estimates approximate solutions of mentioned clusters in real-time in parallel. Additionally, the quality of outcomes was enhanced by swapping the closest clients inside neighbour clusters. The proposed method was tested on Solomon's 56 VRPTW and compared to four heuristics from the literature. Numerical simulations demonstrate that our method was superior for almost all instances compared to other heuristics. Also, results show that compared with other heuristics published in the literature, our approach is an effective tool for solving the VRPTW. Future work can include a better clustering method since it influences outcomes qualities.

References

1. Braca, J., Bramel, J., Posner, B., Simchi-Levi, D.: A computerized approach to the New York city school bus routing problem. IIE Trans. **29**(8), 693–702 (2007). https://doi.org/10.1080/07408179708966379
2. Bräysy, O., Gendreau, M.: Tabu search heuristics for the vehicle routing problem with time windows. TOP **10**(2), 211–237 (2002). https://doi.org/10.1007/BF02579017
3. Christofides, N., Mingozzi, A., Toth, P.: Exact algorithms for the vehicle routing problem, based on spanning tree and shortest path relaxations. Math. Program. **20**, 255–282 (1981). https://doi.org/10.1007/BF01589353
4. Clarke, G., Wright, J.: Scheduling of vehicle routing problem from a central depot to a number of delivery points. Oper. Res. **12**(4), 568–581 (1964). https://doi.org/10.1287/opre.12.4.568
5. CPLEX: IBM ILOG CPLEX Optimization Studio CPLEX User's Manual V 12.7. International Business Machines Corporation (2017). https://www.ibm.com/support/knowledgecenter/SSSA5P_12.7.1/ilog.odms.studio.help/pdf/usrcplex.pdf
6. Dantzig, G.B., Ramser, J.H.: The truck dispatching problem. Management Science **6**(1), 80–91 (October 1959). https://pubsonline.informs.org/doi/10.1287/mnsc.6.1.80
7. Golden, B.L., Wasil, E.A.: Or practice-computerized vehicle routing in the soft drink industry. Oper. Res. **35**(1), 6–17 (1981). https://doi.org/10.1287/opre.35.1.6
8. Gong, Y.J., Zhang, J., Liu, O., Huang, R.Z., Chung, H.S.H., Shi, Y.H.: Optimizing the vehicle routing problem with time windows: a discrete particle swarm optimization approach. IEEE Trans. Syst. Man Cybernetics, Part C (Applications and Reviews) **42**(2), 254–267 (2012). https://doi.org/10.1109/TSMCC.2011.2148712

9. Kallehauge, B., Larsen, J., Madsen, O.B., Solomon, M.M.: Vehicle Routing Problem with Time Windows, pp. 67–98. Springer, US (2005). https://doi.org/10.1007/0-387-25486-2_3

10. Laporte, G.: The vehicle routing problem: an overview of exact and approximate algorithms. Eur. J. Oper. Res. **59**(3), 345–358 (1992). https://doi.org/10.1016/0377-2217(92)90192-C

11. Laporte, G., Toth, P., Vigo, D.: Vehicle routing: historical perspective and recent contributions. EURO J. Transp. Logistics **2**, 1–4 (2013). https://doi.org/10.1007/s13676-013-0020-6

12. Louveaux, F., Lambert, V., Laporte, G.: Designing money collection routes through bank branches. Comput. Oper. Res. **20**(7), 783–791 (1993). https://ur.booksc.me/book/8574920/29e58f

13. Mechti, R., Poujade, S., Roucairol, C., Lemarié, B.: Global and Local Moves in Tabu Search: A Real-Life Mail Collecting Application, pp. 155–174. Springer, US (1999). https://doi.org/10.1007/978-1-4615-5775-3_11

14. Meng-yu, X.: Parallel particle swarm optimization algorithm for vehicle routing problems with time windows. Comput. Eng. Appl. **43**(14), 223–226 (2007)

15. Mor, A., Speranza, M.G.: Vehicle routing problems over time: a survey. 4OR - A Quarterly J. Oper. Res. **18**(2), 129–149 (June 2020). https://doi.org/10.1007/s10288-020-00433-2

16. Nazif, H., Lee, L.S.: Optimised crossover genetic algorithm for capacitated vehicle routing problem. Appl. Math. Model. **36**(5), 2110–2117 (2012). https://doi.org/10.1016/j.apm.2011.08.010

17. Russell, R.A.: Hybrid heuristics for the vehicle routing problem with time windows. Transp. Sci. **29**(2), 156–166 (1995). https://doi.org/10.1287/trsc.29.2.156

18. Savelsbergh, M.W.P.: Local search in routing problems with time windows. Ann. Oper. Res. **4**(1), 285–305 (1985). https://doi.org/10.1007/BF02022044

19. Solomon, M.M.: Algorithms for the vehicle routing and scheduling problems with time window constraints. Oper. Res. **32**(2), 255–265 (1987). https://doi.org/10.1287/opre.35.2.254

20. Tan, K., Lee, L., Ou, K.: Artificial intelligence heuristics in solving vehicle routing problems with time window constraints. Eng. Appl. Artif. Intell. **14**(6), 825–837 (2001). https://doi.org/10.1016/S0952-1976(02)00011-8

21. Toth, P., Vigo, D.: The Vehicle Routing Problem. Society for Industrial and Applied Mathematics, SIAM (2002)

22. Tuba, E., Jovanovic, R., Hrosik, R.C., Alihodzic, A., Tuba, M.: Web intelligence data clustering by bare bone fireworks algorithm combined with k-means. In: Proceedings of the 8th International Conference on Web Intelligence, Mining and Semantics. WIMS 2018. Association for Computing Machinery (2018). https://doi.org/10.1145/3227609.3227650

23. Yu, B., Yang, Z.Z., Yao, B.Z.: A hybrid algorithm for vehicle routing problem with time windows **38**(1), 435–441 (2011). https://doi.org/10.1016/j.eswa.2010.06.082

Topological Dynamic Consistency of Numerical Methods for Dynamical Models in Biosciences

Roumen Anguelov[1,2]([⊠]) [iD]

[1] Department of Mathematics and Applied Mathematics, University of Pretoria,
Pretoria 0028, South Africa
[2] Institute of Mathematics and Informatics, Bulgarian Academy of Sciences,
Sofia 1113, Bulgaria
roumen.anguelov@up.ac.za

Abstract. Qualitative analysis of dynamical systems modelling biological processes makes a valuable contribution to understanding these processes and elucidating relevant properties. These include, invariant sets (including equilibria) and their stability properties, existence of bifurcation thresholds, shape of trajectories, etc. Due to the high level of complexity of the biological systems, the models often involve large number of parameters with unknown or only estimated values. In such setting, qualitative analysis tends to be more important than high precision computations. Hence, the importance of numerical methods which accurately replicate the properties of the model. There are several attempts to capture this property of numerical methods in concepts like qualitative stability, dynamic consistency, structural stability under numerics.

In this paper we present the concept of topological dynamic consistency of numerical methods, which aligns the model and the method in such a way that all properties of the model which are of topological nature, like the mentioned above, are replicated. Further, we present some practical approaches of designing such methods. Let us remark that topological dynamic consistency cannot be derived through the standard tools of numerical analysis based on consistency, stability, order of approximation and convergence. The theory is based on the concepts topological conjugacy of maps, topological equivalence of flows as well as the structural stability of flows and maps.

Keywords: Topological dynamics · Qualitative analysis of numerical methods

1 Introduction

The main goal of this paper is characterising the alignment of properties of discrete and continuous dynamical system modelling the same process. The common

Supported by DST/NRF SARChI Chair in Mathematical Models and Methods in Bioengineering and Biosciences at University of Pretoria.

setting is that the model is constructed as a set of differential equations, that is a continuous dynamical system, with the numerical simulations conducted using a numerical method, that is a discrete dynamical system. Replicating essential properties of the continuous model by the numerical method is particularly relevant to biosciences. These properties include the biologically feasible domain, existence and persistence of the relevant species, periodicity, invariant sets, bifurcation, etc. Considering that, due to the underlying complexity of the biological processes the models typically involve many parameters of unknown or roughly estimated values, any numerical predictions are unlikely to be very reliable, irrespective of the order of accuracy of the numerical method. However, the qualitative accuracy of the simulations provided by the numerical method is always extremely valuable both (i) for better understanding of the modelled processes and (ii) for practical applications, e.g. if the existence of a certain threshold is established, its actual value can be calculated in alternative ways theoretically or experimentally.

The interest in numerical methods preserving specific properties of the continuous model is not new. For example, conservation of Hamiltonian, symplectic systems/volume preservation, reversibility of the flows [12], mimetic discretisation [7]. There has been specifically such interest from in the field of Nonstandard Finite Difference Methods since replicating qualitatively essential properties of the original model is a primary goal of these methods, [19–23]. It was captured in the concept of qualitative stability, [3]. In the particular setting of dynamical systems, the concept of dynamic consistency is widely used to depict the qualitative links between the dynamical system and its numerical method, [24]. Other concepts, e.g. asymptotic stability, [11], have been proposed as well. The concept of elementary stability [21] (preserving equilibria and their asymptotic properties) reflects the basic expectation for a numerical method for continuous dynamical system. Quite often works using these concepts are linked to modelling of biological process, e.g. [1,2,8,9,11].

In the paper [4] an attempt was made to give a definition of dynamic consistency in more precise terms using the topological conjugacy of diffeomorphisms. Considering that the properties of interest in a dynamical system are in essence topological properties, the concept of topological dynamic consistency given in [4] indeed captures well the expectations from qualitative point of view that a numerical method should satisfy. The definition in [4] was given under some restrictive assumptions regarding the continuous model. In this paper we extend the definition of topological dynamic consistency so that it is applicable to a wider class of continuous dynamical systems. The main aim of this paper is to develop appropriate tools and techniques (i) for practical application of this definition to characterise existing numerical methods for dynamical systems as topologically dynamically consistent or not and (ii) for constructing numerical schemes which are topologically dynamically consistent. This work builds on the existence results of topological conjugacy between the flow of a dynamical system and its numerical discretization in works of Garay [10], Li [15–18], Bielecki [6]. We develop a general approach on how the conjugacy, which is established in the mentioned works for sufficiently small step-size of the numerical method, can be extended to a step-size in a given range, thus providing for topological dynamic consistency of the method. The case of gradient systems and the case of dynamical systems admitting a Lyapunov function are specifically discussed.

2 Defining Topological Dynamic Consistency

We consider dynamical systems defined on a set $\Omega \subset \mathbb{R}^n$ via a system of differential equations

$$\dot{x} = f(x). \tag{1}$$

Since our interest in a qualitative link with numerical methods for initial value problems, we consider dynamical systems defined for positive time, referred to as a *forward* dynamical systems. Let (1) define a forward dynamical system on Ω and let $\Phi^t(y)$ denote the solution of (1) with initial condition $x(0) = y$. The operator $\Phi : \mathbb{R} \times \Omega \to \Omega$ is called a semi-flow. Under some mild assumptions on f, which we assume henceforth, the operator Φ is C^1 together with its inverse. The set

$$\{\Phi^t : t \in [0, +\infty)\} \tag{2}$$

is the well known semi-group associated with (1). Recall that the set (2) is a semi-group with respect to composition and we have

$$\Phi^t \cdot \Phi^s = \Phi^{t+s}. \tag{3}$$

A map $F : [0, \overline{h} \times \Omega \to \Omega]$ is called a numerical method of order p for the flow Φ if there exists a constant K such that

$$\|\phi^h(x) - F_h(x)\| < Kh^{p+1}, \; x \in \Omega, \; h \in [0, \overline{h}). \tag{4}$$

The aim of this section is to develop a concept aligning the properties of the set of maps of the numerical method

$$\{F_h : h \in [0, \overline{h})\} \tag{5}$$

and the set of maps (2). The maps Φ^t are all diffeomorphism from Ω to its respective image. Therefore, as also shown in [4], a basic requirement for F_h is to be a diffeomorphism from Ω to $F_h(\Omega)$.

An essential difference between the sets (2) and (5) is that (2) is a semi-group, while (5) is not. More precisely, any composition of operators in (2) is in (2), while the set (5) has no such property. For example, $(F_h)^n$, $n > 1$, does not necessarily belong to the set (5). This difference is reflected in the different definitions for topological equivalence of maps and flows, which we recall below.

Definition 1. *Let $P : \Omega \to P(\Omega) \subseteq \Omega$ and $Q : \Omega \to Q(\Omega) \subset \Omega$ be diffeomorphisms. The diffeomorphisms P and Q are called topologically conjugate (or equivalent) if there exists a homeomeorphism $H : \Omega \to \Omega$ such that*

$$H \cdot P(x) = Q \cdot H(x), \; x \in \Omega. \tag{6}$$

Let $\mathcal{B}$ denote the set of continuous increasing bijections of $[0, +\infty)$ onto itself.

Definition 2. *Let Φ and Ψ be C^1 semi-flows defined on Ω. Φ and Ψ are called topologically equivalent is there exist a homeomorphism $H : \Omega \to \Omega$ and a continuous map $\theta : \Omega \to \mathcal{B}$ such that*

$$H \cdot \Phi^{\theta(x)(t)}(x) = \Psi \cdot H(x), \ t \in [0, +\infty), \ x \in \Omega. \tag{7}$$

The to definitions are illustrated on Fig. 1.

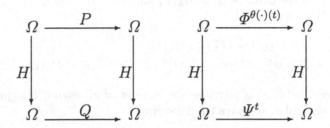

Fig. 1. Conjugacy of maps vs. equivalence of flows.

The conjugacy H in (6) provides equivalence of the actions of P and Q at every point of Ω, that is if $x \overset{P}{\longrightarrow} y$ than $H(x) \overset{Q}{\longrightarrow} H(y)$. In the case of the conjugacy (7) the equivalence is not for the action at a point but rather for orbits with their direction, where the function $\theta(x)$ is essentially a re-scaling of time. Indeed, let $\mathcal{O}_\Phi(x)$ be the orbit of $x \in \Omega$ under the flow Φ, that is

$$\mathcal{O}_\Phi(x) = \{\Phi^t(x) : t \in [0, +\infty)\}.$$

Then using the conjugacy (7) and that $\theta(x)$ increases from 0 to $+\infty$ we have

$$\mathcal{O}_\Psi(H(x)) = \{\Psi^t(H(x)) : t \in [0, +\infty)\} = \{H \cdot \Phi^{\theta(x)(t)}(x) : t \in [0, +\infty)\}$$
$$= H(\{\Phi^t(x) : t \in [0, +\infty)\}) = H(\mathcal{O}_\Phi(x))$$

Taking into account the stated differences, we consider the following definition for qualitative alignment of (2) and (5).

Definition 3. *Given $h \in (0, \overline{h}]$, the numerical method F_h for the flow Φ is said to be topologically dynamically consistent with Φ if there exists a homeomorphism $H_h : \Omega \to \Omega$, $\varepsilon_h > 0$ and a continuous function $\tau_h : \Omega \to [\varepsilon_h, +\infty)$ such that*

$$H_h \cdot \Phi^{\tau_h(x)} = F_h \cdot H_h(x), \ x \in \Omega. \tag{8}$$

Let F_h be topologically dynamically consistent with Φ for $h \in (0, \overline{h}]$. Then we have on Ω the discrete dynamical system

$$y_{k+1} = F_h(y_k). \tag{9}$$

The orbit of $y \in \Omega$ is the set

$$\mathcal{O}_{F_h} = \{(F_h)^m(y) : m = 0, 1, 2, ...\}$$

The conjugacy (8) is an embedding of the discrete dynamical system (9) into the flow Φ defined via (1). More precisely, for any $x \in \Omega$ we have

$$\mathcal{O}_{F_h}(H_h(x)) \subseteq H_h(\mathcal{O}_\Phi(x)),$$

where the sense of direction is also preserved. Indeed, let $x \in \Omega$, $\mathcal{O}_{F_h}(y_0) = \{y_0, y_1, y_2, ...\}$ be the orbit of $y_0 = H_h(x)$ and $x_k = H_h^{-1}(y_k)$, $k = 1, 2, ...$ Then for any $m \in \mathbb{N}$ we have

$$y_m = (F_h)^m(y_0) = (F_h)^m \cdot H_h(x)$$
$$= H_h \cdot \Phi^{\tau_h(x) + \tau_h(x_1) + \tau_h(x_2) + ... + \tau_h(x_{m-1})}(x) \in H_h(\mathcal{O}_\Phi(x)).$$

The mentioned embedding provides for transfer of all asymptotic properties of Φ to F_h. In particular, we have the following

- If x is a fixed point of Φ then $H_h(x)$ is a fixed point of F_h.
- If $\lim_{t \to +\infty} \Phi^t(x) = z$ then $\lim_{m \to +\infty} (F_h)^m(H_h(x)) = H(z)$.
- If $I \subseteq \Omega$ is a positively invariant set of Φ then $H_h(I)$ is a positively invariant set of F_h.
- If $\mathcal{A} \subseteq \Omega$ is an absorbing set of Φ with basin of attraction W, then $H_h(\mathcal{A})$ is an absorbing set of F_h with basin of attraction $H_h(W)$.

$$(10)$$

The inverse implications are not always true. For example, let the flow Φ have a periodic orbit $\mathcal{P}$ with period T. Consider the numerical method $F_h = \Phi^h$. This method, referred to as the *exact scheme*, is clearly topologically dynamically consistent with conjugacy $H_h(x) = x$. Now we may note that any point $x \in \mathcal{P}$ is a fixed point of F_T, while x is not a fixed point of Φ. In the sequel we will consider conditions which provide for the inverse implications in (10) to hold as well.

3   Structural Stability and Topological Dynamic Consistency

The well-posedness according to D'Alambert of a mathematical problem requires that small changes of the data lead to small changes in the solution. In our setting the primary interest is preserving the topological properties of a continuous dynamical system. Hence, in the spirit of the D'Alambert well-posedness, it is reasonable to require that the properties to be replicated to do not change significantly with small change in the flow. If we consider that a change in properties between flows is "insignificant" when they are topologically equivalent, we are lead to the concept of structural stability.

Definition 4. *[13, Definition 2.3.6] A C^1 flow Φ is C^1 structurally stable if any C^1 flow sufficiently close to Φ in the C^1 topology is topologically equivalent to Φ.*

Therefore, in the sequel we consider the construction of topologically dynamically consistent numerical methods for C^1 structurally stable flows.

The following theorem restates the result in [16, Theorem 2] in the terminology used of this paper

Theorem 1. *Let M be a smooth manifold and Ω be a compact subset of M with closure(interior(Ω)) $= \Omega$. For $p \geq 2$, let Φ be a C^{p+1} flow on M which satisfies*

(i) *the closed orbits are dense in the nonwandering set of Φ and this set has a hyperbolic structure;*
(ii) *at every point the stable and the unstable manifolds intersect at a positive angle.*

Let further the restriction of the nonwandering set of Φ to Ω be in the interior of Ω. Then any numerical method F_h for Φ of order p is topologically dynamically consistent with Φ for all sufficiently small h. Moreover, we have that as $h \to 0$ the homeomorphism H_h in (8) converges to identity and $\frac{\tau_h}{h}$ converges to 1, both at a rate of $O(h^p)$.

Remark 1. Condition (i) in Theorem 1 is usually called Axiom A, while condition (ii) is referred to as strong transversality. These conditions are applicable both to flows and to maps. Axiom A and strong transversality are sufficient conditions for structural stability of diffeomorphisms, [25, Theorem 1.1]. The two conditions, where Axiom A refers to the chain-recurrent set of Φ^t rather than the nonwandering set, are sufficient conditions for the flow Φ^t to be structurally stable [26, Theorem A]. In the context of the mentioned qualitative well-posedness, where we expect that Φ^t is structurally stable, conditions (i) and (ii) in Theorem 1 are not surprising.

The following theorem extends Theorem 1 by providing means to quantify the concept of h being sufficiently small.

Theorem 2. *Let Φ satisfy the conditions of Theorem 1 and let F_h be a numerical method for Φ of order $p \geq 2$. If the maps F_h are structurally stable diffeomorfisms on Ω for all $h \in (0, \overline{h}]$, then F_h is topologically dynamically consistent with Φ for all $h \in (0, \overline{h}]$.*

Proof. It follows from Theorem 1 that there exists $\tilde{h}$ such that F_h is topologically dynamically consistent with Φ for all $h \in (0, \tilde{h}]$. Considering that topological dynamic consistency is defined via a conjugacy of the form (8), every diffeomorfism F_h, which is topologically conjugate to $F_{\tilde{h}}$ is topologically dynamically consistent with Φ. The topological conjugacy is an equivalence relation in the set of diffeomorphisms. Then it is enough to show that $\{F_h : h \in [\tilde{h}, \overline{h}]\}$ all belong to the same equivalence class.

Using the given structural stability of F_h, for every $h \in (0, \overline{h}]$ there exists $\delta_h > 0$, such that the maps F_θ, $\theta \in (h - \delta_h, h + \delta_h)$, are topologically conjugate to F_h and, therefore, to each other as well. Considering that $\{(h - \delta_h, h + \delta_h) : h \in (0, \overline{h}]\}$ is an open cover of the compact interval $[\tilde{h}, \overline{h}]$, there exists a finite cover $\{(h_i - \delta_{h_i}, h_i + \delta_{h_i}) : i = 1, ..., k\}$, where without loss of generality we may assume that $h_1 < h_2 < ... < h_k$. Since every two consecutive intervals have a nonempty intersection, the maps in $\{F_h : h \in [\tilde{h}, \overline{h}]\}$ are all conjugate to each other, which completes the proof. $\qquad\qquad\square$

The conditions (i) and (ii) in Theorem 1 are restrictive and also often not easy to establish. Some well known flows satisfying these conditions:

- Gradient flows if the nonwandering set consists of finite number of hyperbolic fixed points and where the transversality condition holds
- Morse-Smale flows [27, Section 8.12]
- In 2D: Flows having finitely many fixed points and periodic orbits, all hyperbolic, and no orbits connecting saddle points, [14, Theorem 2.5].

4   Some Applications of Theorem 2

4.1   Gradient Flows

The flow Φ defined via (1) is called a gradient flow of there exist a map $G : \Omega \to \mathbb{R}$ such that

$$f = \nabla G. \tag{11}$$

Let the flow defined by (1) be a gradient flow with finite number of fixed points and let the transversality condition hold.

Theorem 3. *If a numerical method F_h, $h \in [0, \overline{h}]$ is such that for any $y \in \Omega$ which is not a fixed point of Φ we have*

$$f(F_h(y)) \cdot \frac{dF_h(y)}{dh} > 0, \ h \in [0, \overline{h}], \tag{12}$$

then F_h is topologically dynamically consistent with Φ for all $h \in (0, \overline{h}]$.

Proof. Considering Theorem 2, it is sufficient to show that the diffeomorphism F_h is structurally stable for all $h \in (0, \overline{h}]$.

Let $y \in \Omega$. Consider the function $\varphi(h) = G(F_h(y))$. We have

$$\frac{d\varphi(h)}{dh} = \nabla G(F_h(y)) \cdot \frac{\partial F_h(y)}{\partial h} = f(F_h(y)) \cdot \frac{dF_h(y)}{dh} > 0.$$

Therefore φ is strictly increasing. In particular, this means that

$$G(F_h(y)) > G(F_0(y)) = G(y). \tag{13}$$

Now, we will show that the nonwandering set consist only of the critical points of G. Let y belong to the nonwandering set of F_h. Assume that y is not a fixed point of Φ or equivalently, not a critical point of G. Denote $\delta = G(F_h(y)) - G(y)$. It follows from (13) that $\delta > 0$. Consider the neighborhood of y defined as

$$U = \{x \in \Omega : |G(x) - G(y)| < \frac{\delta}{2} \text{ and } |G(F_h(x)) - G(F_h(y))| < \frac{\delta}{2}\}.$$

Using the defining property of a nonwandering set, there exists $z \in U$ and $m \in \mathbb{N}$ such that $(F_h)^m(z) \in U$. Then we have

$$G(y) + \frac{\delta}{2} > G(F_h)^m(z) \geq G(F_h(z)) \geq G(F_h(y)) - \frac{\delta}{2} = G(y) + \delta - \frac{\delta}{2} = G(y) + \frac{\delta}{2}$$

which is a contradiction. Therefore, every point of the nonwandering set is a fixed point of Φ or, equivalently, a critical point of G. Since the behavior of the orbits around the critical points G are governed by ∇G, the transversality conditions holds as well. Then the required structural stability of F_h follows from [25] (see Remark 1). □

Remark 2. The inequality (12) trivially hold when $h = 0$ as we have

$$f(F_h(y)) \cdot \frac{dF_h(y)}{dh}\bigg|_{h=0} = f(y) \cdot f(y) = \|f(y)\|^2 > 0$$

provided $f(y) \neq 0$. Therefore, $\overline{h} > 0$ with the properties required in Theorem 3 does exist. The inequality (12) provides a practical approach of finding it.

4.2 Flows with a Lyapunov Function

Here we assume that the system (1) admits a Lyapunov function L with the properties

(i) $\dot{L}(x) = \nabla L(x) \cdot f(x) \leq 0$, $x \in \Omega$;
(ii) The set $E = \{x \in \Omega : \dot{L}(x) = 0\}$ consist of fixed points only.

Theorem 4. *If a numerical method F_h, $h \in [0, \overline{h}]$ is such that for any $y \in \Omega \setminus E$ we have*

$$\nabla L(F_h(y)) \cdot \frac{dF_h(y)}{dh} > 0, \ h \in [0, \overline{h}], \tag{14}$$

then F_h is topologically dynamically consistent with Φ for all $h \in (0, \overline{h}]$.

The proof is done in the same way as the proof of Theorem 3 using $\varphi(h) = L(F_h(y))$. Further, similarly to Remark 2, the inequality in the Theorem 4 trivially hold when $h = 0$ as we have

$$\nabla L(F_h(y)).\frac{dF_h(y)}{dh}\bigg|_{h=0} = \nabla L(y).f(y) < 0.$$

Hence, $\overline{h}$ exists and a value for it can be computed by using (14).

5 Conclusion

In this paper we introduced the concept of topological dynamic consistency and we showed that it is relevant for characterising the alignment of properties of continuous and discrete dynamical systems. For numerical methods this concept

- gives in precise terms how the properties of the continuous dynamical systems are replicated in its discretization
- elucidates the qualitative link between systems of ODEs and their numerical methods
- highlights the importance of structural stability of both the flow and its numerics.

It is clear from the technicalities involved in all results, that proving structural stability of the flow and of the numerical method as well as the topological dynamic consistency between the two is usually not easy. We showed for some specific cases how the issue can be dealt with. Future research will be aimed at developing new tools for establishing structural stability and topological dynamic consistency. One possible approach, which was pioneered in [5] is to consider structural stability only in a relevant to the specific model subspace of C^1.

References

1. Al-Kahby, H., Dannan, F., Elaydi, S.: Non-standard discretization methods for some biological models. In: Mickens, R.E. (ed.) Applications of nonstandard finite difference schemes, pp. 155–180. World Scientific, Singapore (2000)
2. Anguelov, R., Dumont, Y., Lubuma, J., Shillor, M.: Dynamically consistent non-standard finite difference schemes for epidemiological models. J. Comput. Appl. Math. **255**, 161–182 (2014)
3. Anguelov, R., Lubuma, J.: Contributions to the mathematics of the nonstandard finite difference method and applications. Numerical Methods for Partial Differential Equations **17**, 518–543 (2001)
4. Anguelov, R., Lubuma, J., Shillor, M.: Topological dynamic consistency of non-standard finite difference schemes for dynamical systems. J. Differ. Equations Appl. **17**(12), 1769–1791 (2011)
5. Anguelov, R., Popova, E.: Topological structure preserving numerical simulations of dynamical models. J. Comput. Appl. Math. **235**(2010), 358–365 (2010)
6. Bielecki, A.: Topological conjugacy of discrete time-map and Euler discrete dynamical systems generated by a gradient flow on a two-dimensional compact manifold. Nonlinear Anal. **51**, 1293–1317 (2002)
7. Bochev, P.B., Hyman, J.M.: Principles of mimetic discretizations of differential operators. In: Arnold, D.N., et al. (eds.) Compatible Spatial Discretizations, pp. 89–120. Springer, New York (2006). https://doi.org/10.1007/0-387-38034-5_5
8. Dimitrov, D.T., Kojouharov, H.V., Chen-Charpentier, B.M.: Reliable finite difference schemes with applications in mathematical biology. In: Mickens, R.E. (ed.) Advances in the applications of nonstandard finite difference schemes, pp. 249–285. World Scientific, Singapore (2005)

9. Dimitrov, D.T., Kojouharov, H.V.: Positive and elementary stable nonstandard numerical methods with applications to predator-prey models. J. Comput. Appl. Math. **189**, 98–108 (2006)
10. Garay, B.M.: On structural stability of ordinary differential equations with respect to discretization methods. Numerical Math. **72**, 449–479 (1996)
11. Gumel, A.B., Patidar, K.C., Spiteri, R.J.: Asymptotically consistent nonstandard finite difference methods for solving mathematical models arrising in population biology. In: Mickens, R.E. (ed.) Advances in the applications of nonstandard finite difference schemes, pp. 513–560. World Scientific, Singapore (2005)
12. Hairer, E., Lubich, C., Wanner, G.: Geometric Numerical Integration: Structure-Preserving Algorithms for Ordinary Differential Equations. Springer, Heidelberg (2002). https://doi.org/10.1007/3-540-30666-8
13. Katok, A., Hasselblatt, B.: Introduction to the Modern Theory of Dynamical Systems. Cambridge University Press, New York (1995)
14. Kuznetsov, Y.A.: Elements of Applied Bifurcation Theory. Springer, Heidelberg (2004). https://doi.org/10.1007/978-1-4757-3978-7
15. Li, M.-C.: Structural stability of Morse-Smale gradient like flows under discretizations. SIAM J. Math. Anal. **28**, 381–388 (1997)
16. Li, M.-C.: Structural stability of flows under numerics. J. Differential Equations **141**, 1–12 (1997)
17. Li, M.-C.: Structural stability for the Euler method. SIAM J. Math. Anal. **30**, 747–755 (1999)
18. Li, M.-C.: Qualitative property between flows and numerical methods. Nonlinear Anal. **59**, 771–787 (2004)
19. Mickens, R.E.: Nonstandard finite difference models of differential equations. World Scientific, Singapore (1994)
20. Mickens, R.E.: Nonstandard finite difference schemes: a status report. In: Teng, Y.C., Shang, E.C., Pao, Y.H., Schultz, M.H., Pierce, A.D. (eds.) Theoretical and Computational Accoustics 97, pp. 419–428. World Scientific, Singapore (1999)
21. Mickens, R.E.: Nonstandard finite difference schemes: methodology and applications. World Scientific, Singapore (2021)
22. Mickens, R.E.: Nonstandard finite difference schemes for differential equations. J. Differential Equations Appl. **8**, 823–847 (2002)
23. Mickens, R.E.: Nonstandard finite difference methods. In: Mickens, R.E. (ed.) Advances in the applications of nonstandard finite difference schemes, pp. 1–9. World Scientific, Singapore (2005)
24. Mickens, R.E.: Dynamic consistency: a fundamental principle for constructing nonstandard finite difference schemes for differential equations. J. Differ. Equations Appl. **11**, 645–653 (2005)
25. Robinson, C.: Structural stability of C^1 diffeomorphisms. J. Differential Equations **22**, 238–265 (1976)
26. Robinson, C.: Structural stability on manifolds with boundary. J. Differential Equations **37**, 1–11 (1980)
27. Robinson, C.: Dynamical Systems: Stability Symbolic Dynamics and Chaos, 2nd edn. CRC Press, Boca Raton (1999)

Parameter Estimation Inspired by Temperature Measurements for a Chemotactic Model of Honeybee Thermoregulation

Atanas Z. Atanasov$^{(\boxtimes)}$, Miglena N. Koleva, and Lubin G. Vulkov

University of Ruse, 8 Studentska Street, 7017 Ruse, Bulgaria
{aatanasov,mkoleva,lvalkov}@uni-ruse.bg

Abstract. The key for the survival of honeybees in winter is in the generation and preservation of heat. A successful study of this process is the modeling based on generalized Keller-Segel model, proposed by R.Bastaansen et al., 2020. The problem is in the form of coupled system of two parabolic equations for the temperature and bee density. The model parameters control the particular population dynamics in the hive. Our goal is to predict the optimal parameters based only on measurements of the temperature with three sensors. We perform the study on two stages. First, we solve an unknown reaction coefficient problem to determine the temperature and density. Then, we solve the next inverse problem for estimation of the parameters in the other parabolic equation, using as measured data the density, obtained by the first inverse problem. Results from numerical experiments are presented.

1 Introduction and Model Problem

In recent years, one of the problems in the worldwide is the decline, even extinction in some regions, of honey bee populations. Some of the major contributing factors are extensive farming and pesticides, pests and diseases, etc. In a number of cases, the cause of the bee mortality has not been identified. Hence, investigating and studying the dynamics of honey bee colonies is of great importance to tackle this problem.

The winter period is critical for bee colonies. The death and weakening of honey bee colonies is because of the cold, starvation and depleting their protein-fat reserves, parasitic mites and so on. During the winter, the bees feed on the stored reserves and remain active, taking care of maintaining the temperature in the hive. They form thermoregulation clusters with higher temperature in the center [4, 10, 20]. There is still a lot the beekeepers do not know about the behavior of the bees in the winter. More research is yet to be done.

In this work we consider coupled reaction-diffusion-transport equations, describing self-thermoregulation process in the bee colony. The model is generalized Keller-Segel problem, introduced in [4], for the bee density $\rho(x,t) \geq 0$

© The Author(s), under exclusive license to Springer Nature Switzerland AG 2023
I. Georgiev et al. (Eds.): NMA 2022, LNCS 13858, pp. 36–47, 2023.
https://doi.org/10.1007/978-3-031-32412-3_4

and the local temperature $T(x,t)$, $(x,t) \in [0,L] \times [0,t_f]$, where $L > 0$ is the edge of the honey bee colony and t_f is the final time

$$\frac{\partial T}{\partial t} = \frac{\partial^2 T}{\partial x^2} + f(T)\rho, \tag{1}$$

$$\frac{\partial \rho}{\partial t} = \frac{\partial^2 \rho}{\partial x^2} - \frac{\partial}{\partial x}\left(\chi(T)\rho\frac{\partial T}{\partial x}\right) - \theta(\rho,T)\rho, \tag{2}$$

$$\frac{\partial T}{\partial x}(0,t) = 0; \quad T(L,t) = T_a < T_\chi, \tag{3}$$

$$\frac{\partial \rho}{\partial x}(0,t) = 0; \quad \left(\frac{\partial \rho}{\partial x} - \chi(T)\rho\frac{\partial T}{\partial x}\right)(L,t) = 0, \tag{4}$$

$$\rho(x,0) = \rho^0(x), \tag{5}$$

$$T(x,0) = T^0(x). \tag{6}$$

Here the function $f(T)\rho$ describes the heat generation by bees and $\chi(T)$ is sign-changing function at T_χ, namely when $T < T_\chi$, bees move towards a higher temperature and when $T > T_\chi$ they move away to lower temperature. Thus the chemotactic coefficient $\chi(T)$ dictate very different dynamics of the problem (1)–(6), compared to the well-studied in the literature generalized Keller-Segel model, where χ has a fixed positive sign.

For the completeness, we give some details and assumptions for the model derivation in [4]. Authors study the dynamic of the winter bee colony, simplifying the model proposed in [20], where the boundary of the honey bee colony is not fixed and the functions f, χ are more complicated, but on the other side they take into account the individual mortality bees rate $\theta(\rho,T) > 0$. The problem (1)- (6) better represents bee colonies with higher mortality.

In [4], the function χ is defined by step-function, since χ switches signs from $\chi(T) > 0$ for small T to $\chi(T) < 0$ for large T. Similarly, given the data in [20], f is defined as a step-function as well

$$f(T) = \begin{cases} f_{\text{low}}, & T < T_f, \\ f_{\text{high}}, & T > T_f, \end{cases} \qquad \chi(T) = \begin{cases} +\chi_1, & T < T_\chi, \\ -\chi_2, & T > T_\chi, \end{cases} \tag{7}$$

Here T_f is the temperature where f changes value and T_χ is the temperature where $\chi(T)$ changes sign ($T_f < T_\chi$), f_{low}, f_{high}, χ_1, $\chi_2 > 0$. The temperature T_χ can be considered the preferred temperature, as bees prefer to move to locations with this temperature.

The bees mortality rate is modeled by the function

$$\theta(T, \rho) = \theta_0 \theta_T(T)\theta_D(\rho)\theta_M(\rho). \tag{8}$$

Here θ_0 is a constant that must fit to the observations, and each multiplier in (8) represent the influence of different factors:

(i) θ_T - effect of the local temperature;
(ii) θ_D - effective refresh rates of heat generating bees;
(iii) θ_M - presence of parasitic mites in the colony.

The local temperature effect indicates that mortality does not increase when the local temperature is above a certain thresholds, $T_\theta > T_a$. If he temperature is too low $(T(x) < T_\theta)$, a bee in that location has to work too hard to generate heat, reducing her lifespan. This effect modeled by the step-function

$$\theta_T(T) = \begin{cases} 1, \text{ if } T < T_\theta, \\ 0, \text{ if } T \geq T_\theta. \end{cases} \tag{9}$$

The second effect arises from the renewal rate of the recovered bees. It is represented by the function

$$\theta_D(\rho) = \frac{\rho}{(\rho_{\text{tot}})^\gamma}, \quad \gamma > 0, \quad \rho_{\text{tot}} = \int_0^L \rho(x)dx, \tag{10}$$

where ρ_{tot} is the colony size and $\gamma > 0$ is some unknown exponent. To heat up the colony, each bee can generate heat by quivering its flight muscles, but only for about 30 min, after which it must recover and replenish its reserves by consuming honey [19]. The bees rotate from the periphery and inside the colony, to a warmer location, so that the recovered bees can take over the duty of warming the colony, and the already exhausted bees can replenish themselves [18]. Therefore, the larger the colony, the longer the recovery time and this affects the mortality rate.

The third effect is connected with the presence of the parasitic mite Varroa destructor in honey bee colonies, who reduces the body weight and protein content of individual bees and shorten their life span [8,15]. Thus, the increasing of the amount of mites per bee increasing the bee mortality as well. Moreover, this fraction increases as the colony size decreases, since mites may jump to another bees when their host bee dies. Let m is the amount of mites in the colony. This effect is modeled as

$$\theta_M(\rho) = 1 + \frac{m}{\rho_{\text{tot}}}. \tag{11}$$

In our previous work [3], we develop positivity preserving second-order numerical method for solving the problem (1)–(11). In [2] we solve inverse problem to find numerically the temperature and density, fitted to the temperature measurements.

In this study, we extend the results in [2] to find numerically the local temperature, bee density and parameters γ, θ_0 fitted to the temperature measurements. In contrast to the bee density, the temperature in the bee hive can be measured. The parameters γ and θ_0 have to be recovered on the base of the observations as well [4]. This motivates our research.

We formulate two inverse problems (IP). First, we solve a coefficient IP1 for (1), (3), (6) to find the temperature and density, matched to the temperature measurements. This problem is used in the first stage of the algorithm and it is proposed in [2]. Next, using the results of IP1 we formulate inverse problem (IP2), based on the linearized convection-diffusion equation for estimation of two important parameters γ and θ_0, that are difficult for observation.

The layout of the current research is arranged as follows. In the next section, the inverse problems are formulated and methodology, involving minimization of least squares cost functionals, using adjoint equations is presented. In Sect. 3, we develop algorithm for numerical study of the inverse problems. Finally, numerical examples are presented to show the efficiency of the proposed approach.

2   Inverse Problems

In this section, we present two inverse problems that employs a information from temperature measurements in the bee hive. The first problem (IP1) is considered in [2] and here we give briefly the problem formulation. Next, using the numerical local temperature and bee density, obtained by IP1, we formulate IP2

2.1   The IP1

We consider the problem for reconstruction of $\rho = \rho(x,t)$ as an unknown reaction coefficient in the parabolic equation (1). The corresponding well-posed *direct problem* (DP1) is (1) with boundary conditions (3) and initial data (6).

Let T be the solution of DP1. For a given temperature measurements, we study the IP1 of identifying the unknown coefficient $\rho = \rho(x,t)$ under the conditions:

- the initial temperature $T_0(x)$ is known;
- given measurements $T(x_m, t_k; \rho) = G_{mk}, \ m = 1, \ldots, M, \ k = 1, \ldots, K.$

We study the problem IP1 for $\rho(x,t) \in A = \{\rho : \rho \geq 0\}$, using additional data G_{mk} in a set of *admissible* solutions A. Then we rewrite the IP1 in a operator form $\mathcal{A}(\rho) = g$, where $\mathcal{A} : A \to G$ is an injective operator, $\rho \in A$, $g \in G$ is an Euclidian space of data $g = \{G_{11}, \ldots, G_{MK}\}$.

The inverse problem $\mathcal{A}(\rho) = g$ is *ill-posed*, namely its solution may not exist and/or its solution is non-unique and/or unstable to errors in measurements, see e.g. [1,5,9].

Further, the IP1 comes down to the minimization problem

$$\rho^* = \operatorname{argmin} J(\rho), \quad J(\rho) = \frac{1}{2} < \mathcal{A}(\rho) - g, \ \mathcal{A}(\rho) - g >,$$

where the quadratic deviation of the model data is expressed by the functional:

$$J(\rho) = \frac{1}{KM} \sum_{k=1}^{K} \sum_{m=1}^{M} (T(x_m, t_k; \rho) - G_{mk})^2. \tag{12}$$

We rewrite $J(\rho)$ in the equivalent form

$$J(\rho) = \frac{1}{KM} \sum_{k=1}^{K} \sum_{m=1}^{M} \int_0^{t_f} \int_0^L (T(x,t;\rho) - G_{mk})^2 \delta(x - x_m) \delta(t - t_k) dx dt, \tag{13}$$

where $\delta(\cdot)$ is the Dirac-delta function and then we apply conjugate gradient method for numerically solving the minimization problem.

The deviation of the temperature, $\delta T(x,t;\rho) = T(x,t;\rho + \delta\rho) - T(x,t;\rho)$ for increment $\delta\rho$, is the solution the following initial boundary value problem (*sensitivity problem*) with accuracy up to second-order term $O(|\delta\rho|^2)$:

$$\frac{\partial \delta T}{\partial t} = \frac{\partial^2 \delta T}{\partial x^2} + f'(T)\rho\delta T + f(T)\delta\rho,$$

$$\frac{\partial \delta T}{\partial x}(0,t) = 0, \quad \delta T(L,t) = 0, \tag{14}$$

$$\delta T(x,0) = 0.$$

The gradient for IP1 without derivation, i.e., the gradient of (13) is given by

$$J'(\rho) = \frac{1}{KM} f(T)Y(x,t), \tag{15}$$

where $Y(x,t)$ satisfies the *adjoint problem*

$$\frac{\partial Y}{\partial t} = -\frac{\partial^2 Y}{\partial x^2} - f'(T)\rho Y$$

$$+ \frac{2}{KM} \sum_{k=1}^{K} \sum_{m=1}^{M} (T(x,t;\rho) - G_{mk})\delta(x - x_m)\delta(t - t_k)dxdt, \tag{16}$$

$$\frac{\partial Y}{\partial x}(0,t) = 0, \quad Y(l,t) = 0, \quad Y(x,t_f) = 0.$$

Further we initiate iteration the process

$$\rho_{l+1} = \rho_l - \alpha_l\beta_l, \tag{17}$$

where l is the number of iteration, α_l is a descent parameter and β_l is the search step size. Let $J'(\rho_l)$ is the gradient of the cost functional at point ρ_l. We determine α_l and β_l in (17), unfolding the gradient method introduced in [5, 7, 12, 14]

$$\alpha_0 = J'(\rho_0), \quad \alpha_l = J'(\rho_l) + \gamma^l J'(\rho_{l-1}), \quad l = 1, 2, \ldots,$$

$$\gamma_0 = 0, \quad \gamma_l = \frac{\int_0^{t_f} \int_0^L [J(\rho_l)]^2 dt dx}{\int_0^{t_f} \int_0^L [J(\rho_{l-1})]^2 dt dx}, \quad l = 1, 2, \ldots, \tag{18}$$

$$\beta_l = \frac{\sum_{k=1}^{K} \sum_{m=1}^{M} [T(x_m, t_k, \rho_l) - G_{mk}] \delta T(x_m, t_k, \rho_l)}{\sum_{k=1}^{K} \sum_{m=1}^{M} [\delta T(x_m, t_k, \rho_l)]^2} \tag{19}$$

The conjugate gradient method for solving the IP1 is performed by the next steps.

Algorithm IP1

1 Choose an initial guess ρ_0 and stopping parameter $\varepsilon_1 > 0$. Suppose that we have ρ_l. We reach the next approximation as follows:
2 Check the stop condition: if $J(\rho_l) < \varepsilon_1$ then ρ_l is an approximate solution of the IP1. Otherwise go to Step 3
3 Solve the direct problem (1), (3), (6) for a given set of the parameters ρ_l and get $T(x_m, t_k; \rho_l)$, $m = 1, \ldots, M$, $k = 1, \ldots, K$.
4 Solve the adjoint problem (16) to get numerically $Y(x, t; \rho_l)$.
5 Compute the gradient of the cost functional $J'(\rho_l)$ by formula (20).
6 Calculate the descent parameter α_l (18).
7 Solve the sensitivity problem (14) to get numerically $\delta T(x, t; \rho_l)$, $\delta \rho = \alpha_l$.
8 Calculate the search step size β_l from (19) for the minimum of the gradient functional $J(\rho)$.
9 Calculate the next approximation by (17) and go to Step 2.

2.2 The IP2

In this section we formulate the second inverse problem for parameter identification for a given measurements ρ^*, obtained by IP1

$$\rho(m^*, k^*) = \rho^*(m^*, k^*), \quad m^* = 1, 2, \ldots M^*, \quad k^* = 1, 2, \ldots K^*,$$

and subject to the solution of DP2, i.e. (2), (4), (5) for known $T(x, t)$ from IP1.

Thus, the IP2 for recovering the parameters θ_0 and γ can be formulated in a variational setting

$$\min_{a \in A_{\text{adm}}} J^*(a), \quad a = (\theta_0, \gamma)$$

where the functional $J^*(a)$ is defined by

$$J^*(a) = \sum_{m^*=1}^{M^*} \left[\sum_{k^*=1}^{K^*} \left(\rho(x_{m^*}, t_{k^*}; a) - \rho_{m^*k^*}^* \right)^2 \right]. \tag{20}$$

Here $\rho(x_{m^*}, t_{k^*}; a)$ is the solution of the problem (2), (4), (5).

To solve the IP2 numerically, we apply Levenberg-Marquardt algorithm [6, 11, 13, 16], which combines gradient-descent and Gauss-Newton methods.

Let rewrite (20) in the form

$$J^*(a) = \mathbf{e}(a)^T \mathbf{e}(a),$$
$$\mathbf{e} = [e_{11}, e_{12}, \ldots, e_{1K^*}, e_{21}, \ldots, e_{2K^*}, \ldots, e_{M^*1}, \ldots, e_{M^*K^*}],$$
$$e_{lq} = \rho(x_{l^*}, t_{q^*}; a) - \rho_{l^*q^*}^*.$$

To minimize the objective functional (20), we consider the iteration process

$$a_{s+1} = a_s + \delta a_{s+1}, \tag{21}$$

where $s = 0, 1, \ldots$ is the iteration number and δa^{s+1} is the solution of the equation

$$[\tilde{J}^T \tilde{J} + \mu I]\delta a_{s+1} = \tilde{J}^T e(a_s). \tag{22}$$

Here I is the unit matrix, μ is the damping factor and $\tilde{J}$ is the Jacobian matrix

$$\tilde{J}(a) = \begin{pmatrix} \dfrac{\partial e_{11}}{\partial \theta_0} & \dfrac{\partial e_{11}}{\partial \gamma} \\ \dfrac{\partial e_{12}}{\partial \theta_0} & \dfrac{\partial e_{12}}{\partial \gamma} \\ \ldots & \ldots \\ \dfrac{\partial e_{M^* K^*}}{\partial \theta_0} & \dfrac{\partial e_{M^* K^*}}{\partial \gamma} \end{pmatrix}.$$

The derivatives in the Jacobian matrix $\tilde{J}$ are determined as follows

$$\begin{aligned} \frac{\partial e_{lq}}{\partial \theta_0} &= \frac{e_{lq}(x_{m^*}, t_{k^*}; \theta_0 + \varepsilon_\theta, \gamma) - e_{lq}(x_{m^*}, t_{k^*}; \theta_0 - \varepsilon_\theta, \gamma)}{2\varepsilon_\theta} + O(\varepsilon_\theta^2), \\ \frac{\partial e_{lq}}{\partial \gamma} &= \frac{e_{lq}(x_{m^*}, t_{k^*}; \theta_0, \gamma + \varepsilon_\gamma) - e_{lq}(x_{m^*}, t_{k^*}; \theta_0, \gamma - \varepsilon_\gamma)}{2\varepsilon_\gamma} + O(\varepsilon_\gamma^2). \end{aligned} \tag{23}$$

The stoping criteria is $\|J^*(a_{s+1}) - J^*(a^s)\| \leq \varepsilon_2$ for small positive ε_2. The iteration procedure executes as follows.

Algorithm IP2

1 Chose a_0 $(s = 0)$ and the accuracy ε_2. We find a_{s+1} by the next steps.
2 Find $\tilde{J}(a_s)$.
3 Compute δa^{s+1} from (22).
4 Calculate the next iteration by (21).
5 Find the objective functional $J^*(a_{s+1})$ by (20).
6 Check the convergence criterion and stop the iteration process if it is achieved, set $a_{s+1} := (\theta_0, \gamma)$ and go to Step 7. Otherwise, $s := s + 1$ and go to Step 2.
7 Find $\rho(x, t)$ from (2), (4) and (5) for known T (from IP1) and restored θ_0, γ.

The damping parameter μ is adjusted at each iteration.

3   Numerical Method

In this section we provide numerical discretizations that are necessary for the numerical realization of Algorithm IP1 and Algorithm IP2 and construct the algorithm for computing the local temperature, bee density, parameters θ_0 and γ, fitted to the temperature measurements.

We consider uniform spatial and temporal meshes

$$\begin{aligned} \overline{\omega}_h &= \{x_i : x_i = ih, \ i = 0, 1, \ldots, N_x, \ x_{N_x} = L\}, \\ \overline{\omega}_\tau &= \{t_n : t_n = n\tau \ n = 0, 1, \ldots, N_t, \ t_{N_t} = t_f\}. \end{aligned}$$

Further, we denote by T_i^n, ρ_i^n the mesh functions at point (x_i, t_n) and use the notations [17]

$$u^{\pm} = \max\{0, \pm u\}, \quad u_{x,i} = \frac{u_{i+1} - u_i}{h}, \quad u_{\overline{x},i} = u_{x,i-1} \quad u_{\overline{x}x,i} = \frac{u_{x,i+1/2} - u_{x,i-1/2}}{h}.$$

As in [2,3], Eqs. (1), (3) are approximated by the standard second-order implicit-explicit finite difference scheme

$$\begin{aligned}
\frac{T_i^{n+1} - T_i^n}{\tau} &= T_{\overline{x}x,i}^{n+1} + f(T_i^n)\rho_i^{n+1}, \quad i = 1, 2, \ldots, N_x - 1, \\
\frac{T_0^{n+1} - T_0^n}{\tau} &= \frac{2}{h}T_{x,0}^{n+1} + f(T_0^n)\rho_0^{n+1}, \quad T_N^{n+1} = T_a.
\end{aligned} \tag{24}$$

The second Eq. (2) with the corresponding boundary conditions (4) is discretized by the explicit-implicit upwind finite difference scheme

$$\frac{\rho_i^{n+1} - \rho_i^n}{\tau_n} - \rho_{\overline{x}x,i}^{n+1} + \theta(\rho_i^n, T_i^{n+1})\rho_i^{n+1}$$

$$+ \frac{F_{i+1/2}(\rho_i^{n+1}, T_i^{n+1}) - F_{i-1/2}(\rho_i^{n+1}, T_i^{n+1})}{h} = 0, \quad i = 1, 2, \ldots, N_x - 1,$$

$$\frac{\rho_N^{n+1} - \rho_N^n}{\tau_n} + 2\frac{\rho_N^{n+1} - \rho_{N-1}^{n+1}}{h^2} + \theta(\rho_N^*, T_N^{n+1})\rho_N^{n+1} = \frac{2F_{N-1/2}(\rho_N^{n+1}, T_N^{n+1})}{h},$$

$$\frac{\rho_0^{n+1} - \rho_0^n}{\tau_n} + 2\frac{\rho_0^{n+1} - \rho_1^{n+1}}{h^2} + \theta(\rho_0^*, T_0^{n+1})\rho_0^{n+1} = -\frac{2F_{1/2}(\rho_0^{n+1}, T_0^{n+1})}{h}, \tag{25}$$

where

$$F_{i+1/2}(\rho_i^{n+1}, T_i^{n+1}) = \chi_{i+1/2}^+\rho_i^{n+1} - \chi_{i+1/2}^-\rho_{i+1}^{n+1},$$

$$F_{i-1/2}(\rho_i^{n+1}, T_i^{n+1}) = \chi_{i-1/2}^+\rho_{i-1}^{n+1} - \chi_{i-1/2}^-\rho_i^{n+1},$$

$$\chi_{i+1/2} = \chi\left(\frac{T_{i+1}^{n+1} + T_i^{n+1}}{2}\right)T_{x,i}^{n+1}, \quad \chi_{i-1/2} = \chi\left(\frac{T_i^{n+1} + T_{i-1}^{n+1}}{2}\right)T_{\overline{x},i}^{n+1}.$$

For known T^{n+1}, the numerical scheme (25) is realized by iteration process, taking ρ^n in the reaction term at old iteration and ρ^{n+1} at new iteration.

To solve the IP1 numerically, we need to discretize the sensitivity and adjoint problems (14), (16). We use the discretizations as in [2]

$$\frac{\delta T_i^{n+1} - \delta T_i^n}{\tau} = \delta T_{\overline{x}x,i}^{n+1} + f'(T_i^{n+1})\rho_i^*\delta T_i^n + f(T_i^{n+1})\delta\rho_i^{n+1}, \quad i = 1, \ldots, N_x - 1,$$

$$\frac{\delta T_0^{n+1} - \delta T_0^n}{\tau} = \frac{2}{h}\delta T_{x,0}^{n+1} + f'(T_0^{n+1})\rho_0^*\delta T_0^n + f(T_0^{n+1})\delta\rho_0^{n+1}, \quad T_N^{n+1} = 0,$$

$$\delta T_i^{N_t} = 0, \quad i = 1, 2, \ldots, N_x - 1. \tag{26}$$

and

$$\frac{Y_i^n - Y_i^{n+1}}{\tau} = -Y_{\overline{x}x,i}^n - f'(T_i^n)\rho_i^* Y_i^{n+1}$$

$$+ \frac{2}{\tau h} \sum_{k=1}^{K} \sum_{m=1}^{M} (T_i^n - G_{mk}) \, \mathbf{1}_{km}(i,j), \quad i = 1, \ldots, N_x - 1,$$

$$\frac{Y_0^n - Y_0^{n+1}}{\tau} = -\frac{2}{h} Y_{x,0}^n - f'(T_0^n)\rho_0^* Y_o^{n+1} + \frac{2}{\tau h} \sum_{k=1}^{K} \sum_{m=1}^{M} (T_i^n - G_{mk})^2 \, \mathbf{1}_{km}(i,j),$$

$$Y_N^n = 0, \quad Y_i^{N_t} = 0, \quad i = 1, 2, \ldots, N_x - 1, \tag{27}$$

where $\mathbf{1}_{km}(i,j)$ is the indicator function. We find the derivative $f'(T)$, by smoothing the function f:

$$f(T) = s(T) f_{\text{low}} + (1 - s(T)) f_{\text{high}}, \quad s(x) = 0.5 - 0.5 \tanh \frac{T - Tf}{\epsilon}, \quad 0 < \epsilon < 1.$$

The double integrals in (18) is approximated by the second order quadrature

$$\int_0^{t_f} \int_0^L g(x,t) dt dx \approx \frac{h\tau}{4} \sum_{n=0}^{N_t-1} \sum_{i=0}^{N_x-1} \Big(g(x_i, t_j) + g(x_{i+1}, t_j) $$
$$+ g(x_i, t_{j+1}) + g(x_{i+1}, t_{j+1}) \Big). \tag{28}$$

We realize the numerical method as follows

Algorithm
Solving IP1: Compute ρ^*, T

1 Require ρ_0, T_0, $\varepsilon_1 > 0$, $\epsilon > 0$.
2 Set $l = 0$ and compute $J(\rho_0)$ by (12).
3 While $J(\rho_l) > \varepsilon$ and $(J(\rho_{l-1}) > J(\rho_l), l > 0)$ perform the following steps
  4.1 Find $T(x,t;\rho_l)$ from direct problem (24)
  4.2 Compute $J(\rho_l)$, using (12).
  4.3 Compute $Y(x,t;\rho_l)$ from adjoint problem (27) for $T(x,t;\rho_l), G_{MK}, \rho_l)$.
  4.4 Compute $J'(\rho_l)$ from (15).
  4.5 Find descent parameter α_l from (18), (28) and state $\delta\rho^{n+1} = \alpha_l$.
  4.6 Compute $\delta T(x,t;\rho_l)$ from sensitivity problem (26) for $(T(x,t;\rho_l), \rho_l)$.
  4.7 Calculate β_l from (19).
  4.8 Set $\rho_{l+1} = \rho_l - \alpha_l \beta_l$.
5 $\rho^* := \rho_{l+1}$ and find $T := T(x,t;\rho_{l+1})$, solving direct problem (24).

Solving IP2: Compute ρ, θ_0, γ.

1 Require a_0 $(s = 0)$, $\varepsilon_2 > 0$, $\varepsilon_\theta > 0$, $\varepsilon_\gamma > 0$, T. We find a_{s+1} as follows.
2 Find $\tilde{J}(a_s)$ from (23) and (25) for $a_s = (\theta_0 \pm \varepsilon_\theta, \gamma)$ and $a_s = (\theta_0, \gamma \pm \varepsilon_\gamma)$.
3 Compute δa^{s+1} from (22).
4 Calculate the next iteration by (21)

5 Find the objective functional $J^*(a_{s+1})$ by (20).

6 Check the convergence criterion $\|J^*(a_{s+1}) - J^*(a^s)\| \leq \varepsilon_2$. Stop the iteration process if it is achieved, set $a_{s+1} := (\theta_0, \gamma)$ and go to Step 7. Otherwise, $s := s+1$ and go to Step 2.

7 Find ρ from (2), (4) and (5) for known T (from IP1) and restored θ_0, γ.

4 Numerical Simulations

In this section we present results from numerical tests, obtained by Algorithm. We consider the following model parameters $T_\theta = 21$, $T_\chi = 25$ $\chi_1 = \chi_2 = 1$, $f_{\text{high}} = 0.6$, $f_{\text{low}} = 3$, $T_f = 15$, $\gamma = 1$, $m = 10$, $\theta_0 = 4.10^{-3}$ [4].

In order to save computational time, we rescale the problem (1)–(4) [2,3], introducing new variables $x := x/L$, $t := t/L^2$. As a result, the problem is defined in $x \in [0,1]$, $t \in [0, t_f/L^2]$ and $\rho := L\rho$, $\theta_0 := L\theta_0$, $f := Lf$. Further all computations will be performed with the rescaled problem, but the data are given for the original problem (1)–(4).

We consider a test example with exact solution, adding residual function in the right-hand side of the equations. The exact bee density and temperature are $T(x,t) = e^{-t/(3t_f)}(T_a + 40(1 - (x/L)^4))$, $\rho(x,t) = e^{-t/(t_f)}\rho^0(x)$, where $\rho^0(x)$ is the steady-state solution, derived in [4].

We give relative errors $\mathcal{E}_\theta = |\theta_0 - \theta_{0,\text{res}}|/\theta_0$, $\mathcal{E}_\rho = |\gamma - \gamma_{\text{res}}|/\gamma$,

$$\mathcal{E} = \frac{\max\limits_{0 \leq i \leq N_x} \max\limits_{0 \leq n \leq N_t} |E_i^n|}{\max\limits_{0 \leq i \leq N_x} \max\limits_{0 \leq n \leq N_t} |V(x_i, t_n)|}, \qquad \mathcal{E}_2 = \frac{\left(\sum\limits_{i=0}^{N_x} \sum\limits_{n=0}^{N_y} (E_i^n)^2\right)^{1/2}}{\left(\sum\limits_{i=0}^{N_x} \sum\limits_{n=0}^{N_y} (V(x_i, t_n))^2\right)^{1/2}},$$

where $E_i^n = V(x_i, t_n) - V_i^n$, $V = \{T, \rho\}$ and $\theta_{0,\text{res}}$, γ_{res} are restored values of θ_0 and γ, respectively.

In reality, the temperature is measured with sensors, positioned 4–5 cm in aline distance apart, that can provide information at very small time intervals [2,3]. For the numerical test, we utilize perturbed data, adding a Gaussian noise to the exact solution and parameters θ_0, γ. Let $M = M^* = 3$, $K = K^* = 27$, $T_a = -9\,^\circ\text{C}$, $L = 10$ cm, $t_f = 100$, $\epsilon = 0.5$, $\varepsilon_1 \approx 10^{-3}$, $\varepsilon_2 = \approx 10^{-7}$, $\varepsilon_\theta = 10^{-5}$, $\varepsilon_\gamma = 10^{-3}$, $N = 80$, $\tau = h^2$.

In Table 1 we give the results, obtained for perturbations p_T, p_ρ, p_θ and p_γ of the local temperature, density, parameters and θ_0, γ. We observe optimal precision of the local temperature, density and parameters, fitted to the temperature measurements. Bigger deviation of the input data, leads to an increase in error.

On Fig. 1 we depict numerical temperature and density, obtained by Algorithm, $p_\rho = 2$, $p_T = 2$, $p_\theta = 4$, $p_\gamma = 3$.

Table 1. Relative errors of the restored solution and parameters.

p_ρ	p_T	p_θ	p_γ	Density $\mathcal{E}$	Density $\mathcal{E}_2$	Temperature $\mathcal{E}$	Temperature $\mathcal{E}_2$	Parameters $\mathcal{E}_\theta$	Parameters $\mathcal{E}_\rho$
5%	2%	5%	1%	2.052e−2	1.554e−2	1.117e−3	5.888e−4	3.214e−2	5.706e−4
2%	2%	4%	2%	1.429e−2	1.438e−2	9.565e−4	5.541e−4	1.372e−2	1.274e−3
5%	5%	3%	3%	2.052e−2	2.279e−2	2.131e−3	1.285e−3	9.432e−3	2.846e−3

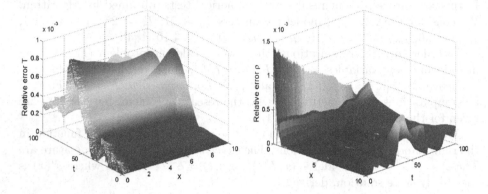

Fig. 1. Relative error $E_i^n / \max\limits_{0 \leq i \leq N_x} \max\limits_{0 \leq n \leq N_t} |V(x_i, t_n)|$ of the recovered temperature (*left*) and bee density (*right*), $p_T = 2\%$, $p_\rho = 2\%$.

Conclusions

Papers [3,4,10,12,20] analyze mathematical models that explored the thermoregulation of the honeybees in the hive. But only the paper [2] and the present research are aimed to predict the optimal parameters of the processes development. Relying on the fact that the beekeepers are able to measure the temperature, we predict the thermoregulation process.

The research might be expanded to 2D case, as well as for more complicated problems with free boundary of the honey bee colony. Rigorous theoretical analysis is also planed in our future work.

Acknowledgements. This work is supported by the Bulgarian National Science Fund under the Project KP-06-PN 46-7 Design and research of fundamental technologies and methods for precision apiculture.

References

1. Alifanov, O.M., Artioukhine, E.A., Rumyantsev, S.V.: Extreme Methods for Solving Ill-posed problems with Applications to Inverse Heat Transfer Problems, Begell House, New York - Wallingford (U.K.) (1995)

2. Atanasov, A.Z., Koleva, M.N., Vulkov, L.G.: Numerical optimization identification of a Keller-Segel model for thermoregulation in honey bee colonies in winter, accepted in Springer book series CCIS (2023)
3. Atanasov, A.Z., Koleva, M.N., Vulkov, L.G.: Numerical analysis of thermoregulation in honey bee colonies in winter based on sign-changing chemotactic coefficient model, accepted in Springer Proceedings in Mathematics & Statistics (2023)
4. Bastaansen, R., Doelman, A., van Langevede, F., Rottschafer, V.: Modeling honey bee colonies in winter using a Keller-Segel model with a sign-changing chemotactic coefficient. SIAM J. Appl. Math. **80**(20), 839–863 (2020)
5. Chavent, G.: Nonlinear Least Squares for Inverse Problems: Theoretical Foundation and Step-by Guide for Applications. Springer (2009). https://doi.org/10.1007/978-90-481-2785-6
6. Cui, M., Yang, K., Xu, X., Wang, S., Gao, X.: A modified Levenberg-Marquardt algorithm for simultaneous estimation of multi-parameters of boundary heat flux by solving transient nonlinear inverse heat conduction problems. Int. J. Heat Mass Transf. **97**, 908–916 (2016)
7. Fakhraie, M., Shidfar, A., Garshasbi, M.: A computational procedure for estimation of an unknown coefficient in an inverse boundary value problem. Appl. Math. Comput. **187**, 1120–1125 (2007)
8. Guzman-Novoa, E., Eccles, C.Y., McGowan, J., Kelly, P.G., Correa- Bentez, A.: Varroa destructor is the main culprit for the death and reduced populations of overwintered honey bee (Apis mellifera) colonies in Ontario. Canada, Apidologie **41**, 443–450 (2010)
9. Hasanov, A., Romanov, V.: Introduction to the Inverse Problems for Differential Equations. Springer, N.Y. (2017). https://doi.org/10.1007/978-3-319-62797-7
10. Jarimi, J., Tapia-Brito, E., Riffat, S.: A review on thermoregulation techniques in honey bees (Apis Mellifera) beehive microclimate and its similarities to the heating and cooling management in buildings. Future Cities Environ. **6**(1), 7, 1–8 (2020)
11. Kwak, Y., Hwang, J., Yoo, C.: A new damping strategy of Levenberg-Marquardt algorithm for multilayer perceptrons. Neural Network World **21**(4), 327–340 (2011)
12. Lemke, M., Lamprecht, A.: A model of heat production and thermoregulation in winter clusters of honey bee using differential heat conduction equations. J. Theor. Biol. **142**(2), 261–0273 (1990)
13. Levenberg, K.: A method for the solution of certain non-linear problems in Least-Squares. Quarterly Appl. Math. **2**, 164–168 (1944)
14. Lesnic, D.: Inverse Problems with Applications in Science and Engineering. CRC Press, London (2020)
15. Ratti, V., Kevan, P.G., Eberl, H.J.: A mathematical model of forager loss in honeybee colonies infested with varroa destructor and the acute bee paralysis virus. Bul. Math. Biol. **79**, 1218–1253 (2017)
16. Marquardt, D.: An algorithm for least-squares estimation of non- linear parameters. SIAM J. Appl. Math. **11**(2), 431–441 (1963)
17. Samarskii A.A.: The Theory of Difference Schemes. Marcel Dekker Inc. (2001)
18. Stabentheiner, A., Kovac, H., Brodschneider, R.: Honeybee colony thermoregulation regulatory mechanisms and contribution of individuals in dependence on age, location and thermal stress. PLoS ONE **5**(1), e8967 (2010)
19. Tautz, J.: The Buzz About Bees: Biology of a Superorganism. Springer, Berlin (2008). https://doi.org/10.1007/978-3-540-78729-7
20. Watmough, J., Camazine, S.: Self-organized thermoregulation of honeybee clusters. J. Theor. Biol. **176**(3), 391–402 (1995)

On the Consistency and Convergence of Repeated Richardson Extrapolation

Teshome Bayleyegn[1] , István Faragó[1,2,3(✉)] , and Ágnes Havasi[1,2]

[1] ELTE Eötvös Loránd University, Pázmány Péter s. 1/C, Budapest 1117, Hungary
faragois@gmail.com, agnes.havasi@ttk.elte.hu
[2] MTA-ELTE Numerical Analysis and Large Networks Research Group,
Pázmány Péter s. 1/C, Budapest 1117, Hungary
[3] Institute of Mathematics, Budapest University of Technology and Economics,
Egry J. u. 1., Budapest 1111, Hungary

Abstract. Richardson extrapolation is a sequence acceleration method, which has long been used to enhance the accuracy of time integration methods for solving differential equations. Its classical version is based on a suitable linear combination of numerical solutions obtained by the same numerical method with two different discretization parameters. We present the principle of Richardson extrapolation, and introduce a possible generalization of this method called repeated Richardson extrapolation (RRE). The consistency and convergence of the new method obtained by combining certain Runge-Kutta methods with RRE are analysed and illustrated with numerical experiments.

Keywords: explicit Runge–Kutta methods · consistency · convergence · Repeated Richardson extrapolation

1 Introduction

Numerous changes are occurring in real life due to both natural and artificial factors. The majority of the derived models are often too large or complex to be solved by traditional analytical or numerical approaches [5]. Accuracy is an important consideration along with the existing analytical and numerical methods' limited efficiency in problem solving. Richardson extrapolation is widely used to increase the order of accuracy of numerical results and control the step size [3,6,7,9]. This procedure is applied to the solutions obtained on two different grids with time step sizes h and $h/2$ by the same numerical method, and the two solution are combined linearly [2,6,7]. This original version of the method is called classical Richardson extrapolation (CRE) and it has a natural generalization called repeated Richardson extrapolation (RRE) [10–12]. In this paper, the consistency and convergence of RRE is analysed when the underlying method is some explicit Runge–Kutta method.

The paper is organized as follows. In Chap. 2 we present the classical and repeated Richardson extrapolations. In Sect. 3 the family of Runge-Kutta techniques and their well-known order conditions for $p \leq 4$ are reviewed, and

I. Georgiev et al. (Eds.): NMA 2022, LNCS 13858, pp. 48–58, 2023.
https://doi.org/10.1007/978-3-031-32412-3_5

we demonstrate that applying the active Richardson extrapolation to any Runge–Kutta method of order $p = 1, 2$ or 3 raises the order of consistency by one. In Sect. 4, we show that any Runge-Kutta method with order $p = 1$ paired with the repeated Richardson extrapolation results in an improvement of two in the order of consistency. In Sect. 5, we explore the convergence property of RRE when combined with any ERK method, and give an illustration of our findings.

2   The Method of Classical and Repeated Richardson Extrapolation

Consider the Cauchy problem for a system of ODE's (1)

$$y'(t) = f(t, y(t)) \tag{1}$$

$$y(0) = y_0, \tag{2}$$

where $f : \mathbb{R} \times \mathbb{R}^d \to \mathbb{R}^d$ and $y \colon \mathbb{R} \to \mathbb{R}^d$ and $y_0 \in \mathbb{R}^d$. We divide $[0, T]$ into N subintervals of length h, and define the following two meshes on $[0, T]$:

$$\Omega_h := \{t_n = nh : n = 0, 1, \ldots, N_t\}, \text{ where } N_t = T/h, \tag{3}$$

$$\Omega_{h/2} := \{t_k = kh/2 : k = 0, 1, \ldots, 2N_t\}. \tag{4}$$

We use the same one-step numerical method of order p to solve the problem on both meshes. The numerical solutions at time t_n will be denoted by $z(t_n)$ and $w(t_n)$.

If the exact solution $y(t)$ is $p + 1$ times continuously differentiable, then

$$y(t_n) - z(t_n) = K \cdot h^p + \mathcal{O}(h^{p+1}), \text{ and} \tag{5}$$

$$y(t_n) - w(t_n) = K \cdot (h/2)^p + \mathcal{O}(h^{p+1}) \tag{6}$$

By eliminating the p-th order terms, we get the following linear combination, defining the active CRE method

$$y_n := \frac{2^p w_n - z_n}{2^p - 1}. \tag{7}$$

Consider the Cauchy problem (1) and assume that a numerical method is applied to solve (1), which is consistent in order p. We divide $[0, T]$ into N sub-intervals of length h, and define three meshes on $[0, T]$:

$$\Omega_h^0 := \{t_n = nh : n = 0, 1, \ldots, N\}, \tag{8}$$

$$\Omega_h^1 := \{t_n = n\frac{h}{2} : n = 0, 1, \ldots, 2N\}, \tag{9}$$

and

$$\Omega_h^2 = \{t_n = \frac{nh}{4}, i = 0, 1, \ldots, 4N\} \tag{10}$$

Repeated Richardson extrapolation (RRE) is a technique that is created by linearly combining the numerical results from the aforementioned three grids (8),(9) and (10) [9,11].

$$y_{\text{RRE},n} := \frac{2^{2p+1} z_n^{[2]} - 3 \cdot 2^{p+1} z_n^{[1]} + z_n^{[0]}}{2^{2p+1} - 3 \cdot 2^{p-1} + 1} \tag{11}$$

In order to attain even higher accuracy, the repeated Richardson extrapolation is a proposed generalization of the Richardson extrapolation concept.

3   Runge–Kutta Methods and Their Order Conditions

Ordinary differential equations can be solved using the family of Runge-Kutta methods, which have the general form

$$y_{n+1} = y_n + h \sum_{i=1}^{s} b_i k_i \tag{12}$$

with [8]

$$k_i = f\left(t_n + c_i h, y_n + h \sum_{j=1}^{s} a_{ij} k_j \right), i = 1, 2, ..., s, \tag{13}$$

where b_i, c_i and $a_{ij} \in \mathbb{R}$ are given constants, and s is the number of stages. The Butcher tableau quickly identifies this technique:

$$\frac{c \| A}{\| b^T} \tag{14}$$

where $c \in \mathbb{R}^{s \times 1}$, $b \in \mathbb{R}^{s \times 1}$ and the matrix $A = [a_{ij}]_{i,j=1}^s$. In the case of an explicit Runge-Kutta technique, the matrix A is strictly lower triangular, and lower triangular for diagonally implicit Runge-Kutta methods. Assume that the row-sum condition holds [4], i.e.,

$$c_i = \sum_{j=1}^{s} a_{ij}, \ i = 1, 2, \ldots, s. \tag{15}$$

In addition to the row-sum requirement, Table 1 also provides the conditions that are sufficient for order of consistency up to $p = 4$. In the following we use the notations

$$\mathbf{b}^T = \left[\left(\frac{1+\rho}{2} \right) b^T, \left(\frac{1+\rho}{2} \right) b^T, -\rho b^T \right] \in \mathbb{R}^{1 \times 3s},$$

$$\mathbf{c} := \begin{bmatrix} \frac{1}{2} c \\ \frac{1}{2}(e + c) \\ c \end{bmatrix} \in \mathbb{R}^{3s \times 1}, \ \mathbf{A} = \begin{bmatrix} \frac{1}{2} A & 0 & 0 \\ \frac{1}{2} e b^T & \frac{1}{2} A & 0 \\ 0 & 0 & A \end{bmatrix} \in \mathbb{R}^{3s \times 3s},$$

In [1] based on the order conditions, we proved that the active Richardson extrapolation increases the order of consistency by one for RK methods of orders $p = 1, 2$ and 3.

Table 1. Conditions for orders of consistency $p = 1, 2, 3$ and 4 of Runge–Kutta methods.

Order p	Conditions
1	$b^T e = 1$
2	$b^T c = \frac{1}{2}$
3	$b^T c^2 = \frac{1}{3}$, $b^T A c = \frac{1}{6}$
4	$b^T c^3 = \frac{1}{4}$, $b^T c A c = \frac{1}{8}$, $b^T A c^2 = \frac{1}{12}$, $b^T A^2 c = \frac{1}{24}$

4   Increasing the Order by RRE

We solve the problem (1) on the three meshes (8), (9) and (10) with Runge–Kutta method (12) of order p.

Step 1: First we take a step of length h, i.e. on the mesh

$$\Omega_h = \{t_n = nh, i = 0, 1, \ldots, N\} \tag{16}$$

with the general RK method:

$$z_{n+1} = y_n + h \sum_{i=1}^{s} b_i k_i, \tag{17}$$

where

$$k_1 = f(t_n, y_n)$$

$$k_i = f\left(t_n + c_i h, y_n + h \sum_{j=1}^{i-1} a_{ij} k_j\right), \quad i = 2, \ldots, s.$$

Step 2: Take two steps of length $h/2$. First we take a step of length h on the mesh

$$\Omega_{h/2} = \left\{t_n = \frac{nh}{2}, i = 0, 1, \ldots, 2N\right\} \tag{18}$$

with the first step

$$w_{n+\frac{1}{2}} = y_n + \frac{h}{2} \sum_{i=1}^{s} b_i k_i^1,$$

where

$$k_1^1 = f(t_n, y_n)$$

$$k_i^1 = f\left(t_n + c_i \frac{h}{2}, y_n + \frac{h}{2} \sum_{j=1}^{i-1} a_{ij} k_j^1\right), \quad i = 2, \ldots, s,$$

and the second step is

$$w_{n+1} = w_{n+\frac{1}{2}} + \frac{h}{2} \sum_{i=1}^{s} b_i k_i^2,$$

where

$$k_1^2 = f\left(t_n + \frac{h}{2}, w_{n+\frac{1}{2}}\right)$$

$$k_i^2 = f\left(t_n + \frac{h}{2} + \frac{c_i h}{2}, w_{n+\frac{1}{2}} + \frac{h}{2}\sum_{j=1}^{i-1} a_{ij}k_j^2\right)$$

$$= f\left(t_n + \frac{h}{2}(1+c_i), y_n + \frac{h}{2}\left(b^T K^1 + AK^2\right)\right), \quad i = 2,\ldots,s,$$

$$w_{n+1} = y_n + \frac{h}{2}\sum_{i=1}^{s} b_i k_i^1 + \frac{h}{2}\sum_{i=1}^{s} b_i k_i^2.$$

Step 3: From the initial value y_n we take four steps of length $h/4$ and we have a set

$$\Omega_{h/4} = \{t_n = \frac{nh}{4}, \ n = 0, 1, \ldots, 4N\} \tag{19}$$

with the first step

$$q_{n+\frac{1}{4}} = y_n + \frac{h}{4}\sum_{i=1}^{s} b_i k_{i1},$$

where

$$k_{11} = f\left(t_n, y_n\right)$$

$$k_{i1} = f\left(t_n + c_i\frac{h}{4}, y_n + \frac{h}{4}\sum_{j=1}^{i-1} a_{ij}k_{j1}\right), \quad i = 2,\ldots,s,$$

The second step

$$q_{n+\frac{1}{2}} = q_{n+\frac{1}{4}} + \frac{h}{4}\sum_{i=1}^{s} b_i k_{i1}^1.$$

where

$$k_{11}^1 = f\left(t_n + \frac{h}{4}, \ q_{n+\frac{1}{4}}\right)$$

$$k_{i1}^1 = f\left(t_n + \frac{h}{4} + \frac{c_i h}{4}, \ q_{n+\frac{1}{4}} + \frac{h}{4}\sum_{j=1}^{i-1} a_{ij}k_{j1}^1\right), \quad i = 2,\ldots,s.$$

The result of the third step is

$$q_{n+\frac{3}{4}} = q_{n+\frac{1}{2}} + \frac{h}{4}\sum_{i=1}^{s} b_i k_{i1}^2.$$

where

$$k_{11}^2 = f\left(t_n + \frac{h}{2}, \ q_{n+\frac{1}{2}}\right)$$

$$k_{i1}^2 = f\left(t_n + \frac{h}{2} + \frac{c_i h}{4}, \; q_{n+\frac{1}{2}} + \frac{h}{4}\sum_{j=1}^{i-1} a_{ij} k_{j1}^2\right), \; i = 2, \ldots, s,$$

and the result of the fourth step:

$$q_{n+1} = q_{n+\frac{3}{4}} + \frac{h}{4}\sum_{i=1}^{s} b_i k_{i1}^3,$$

where

$$k_{11}^3 = f\left(t_n + \frac{3h}{4}, \; q_{n+\frac{3}{4}}\right)$$

$$k_{i1}^3 = f\left(t_n + \frac{3h}{4} + \frac{c_i h}{4}, \; q_{n+\frac{3}{4}} + \frac{h}{4}\sum_{j=1}^{i-1} a_{ij} k_{j1}^3\right), \; i = 2, \ldots, s.$$

$$q_{n+1} = y_n + \frac{h}{4}\sum_{i=1}^{s} b_i k_{i1} + \frac{h}{4}\sum_{i=1}^{s} b_i k_{i1}^1 + \frac{h}{4}\sum_{i=1}^{s} b_i k_{i1}^2 + \frac{h}{4}\sum_{i=1}^{s} b_i k_{i1}^3.$$

Step 4: We combine the numerical solutions z_{n+1}, w_{n+1} and q_{n+1} using repeated Richardson extrapolation linearly (c.f. 11).

$$y_{RRE,n} := \frac{2^{2p+1} q(t_{n+1}) - 3 \cdot 2^p w(t_{n+1}) + z(t_{n+1})}{2^{2p+1} - 3 \cdot 2^p + 1}$$

Then by simple calculation we get

$$y_{n+1} = y_n + h\left(\frac{2^{2p}}{2^{2p+2} - 3 \cdot 2^{p+1} + 2}\right) b^T R^1 + h\left(\frac{2^{2p}}{2^{2p+2} - 3 \cdot 2^{p+1} + 2}\right) b^T R^2$$

$$+ h\left(\frac{2^{2p}}{2^{2p+2} - 3 \cdot 2^{p+1} + 2}\right) b^T R^3 + h\left(\frac{2^{2p}}{2^{2p+2} - 3 \cdot 2^{p+1} + 2}\right) b^T R^4$$

$$- hb^T \frac{3 \cdot 2^p}{2^{2p+2} - 3 \cdot 2^{p+1} + 2}\left(K^1 + K^2\right) + hb^T\left(\frac{1}{2^{2p+1} - 3 \cdot 2^p + 1}\right) K$$

with

$$K = \begin{pmatrix} k_1 \\ k_2 \\ \cdot \\ \cdot \\ \cdot \\ k_s \end{pmatrix}, K^1 = \begin{pmatrix} k_1^1 \\ k_2^1 \\ \cdot \\ \cdot \\ \cdot \\ k_s^1 \end{pmatrix}, K^2 = \begin{pmatrix} k_1^2 \\ k_2^2 \\ \cdot \\ \cdot \\ \cdot \\ k_s^2 \end{pmatrix} \text{ and } b^T = (b_1, b_2, \ldots, b_s)^T, \; e = (1, 1, \ldots, 1)^T \in \mathbb{R}^s$$

$$R^1 = \begin{pmatrix} k_{11} \\ k_{21} \\ \cdot \\ \cdot \\ \cdot \\ k_{s1} \end{pmatrix}, R^2 = \begin{pmatrix} k_{11}^1 \\ k_{21}^1 \\ \cdot \\ \cdot \\ \cdot \\ k_{s1}^1 \end{pmatrix} R^3 = \begin{pmatrix} k_{11}^2 \\ k_{21}^2 \\ \cdot \\ \cdot \\ \cdot \\ k_{s1}^2 \end{pmatrix} \text{ and } R^4 = \begin{pmatrix} k_{11}^3 \\ k_{21}^3 \\ \cdot \\ \cdot \\ \cdot \\ k_{s1}^3 \end{pmatrix}$$

By the notations

$$M = \frac{2^{2p}}{2^{2p+2} - 3 \cdot 2^{p+1} + 2}, \quad Q = \frac{3 \cdot 2^p}{2^{2p+2} - 3 \cdot 2^{p+1} + 2}, \quad N = \frac{1}{2^{2p+1} - 3 \cdot 2^p + 1},$$

we have

$$y_{n+1} = y_n + hb^T \left(MR^1 + MR^2 + MR^3 + MR^4 - Q\,K^1 - Q\,K^2 + NK \right)$$

Table 2. The Butcher tableau of RK+RRE.

$\frac{1}{4}c$	$\frac{1}{4}A$	0	0	0	0	0	0
$\frac{1}{4}(e+c)$	$\frac{1}{4}eb^T$	$\frac{1}{4}A$	0	0	0	0	0
$\frac{1}{2}e + \frac{1}{4}c$	$\frac{1}{4}eb^T$	$\frac{1}{4}eb^T$	$\frac{1}{4}A$	0	0	0	0
$\frac{3}{4}e + \frac{1}{4}c$	$\frac{1}{4}eb^T$	$\frac{1}{4}eb^T$	$\frac{1}{4}eb^T$	$\frac{1}{4}A$	0	0	0
$\frac{1}{2}c$	0	0	0	0	$\frac{1}{2}A$	0	0
$\frac{1}{2}(e+c)$	0	0	0	0	$\frac{1}{2}eb^T$	$\frac{1}{2}A$	0
c	0	0	0	0	0	0	A
	Mb^T	Mb^T	Mb^T	Mb^T	$-\,Qb^T$	$-\,Qb^T$	Nb^T

Example 1. Let us check the Butcher tableau under Table 2 by choosing first order Runge–Kutta method (explicit Euler (EE)) as underlying method (Table 3):

Table 3. The Butcher tableau of EE + RRE.

0	0	0	0	0	0	0	0
$\frac{1}{4}$	$\frac{1}{4}$	0	0	0	0	0	0
$\frac{1}{2}$	$\frac{1}{4}$	$\frac{1}{4}$	0	0	0	0	0
$\frac{3}{4}$	$\frac{1}{4}$	$\frac{1}{4}$	$\frac{1}{4}$	0	0	0	0
0	0	0	0	0	0	0	0
$\frac{1}{2}$	0	0	0	0	$\frac{1}{2}$	0	0
0	0	0	0	0	0	0	0
	$\frac{2}{3}$	$\frac{2}{3}$	$\frac{2}{3}$	$\frac{2}{3}$	-1	-1	$\frac{1}{3}$

$$y_{n+1} = y_n + hf(t_n, y_n), \tag{20}$$

Clearly, here **A** is a strictly lower triangular matrix and this shows that the explicit Euler method combined with RRE is again an explicit Runge–Kutta method. We verified that EE + RRE satisfies every order condition of a 3rd order method, but not all order conditions of 4th order methods. So RRE enhanced the order of the explicit Euler method by two.

4.1   Order Increase from 1 to 3

Proposition 1. *Assume that the underlying explicit Runge–Kutta method (14) is first order consistent. Then its combination with the repeated Richardson extrapolation yields a third order consistent method.*

Proof 1. Since $p = 1$, therefore $M = \frac{2}{3}$, $Q = 1$ and $N = \frac{1}{3}$. We have to verify the conditions, given in Table 1:

$$\mathbf{b}^T \mathbf{e} = 1, \tag{21}$$

$$\mathbf{b}^T \mathbf{c} = \frac{1}{2}, \tag{22}$$

$$\mathbf{b}^T \mathbf{c}^2 = \frac{1}{3}, \tag{23}$$

$$\mathbf{b}^T \mathbf{A} \mathbf{c} = \frac{1}{6}. \tag{24}$$

Checking (21):

$$\mathbf{b}^T \mathbf{e} = M b^T e + M b^T e + M b^T e + M b^T e - Q b^T e - Q b^T e + N b^T e = b^T e = 1$$

Checking (22):

$$\mathbf{b}^T \mathbf{c} = \left(M b^T \ M b^T \ M b^T \ M b^T \ -Q b^T \ -Q b^T \ N b^T \right) \cdot \begin{bmatrix} \frac{1}{4} c \\ \frac{1}{4}(e+c) \\ \frac{e}{2} + \frac{c}{4} \\ \frac{3e}{4} + \frac{c}{4} \\ \frac{c}{2} \\ \frac{1}{2}(e+c) \\ c \end{bmatrix}$$

$$= \left(\frac{4 - 3 - 3 + 2}{6} \right) b^T c + \left(\frac{1 + 2 + 3 - 3}{6} \right) b^T e = \frac{3}{6} b^T e = \frac{1}{2}$$

Checking (23):

$$\mathbf{b}^T \mathbf{c}^2 = M b^T \left(\frac{1}{4} c \right)^2 + M b^T \left(\frac{1}{4}(e+c) \right)^2 + M b^T \left(\frac{e}{2} + \frac{c}{4} \right)^2 + M b^T \left(\frac{3e}{4} + \frac{c}{4} \right)^2 - Q b^T (\frac{c}{2})^2$$

$$- Q b^T (\frac{e+c}{2})^2 + N b^T c^2$$

Since $p = 1$, therefore $M = \frac{2}{3}$, $Q = 1$ and $N = \frac{1}{3}$, and we have

$$\mathbf{b}^T \mathbf{c}^2 = \frac{2}{3} b^T \left(\frac{1}{4} c \right)^2 + \frac{2}{3} b^T \left(\frac{1}{4}(e+c) \right)^2 + \frac{2}{3} b^T \left(\frac{e}{2} + \frac{c}{4} \right)^2 + \frac{2}{3} b^T \left(\frac{3e}{4} + \frac{c}{4} \right)^2$$

$$- b^T (\frac{c}{2})^2 - b^T (\frac{e+c}{2})^2 + \frac{1}{3} b^T c^2 = \frac{8}{24} b^T e^2 = \frac{1}{3}$$

Checking (24):

$$\mathbf{b}^T \mathbf{A} \mathbf{c} =$$

$$(Mb^T \;\; Mb^T \;\; Mb^T \;\; Mb^T \;\; -Qb^T \;\; -Qb^T \;\; Nb^T) \begin{bmatrix} \frac{A}{4} & 0 & 0 & 0 & 0 & 0 & 0 \\ \frac{1}{4}eb^T & \frac{A}{4} & 0 & 0 & 0 & 0 & 0 \\ \frac{1}{4}eb^T & \frac{1}{4}eb^T & \frac{A}{4} & 0 & 0 & 0 & 0 \\ \frac{1}{4}eb^T & \frac{1}{4}eb^T & \frac{1}{4}eb^T & \frac{A}{4} & 0 & 0 & 0 \\ 0 & 0 & 0 & 0 & \frac{1}{2}A & 0 & 0 \\ 0 & 0 & 0 & 0 & \frac{1}{2}eb^T & \frac{1}{2}A & 0 \\ 0 & 0 & 0 & 0 & 0 & 0 & A \end{bmatrix} \begin{bmatrix} \frac{1}{4}c \\ \frac{1}{4}(e+c) \\ \frac{e}{2}+\frac{c}{4} \\ \frac{3e}{4}+\frac{c}{4} \\ \frac{c}{2} \\ \frac{1}{2}(e+c) \\ c \end{bmatrix}$$

$$= \frac{1}{6}$$

□

5   Convergence Analysis for RRE

The matrix A is strictly lower triangular in the Butcher tableau of an explicit Runge–Kutta method (ERK), and it was shown that the matrix $\mathbf{A}$ in the Butcher tableau of ERK + RRE is also strictly lower triangular. Hence, ERK + RRE yields an ERK method. Its order of consistency has been shown to increase by two if $p = 1$. It is known that in case the right-hand side function f has the Lipschitz property in its second argument, a pth order consistent method will also be pth order convergent. Therefore, we have the following conclusion.

Theorem 1. *The first order consistent ERK method combined with the RRE is convergent in order 3 if the right-hand side function f has the Lipschitz property in its second argument.*

Example 1. We considered the scalar problem

$$\begin{cases} y' = -2t \sin y, & t \in [0,1] \\ y(0) = 1. \end{cases} \tag{25}$$

The exact solution is $y(t) = 2\cot^{-1}\left(e^{t^2}\cot(\frac{1}{2})\right)$. The global errors were calculated in absolute value at the end of the time interval $[0,1]$. One can see the expected increase in the order of accuracy by two when EE is combined with RRE (Table 4).

Table 4. Global errors at $t = 1$ in absolute value using the explicit Euler method as underlying method.

H	EE	Order	EE+CRE	Order	EE+MRE	Order	EE+RRE	Order
0.1	1.99e-02		7.8397e-04		1.8774e-05		1.2348e-05	
		1.0820		2.1059		3.1410		3.1755
0.05	9.40e-03		1.8212e-04		2.1282e-06		1.3667e-06	
		1.0627		2.0511		3.0715		3.0879
0.025	4.50e-03		4.3945e-05		2.5317e-07		1.6074e-07	
		1.0324		2.0251		3.0360		3.0439
0.0125	2.20e-03		1.0797e-05		3.0867e-08		1.9490e-08	

6 Conclusion

To improve the accuracy of any convergent time-integration method, Richardson extrapolation and its variants (CRE and RRE) are applied. For RK methods with order $p = 1$, we proved that the RRE increases the order of consistency by two. Moreover, the convergence order is also increased by two when the right-hand side function f possesses the Lipschitz property in its second argument.

Acknowledgements. We appreciate Zahari Zlatev's useful advice during our investigation of the RRE method. The project has been supported by the Hungarian Scientific Research Fund OTKA SNN125119 and also OTKA K137699.

References

1. Bayleyogn, T., Faragó, I., Havasi, Á.: On the Consistency sOrder of Runge-Kutta Methods Combined with Active Richardson Extrapolation. In: Lirkov, I., Margenov, S. (eds) Large-Scale Scientific Computing. LSSC 2021. Lecture Notes in Computer Science, vol 13127. Springer, Cham (2022). 10.1007/978-3-030-97549-4_11

2. Faragó, I., Havasi, Á., Zlatev, Z. : Efficient Implementation Of Stable Richardson Extrapolation Algorithms. Comput. Math. Appl. **60**(8), 2309–2325 (2010) https://doi.org/10.1016/j.camwa.2010.08.025

3. Franke, J., Frank, W.: Application of generalized Richardson extrapolation to the computation of the flow across an asymmetric street intersection, J. Wind Eng. Ind. Aerodyn.. 4th International Symposium on Computational Wind Engineering, 96, 1616–1628 (2008)

4. Iserles, A.: A First Course in the Numerical Analysis of Differential Equations, Cambridge University Press, USA (1996)

5. Lambert, J.D.: Numerical Methods for Ordinary Differential Equations. Wiley, New York (1991)

6. Richardson, L.F.: The approximate arithmetical solution by finite differences of physical problems including differential equations, with an application to the stresses in a masonry dam. Philos. Trans. R. Soc. Lond. Ser. A **210**, 307–357 (1911)

7. Richardson, L.F.: The deferred approach to the limit, I. Single Lattice, Philos. Trans. R. Soc. A **226**, 299–349 (1927)
8. Süli, E., Mayers, D.: An introduction to numerical analysis, Cambridge University Press,New York (2003)
9. Zlatev, Z., Dimov, I., Faragó, I., Havasi, Á.: Richardson Extrapolation - practical aspects and applications. De Gruyter, Boston (2017)
10. Zlatev, Z., Dimov, I., Faragó, I., Georgiev, K., Havasi, Á.: Explicit Runge-Kutta methods combined with advanced versions of the richardson extrapolation. Comput. Methods Appl. Math. (2019). https://doi.org/10.1515/cmam-2019-0016
11. Zlatev, Z., Dimov, I., Faragó, I., Georgiev, K., Havasi, Á.: Stability properties of repeated Richardson extrapolation applied together with some implicit Runge-Kutta methods. In: Dimov, Ivan; Faragó, István; Vulkov, Lubin (eds.) Finite Difference Methods : Theory and Applications Cham, Svájc : Springer Nature Switzerland AG, pp. 114–125, 12 p. (2019)
12. Zlatev, Z., Dimov, I., Faragó, I., Georgiev, K., Havasi, Á.: Absolute Stability and Implementation of the Two-Times Repeated Richardson Extrapolation Together with Explicit Runge-Kutta Methods. In: Dimov, I., Faragó, I., Vulkov, L. (eds.) FDM 2018. LNCS, vol. 11386, pp. 678–686. Springer, Cham (2019). https://doi.org/10.1007/978-3-030-11539-5_80

Convergence Analysis of a Finite Volume Scheme for a Distributed Order Diffusion Equation

Fayssal Benkhaldoun[1] and Abdallah Bradji[1,2]([✉])

[1] LAGA, Sorbonne Paris Nord University, Villetaneuse, Paris, France
fayssal@math.univ-paris13.fr, abdallah.bradji@gmail.com,
abdallah.bradji@etu.univ-amu.fr
[2] Department of Mathematics, Faculty of Sciences, Badji Mokhtar-Annaba
University, Annaba, Algeria
abdallah.bradji@univ-annaba.dz
https://www.math.univ-paris13.fr/~fayssal/,
https://www.i2m.univ-amu.fr/perso/abdallah.bradji/

Abstract. We consider a distributed order diffusion equation with space-dependent conductivity. The distributed order operator is defined via an integral of the usual fractional Caputo derivative multiplied by a weight function ω, i.e. $\mathbb{D}_t^\omega u(t) = \int_0^1 \omega(\alpha)\partial_t^\alpha u(t)d\alpha$, where ∂_t^α is the Caputo derivative of order α given by $\partial_t^\alpha u(t) = \dfrac{1}{\Gamma(1-\alpha)} \int_0^t (t-s)^{-\alpha}u_s(s)ds$.

We establish a new fully discrete finite volume scheme in which the discretization in space is performed using the finite volume method developed in [9] whereas the discretization of the distributed order operator $\mathbb{D}_t^\omega u$ is given by an approximation of the integral, over the unit interval, using the known Mid Point rule and the approximation of the Caputo derivative $\partial_t^\alpha u$ is defined by the known $L1$-formula on the uniform temporal mesh.

We prove rigorously new error estimates in $L^\infty(L^2)$ and $L^2(H^1)$–discrete norms. These error estimates are obtained thanks to a new well developed discrete *a priori* estimate and also to the fact that the full discretization of the distributed-order fractional derivative leads to multi-term fractional order derivatives but the number of these terms is varying accordingly with the approximation of the integral over $(0,1)$.

This note is a continuation of our previous work [6] which dealt with the Gradient Discretization method (GDM) for time fractional-diffusion equation in which the fractional order derivative is fixed and it is given in the Caputo sense (without consideration the distributed-order fractional derivative) and conductivity is equal to one.

Supported by MCS team (LAGA Laboratory) of the "Université Sorbonne- Paris Nord".

I. Georgiev et al. (Eds.): NMA 2022, LNCS 13858, pp. 59–72, 2023.
https://doi.org/10.1007/978-3-031-32412-3_6

Keywords: distributed order · space-dependent conductivity · finite volumes · error estimates

MSC2010: 65M08 · 65M12

1   Introduction

We consider the following distributed order diffusion equation with space-dependent conductivity:

$$\mathbb{D}_t^\omega u(\boldsymbol{x},t) - \nabla \cdot (\kappa \nabla u)(\boldsymbol{x},t) = f(\boldsymbol{x},t), \quad (\boldsymbol{x},t) \in \Omega \times (0,T), \tag{1}$$

where Ω is an open bounded polyhedral subset in $\mathbb{R}^d$, with $d \in \mathbb{N}^* = \mathbb{N} \setminus \{0\}$, $T > 0$, and κ, ω, and f are given functions defined respectively on Ω, $(0,1)$, and $\Omega \times (0,T)$.

The distributed order operator $\mathbb{D}_t^\omega$ is defined by

$$\mathbb{D}_t^\omega \varphi(t) = \int_0^1 \omega(\alpha) \partial_t^\alpha \varphi(t) d\alpha \tag{2}$$

with ∂_t^α is the Caputo fractional derivative of order α given by

$$\partial_t^\alpha \varphi(t) = \frac{1}{\Gamma(1-\alpha)} \int_0^t (t-s)^{-\alpha} \varphi' ds, \quad \text{when} \quad 0 < \alpha < 1 \tag{3.a}$$

and

$$\partial_t^\alpha \varphi(t) = \varphi'(t), \quad \text{when} \quad \alpha = 1. \tag{3.b}$$

Initial condition is given by, for a given function u^0 defined on Ω

$$u(0) = u^0. \tag{3}$$

Homogeneous Dirichlet boundary conditions are given by

$$u(\boldsymbol{x},t) = 0, \quad (\boldsymbol{x},t) \in \partial\Omega \times (0,T). \tag{4}$$

For the sake of simplicity of the present note, we assume the following hypotheses. These hypotheses (which can be weakened) are needed to define a finite volume scheme and to analyse its convergence.

Assumption 1. (Assumption on the data of the problem (1)–(4)). *We assume the following assumptions:*

– *The conductivity κ is satisfying $\kappa \in \mathcal{C}^1(\overline{\Omega})$ and for some given $\kappa_0 > 0$*

$$\kappa(\boldsymbol{x}) > \kappa_0 > 0, \quad \forall \boldsymbol{x} \in \Omega. \tag{5}$$

– *The function (source term) f is satisfying $f \in \mathcal{C}\left([0,T]; \mathcal{C}(\overline{\Omega})\right)$.*

- *The initial data function u^0 is satisfying $u^0 \in H^2(\Omega)$.*
- *The weight function ω is satisfying $\omega \in \mathcal{C}([0,1])$ and for some given $\omega_0 > 0$*

$$\omega(\alpha) > \omega_0, \quad \forall \alpha \in [0,1]. \tag{6}$$

Fractional differential equations became one of the main tools to describe properties of various real materials, e.g. polymers, and to model anomalous diffusion phenomena in some heterogeneous aquifers and complex viscoelastic materials, see [15,17] and references therein. Time-fractional derivatives can be used to model time delays in some diffusion process. Distributed-order fractional derivative is a good tool for modeling a mixture delay sources, see [16,19].

From mathematical point of view, due to the presence of the integral (2) and consequently there is a variation of the order of the derivatives in $(0,1)$ (the fractional order derivative is not fixed but it is a variable in $(0,1)$), some additional considerable efforts should be paid with respect to fixed fractional order derivative as in (3.a). This is one of the main reasons of the existing of a considerable literature dealing with distributed order diffusion problems in addition to the one which treats only fixed-order fractional derivatives. As in the existing literature, we will remark that the convergence order in time is moving from $k^{2-\alpha}$ when considering the fixed order time fractional diffusion equation $\partial_t^\alpha u - \Delta u = f$ to $k^{1+\rho/2} + \sigma^2$ when considering the distributed order diffusion equation (1), see Theorem 1 below or for instance [19, Theorem 4.4]. This means that the order $k^{2-\alpha}$ is reduced to $k^{1+\rho/2}$ because of the approximation of the integral, defined over the unit interval, involved in (2). This fact (among other ones) shows how it is important to study extensively distributed order diffusion equation.

The aim if this note is to apply the finite volume method (called SUSHI-Scheme using Stabilization and Hybrid Interfaces), developed originally in [9] for the discretization of heterogeneous and anisotropic diffusion problems, to approximate the distributed order diffusion problem (1)–(4) in which the conductivity is space-dependent. We first shall establish a numerical scheme for the problem (1)–(4) in which the discretization in space is given by SUSHI whereas the discretization in time is perform using the Mid-Point rule and the $L1$ formula. We present a new convergence analysis approach which is based on new developed *a priori* estimates on the discrete solution in $L^\infty(L^2)$ and $L^2(H^1)$-discrete norms. This analysis will be extended to $L^\infty(H^1)$-discrete norm in future research papers. Indeed, the convergence analysis in $L^2(H^1)$-discrete norm seems new, in the area of the numerical methods devoted for distributed order diffusion problems, see for instance [8,11,12,19] in which there are no convergence results in the energy norm with respect to space direction, i.e. the convergence in space is given either in L^2 or L^∞.

This contribution is a continuation of some of our previous works which dealt with the numerical methods for time fractional partial differential equations in which the fractional order derivative is **fixed** and it is given in the Caputo sense (3.a), see [5–7].

2   Space Discretization

We use as discretization in space the following non-conforming mesh developed originally for elliptic equation in [9]. We recall its definition here for the sake of completeness.

Definition 1. (Space discretization, cf. [9]). *Let Ω be a polyhedral open bounded subset of $\mathbb{R}^d$, where $d \in \mathbb{N} \setminus \{0\}$, and $\partial\Omega = \overline{\Omega} \setminus \Omega$ its boundary. A discretization of Ω, denoted by $\mathcal{D}$, is defined as the triplet $\mathcal{D} = (\mathcal{M}, \mathcal{E}, \mathcal{P})$, where:*

1. *$\mathcal{M}$ is a finite family of non empty connected open disjoint subsets of Ω (the "control volumes") such that $\overline{\Omega} = \cup_{K \in \mathcal{M}} \overline{K}$. For any $K \in \mathcal{M}$, let $\partial K = \overline{K} \setminus K$ be the boundary of K; let $\mathrm{m}(K) > 0$ denote the measure of K and h_K denote the diameter of K.*
2. *$\mathcal{E}$ is a finite family of disjoint subsets of $\overline{\Omega}$ (the "edges" of the mesh), such that, for all $\sigma \in \mathcal{E}$, σ is a non empty open subset of a hyperplane of $\mathbb{R}^d$, whose $(d-1)$–dimensional measure is strictly positive. We also assume that, for all $K \in \mathcal{M}$, there exists a subset $\mathcal{E}_K$ of $\mathcal{E}$ such that $\partial K = \cup_{\sigma \in \mathcal{E}_K} \overline{\sigma}$. For any $\sigma \in \mathcal{E}$, we denote by $\mathcal{M}_\sigma = \{K, \sigma \in \mathcal{E}_K\}$. We then assume that, for any $\sigma \in \mathcal{E}$, either $\mathcal{M}_\sigma$ has exactly one element and then $\sigma \subset \partial\Omega$ (the set of these interfaces, called boundary interfaces, denoted by $\mathcal{E}_{\text{ext}}$) or $\mathcal{M}_\sigma$ has exactly two elements (the set of these interfaces, called interior interfaces, denoted by $\mathcal{E}_{\text{int}}$). For all $\sigma \in \mathcal{E}$, we denote by $\boldsymbol{x}_\sigma$ the barycentre of σ. For all $K \in \mathcal{M}$ and $\sigma \in \mathcal{E}$, we denote by $\boldsymbol{n}_{K,\sigma}$ the unit vector normal to σ outward to K.*
3. *$\mathcal{P}$ is a family of points of Ω indexed by $\mathcal{M}$, denoted by $\mathcal{P} = (\boldsymbol{x}_K)_{K \in \mathcal{M}}$, such that for all $K \in \mathcal{M}$, $\boldsymbol{x}_K \in K$ and K is assumed to be $\boldsymbol{x}_K$–star-shaped, which means that for all $\boldsymbol{x} \in K$, the property $[\boldsymbol{x}_K, \boldsymbol{x}] \subset K$ holds. Denoting by $d_{K,\sigma}$ the Euclidean distance between $\boldsymbol{x}_K$ and the hyperplane including σ, one assumes that $d_{K,\sigma} > 0$. We then denote by $\mathcal{D}_{K,\sigma}$ the cone with vertex $\boldsymbol{x}_K$ and basis σ.*

We define the discrete space $\mathcal{X}_{\mathcal{D},0}$ as the set of all $v = \left((v_K)_{K \in \mathcal{M}}, (v_\sigma)_{\sigma \in \mathcal{E}}\right)$, where $v_K, v_\sigma \in \mathbb{R}$ and $v_\sigma = 0$ for all $\sigma \in \mathcal{E}_{\text{ext}}$. Let $H_\mathcal{M}(\Omega) \subset L^2(\Omega)$ be the space of functions which are constant on each control volume K of the mesh $\mathcal{M}$. For all $v \in \mathcal{X}_\mathcal{D}$, we denote by $\Pi_\mathcal{M} v \in H_\mathcal{M}(\Omega)$ the function defined by $\Pi_\mathcal{M} v(\boldsymbol{x}) = v_K$, for a.e. $\boldsymbol{x} \in K$, for all $K \in \mathcal{M}$. To analyze the convergence, we consider the size of the discretization $\mathcal{D}$ defined by $h_\mathcal{D} = \sup \{\mathrm{diam}(K), K \in \mathcal{M}\}$. We define the regularity of the mesh using the parameter $\theta_\mathcal{D}$ given by

$$\theta_\mathcal{D} = \max \left(\max_{\sigma \in \mathcal{E}_{\text{int}}, K, L \in \mathcal{M}} \frac{d_{K,\sigma}}{d_{L,\sigma}}, \max_{K \in \mathcal{M}, \sigma \in \mathcal{E}_K} \frac{h_K}{d_{K,\sigma}} \right). \tag{7}$$

The scheme we shall present is based on the discrete gradient of [9]. For $u \in \mathcal{X}_\mathcal{D}$, we define, for all $K \in \mathcal{M}$

$$\nabla_\mathcal{D} u(\boldsymbol{x}) = \nabla_K u + \left(\frac{\sqrt{d}}{d_{K,\sigma}} (u_\sigma - u_K - \nabla_K u \cdot (\boldsymbol{x}_\sigma - \boldsymbol{x}_K)) \right) \boldsymbol{n}_{K,\sigma}, \quad \text{a.e. } \boldsymbol{x} \in \mathcal{D}_{K,\sigma}, \tag{8}$$

where $\nabla_K u = \dfrac{1}{m(K)} \displaystyle\sum_{\sigma \in \mathcal{E}_K} m(\sigma)\, (u_\sigma - u_K)\, \mathbf{n}_{K,\sigma}$.

We define the bilinear form $\langle \cdot, \cdot \rangle_\kappa$ defined on $\mathcal{X}_{\mathcal{D},0} \times \mathcal{X}_{\mathcal{D},0}$ by

$$\langle u, v \rangle_\kappa = \int_\Omega \kappa(\boldsymbol{x}) \nabla_\mathcal{D} u(\boldsymbol{x}) \cdot \nabla_\mathcal{D} v(\boldsymbol{x}) d\boldsymbol{x}. \tag{9}$$

3   Approximation of the Distributed Order Operator and Some Preliminaries

To discretize the distributed order operator (2), we have to perform two approximations:

1. **Approximation of the integral of (2).** The integral (2) defined over $\alpha \in (0,1)$ can be discretized using various known quadrature formulas, e.g. composite Trapezoid and Simpson formulas. For the sake of simplicity, we only consider in this note the Mid-Point formula on the uniform mesh. We will address a general case, in which the discretization of the integral defined over $\alpha \in (0,1)$ is performed using a "generic approximation" on general mesh, in a future paper.

For a uniform mesh with constant step $\rho = 1/q$, with $q \in \mathbb{N} \setminus \{0\}$, we consider the Mid-Point formula as approximation of the integral $\displaystyle\int_0^1 \omega(\alpha)\psi(\alpha)d\alpha$ given by

$$\int_0^1 \omega(\alpha)\psi(\alpha)d\alpha \approx \frac{1}{q} \sum_{i=0}^{q-1} \omega(\alpha_i)\psi(\alpha_i), \tag{10}$$

where α_i are the mid points of the interval $(\xi_i, \xi_{i+1}) = (\dfrac{i}{q}, \dfrac{i+1}{q})$, i.e., for $i \in [\![0, q-1]\!]$

$$\alpha_i = \frac{\xi_i + \xi_{i+1}}{2} = \xi_i + \frac{1}{2q} = \frac{i}{q} + \frac{1}{2q} = \frac{1}{q}\left(i + \frac{1}{2}\right).$$

By replacing ψ with $\partial_t^\alpha \varphi(t)$ for a fixed t in (10), we get the following approximation for the distributed-order fractional derivative (2)

$$\mathbb{D}_t^\omega \varphi(t) \approx \frac{1}{q} \sum_{i=0}^{q-1} \omega(\alpha_i)\partial_t^{\alpha_i} \varphi(t). \tag{11}$$

Thanks to this approximation, it remains only to approximate the terms $\partial_t^{\alpha_i} \varphi(t)$, for each $i \in [\![0, q-1]\!]$, which appear on the r.h.s. (right hand side) of (11) in order to get a full discretization for the distributed order derivative (2)–(3.a).

2. Approximation of the Fractional Caputo derivatives of r.h.s. of (11).

The approximation of Caputo derivatives of the r.h.s. of (11) can be performed using the known formula $L1$ and $L2-1_\sigma$ on general temporal meshes, see more details on these formulas in [1,14,15,20]. For the sake of clarity, we focus in this note on the use of $L1$-formula on the uniform temporal mesh as approximation for (3.a) and address a general "framework" in a near future work. The basic idea of $L1$-formula is to approximate u involved in (3.a) using piece-wise linear functions: on each sub-interval arising after discretization, the function u is "considered" as a linear function and consequently, its derivative on this sub-interval is constant. To be more precise, let us first consider the uniform mesh $t_n = nk$, for $n \in [\![0, N+1]\!]$, where k is the constant step $k = \dfrac{T}{N+1}$ and $N \in \mathbb{N} \setminus \{0\}$ is given. Accordingly with the stated principle of $L1$-formula, we approximate Caputo derivatives of the r.h.s. of (11) as

$$\partial_t^{\alpha_i} \varphi(t_{n+1}) = \frac{1}{\Gamma(1-\alpha_i)} \sum_{j=0}^{n} \int_{t_j}^{t_{j+1}} (t-s)^{-\alpha_i} \varphi'(s)ds \approx \sum_{j=0}^{n} k\lambda_j^{n+1,\alpha_i} \partial^1 u(t_{j+1}), \quad (12)$$

where ∂^1 is the discrete first time derivative given by $\partial^1 v^{n+1} = \dfrac{v^{n+1} - v^n}{k}$ and

$$\lambda_j^{n+1,\alpha_i} = \frac{(t_{n+1} - t_j)^{1-\alpha_i} - (t_{n+1} - t_{j+1})^{1-\alpha_i}}{k\Gamma(2-\alpha_i)}. \quad (13)$$

3. Full approximation for the distributed order operator (2).

Thanks to (11) and (12), we have the following approximation for the distributed order operator (2):

$$\mathbb{D}_t^\omega \varphi(t_{n+1}) \approx \frac{1}{q} \sum_{i=0}^{q-1} \omega(\alpha_i) \sum_{j=0}^{n} k\lambda_j^{n+1,\alpha_i} \partial^1 \varphi(t_{j+1})$$

$$= \sum_{j=0}^{n} kP_j^{q,n+1} \partial^1 \varphi(t_{j+1}), \quad (14)$$

where

$$P_j^{q,n+1} = \frac{1}{q} \sum_{i=0}^{q-1} \omega(\alpha_i)\lambda_j^{n+1,\alpha_i}. \quad (15)$$

Throughout this paper, the letter C stands for a positive constant independent of the parameters of discretizations.

The approximations (10) and (12) generate the following errors:

– For a function $\psi \in \mathcal{C}([0,1])$, we define the error

$$\mathbb{T}_1(\psi) = \int_0^1 \omega(\alpha)\psi(\alpha)d\alpha - \frac{1}{q} \sum_{i=0}^{q-1} \omega(\alpha_i)\psi(\alpha_i). \quad (16)$$

The following error estimate is known for the Mid-Point rule (it can be justified using Taylor's expansion with integral remainder), for a function $\psi \in H^2(0,1)$ (this implies, thanks to Sobolev embeddings, $\psi \in C^1[0,1]$) and under the assumption $\omega \in C^2([0,1])$, see also [10, Page 198]:

$$|\mathbb{T}_1(\psi)| \leq C\rho^2 |\psi|_{H^2(0,1)}. \tag{17}$$

– For any function $\varphi \in C^1([0,T])$ and for any $0 < \beta < 1$, we define the error

$$\mathbb{T}_2^{n+1,\beta}(\varphi) = \partial_t^\beta \varphi(t_{n+1}) - \sum_{j=0}^{n} k\lambda_j^{n+1,\beta} \partial^1 u(t_{j+1}). \tag{18}$$

It is proved in [20, Lemma 2.1, Page 197] that, for $\varphi \in C^2([0,T])$

$$|\mathbb{T}_2^{n+1,\beta}(\varphi)| \leq Ck^{2-\beta} \|\varphi''\|_{C([0,T])}. \tag{19}$$

We will need to use the following technical lemma which can be found for instance in [6].

Lemma 1. (Properties of the coefficients λ_j^{n+1,α_i}, see [6]). *Let λ_j^{n+1,α_i} be defined as in (13). Then, the following properties hold:*

– *Increasing property, for all $i \in [\![0, q-1]\!]$*

$$\frac{k^{-\alpha_i}}{\Gamma(2-\alpha_i)} = \lambda_n^{n+1,\alpha_i} > \lambda_{n-1}^{n+1,\alpha_i} > \ldots > \lambda_0^{n+1,\alpha_i} \geq \lambda_0 = \frac{T^{-\alpha_i}}{\Gamma(1-\alpha_i)}. \tag{20}$$

– *Estimate on the sum $\sum_{j=0}^{n} k\lambda_j^{n+1,\alpha_i}$. For all $i \in [\![0, q-1]\!]$*

$$\sum_{j=0}^{n} k\lambda_j^{n+1,\alpha_i} \leq \frac{T^{1-\alpha_i}}{\Gamma(2-\alpha_i)}. \tag{21}$$

4 Statement of the Main Results of This Work

This section is devoted to the main results of this note. We first provide the formulation of a finite volume scheme approximating the problem (1)–(4). We then prove its unique solvability, stability, and its convergence. These results are proved thanks to a new discrete a priori estimate.

4.1 Formulation of a New Finite Volume Scheme for the Problem (1)–(4)

Using the approximation (14) together with the principles of the finite volume method developed in [9], we suggest the following scheme as approximation for the problem (1)–(4):

– Discretization of initial condition (3) with Dirichlet boundary conditions (4): Find $u_{\mathcal{D}}^0 \in \mathcal{X}_{\mathcal{D},0}$ such that

$$\langle u_{\mathcal{D}}^0, v \rangle_{\kappa} = - \left(\nabla \cdot (\kappa \nabla u^0), \Pi_{\mathcal{M}} v \right)_{L^2(\Omega)}, \quad \forall v \in \mathcal{X}_{\mathcal{D},0}. \tag{22}$$

– Discretization of the distributed order diffusion equation (1) with Dirichlet boundary conditions (4): For any $n \in [\![0, N]\!]$, find $u_{\mathcal{D}}^{n+1} \in \mathcal{X}_{\mathcal{D},0}$ such that, for all $v \in \mathcal{X}_{\mathcal{D},0}$

$$\sum_{j=0}^{n} k P_j^{q,n+1} \left(\partial^1 \Pi_{\mathcal{M}} u_{\mathcal{D}}^{j+1}, \Pi_{\mathcal{M}} v \right)_{L^2(\Omega)} + \langle u_{\mathcal{D}}^{n+1}, v \rangle_{\kappa} = (f(t_{n+1}), \Pi_{\mathcal{M}} v)_{L^2(\Omega)}. \tag{23}$$

Thanks to Assumption 1, the numerical scheme (22)–(23) is meaningful.

4.2 Unique Solvability, Stability, and Convergence Rate for Scheme (22)–(23)

In order to provide an error estimate with a convergence rate, we need the following hypotheses on the exact solution u of the problem (1)–(4) and on the weight function ω. Such hypotheses are similar to the ones assumed in [8, Theorem 2, Page 426] and [11, Theorem 2, Page 930] and they are not needed nor to the unique solvability neither to the stability of scheme (22)–(23):

Assumption 2. (Assumptions on u and ω). *We assume the following assumptions on the exact solution of (1)–(4) and on the weight function ω:*

– $u \in \mathcal{C}^2([0, T]; \mathcal{C}^2(\overline{\Omega}))$.
– *For any $t \in [0, T]$, the function $\theta : \alpha \in (0,1) \mapsto \partial_t^\alpha u(t)$ is twice differentiable and θ'' is bounded uniformly with respect to t.*
– *The function ω satisfies $\omega \in \mathcal{C}^2([0, 1])$.*

The following theorem provides unique solvability and convergence rate of scheme (22)–(23):

Theorem 1. (Error estimates for scheme (22)–(23)). *Under Assumption 1, let Ω be a polyhedral open bounded subset of $\mathbb{R}^d$, where $d \in \mathbb{N} \setminus \{0\}$. Let $\mathcal{D} = (\mathcal{M}, \mathcal{E}, \mathcal{P})$ be a discretization in the sense of Definition 1. Assume that $\theta_{\mathcal{D}}$ given by (7) satisfies $\theta \geq \theta_{\mathcal{D}}$, for some given positive number θ. Let $\nabla_{\mathcal{D}}$ be the discrete gradient given by (8). Let $\langle \cdot, \cdot \rangle_{\kappa}$ be the bilinear form $\mathcal{X}_{\mathcal{D},0} \times \mathcal{X}_{\mathcal{D},0}$ defined by (9). Consider a discretization of the interval $[0,1]$, in which the fractional orders are varying, by the equidistant mesh points $0 = \xi_0 < \ldots < \xi_q = 1$ with $\xi_i = i/q$ and $q \in \mathbb{N} \setminus \{0\}$ is given. Let us set $\rho = 1/q$. Let $\alpha_0 < \ldots < \alpha_{q-1}$ be the mid-points of the sub-intervals $[0, \xi_1], \ldots, [\xi_{q-1}, \xi_q]$, i.e. $\alpha_i = \dfrac{1}{q} \left(i + \dfrac{1}{2} \right)$. The discretization of $[0, T]$, where the time is varying, is performed using uniform mesh given by $t_n = nk$ with $k = \dfrac{T}{N+1}$ and $N \in \mathbb{N} \setminus \{0\}$. Then, there exists a unique solution $(u_{\mathcal{D}}^n)_{n=0}^{N+1} \in \mathcal{X}_{\mathcal{D},0}^{N+2}$ for the finite volume scheme (22)–(23).*

If we assume in addition that the Assumption 2 is satisfied, then the following $L^2(H_0^1)$ *and* $L^\infty(L^2)$ *error estimates hold:*

$$\left(\sum_{n=0}^{N} k \|\nabla_{\mathcal{D}} u_{\mathcal{D}}^n - \nabla u(t_n)\|_{L^2(\Omega)^d}^2 \right)^{\frac{1}{2}} \leq C \left(h_{\mathcal{D}} + k^{1+\rho/2} + \rho^2 \right) \qquad (24)$$

and

$$\max_{n=0}^{N+1} \|\Pi_{\mathcal{M}} u_{\mathcal{D}}^n - u(t_n)\|_{L^2(\Omega)} \leq C \left(h_{\mathcal{D}} + k^{1+\rho/2} + \rho^2 \right). \qquad (25)$$

To prove Theorem 1, we need some lemmata and an *a priori* estimate. This *a priori* estimate will be set as Theorem 2 for its own importance. Theorem 2 will also serve to prove the stability of scheme (22)–(23). Theorem 2 is an extension of [6, Theorem 402, Page 504], which dealt with numerical schemes for time fractional PDEs in which the fractional order derivative is fixed and given by the Caputo derivative formula (3.a) and the conductivity is equal to one, to distributed-order fractional derivative and space-dependent conductivity.

Theorem 2. (New discrete a priori estimates for the discrete problem, extension of [6]). *Under the same hypotheses of Theorem 1, except Assumption 2, we assume that there exists* $(\eta_{\mathcal{D}}^n)_{n=0}^{N+1} \in (\mathcal{X}_{\mathcal{D},0})^{N+2}$ *such that for all* $n \in [\![0, N]\!]$ *and for all* $v \in \mathcal{X}_{\mathcal{D},0}$

$$\sum_{j=0}^{n} k P_j^{q,n+1} \left(\partial^1 \Pi_{\mathcal{M}} \eta_{\mathcal{D}}^{j+1}, \Pi_{\mathcal{M}} v \right)_{L^2(\Omega)} + \langle \eta_{\mathcal{D}}^{n+1}, v \rangle_{\kappa} = \left(S^{n+1}, \Pi_{\mathcal{M}} v \right)_{L^2(\Omega)} (26)$$

with, for all $n \in [\![0, N]\!]$, $S^{n+1} \in L^2(\Omega)$. *Let us denote* $S = \max_{n=0}^{N} \|S^{n+1}\|_{L^2(\Omega)}$.
Then, the following $L^2(H_0^1)$ *and* $L^\infty(L^2)$*–estimates hold:*

$$\max_{n=0}^{N+1} \|\Pi_{\mathcal{M}} \eta_{\mathcal{D}}^n\|_{L^2(\Omega)} + \left(\sum_{n=0}^{N+1} k \|\nabla_{\mathcal{D}} \eta_{\mathcal{D}}^n\|_{L^2(\Omega)^d}^2 \right)^{\frac{1}{2}}$$
$$\leq C \left(S + \|\Pi_{\mathcal{M}} \eta_{\mathcal{D}}^0\|_{L^2(\Omega)} \right). \qquad (27)$$

An Outline on the Proof. The proof of Theorem 2 follows the techniques of [6, Theorem 402, Page 504] combined with the following property which can be checked using Lemma 1 and Lemma 3 below, for some $P_0 > 0$ independent of the parameters of discretizations

$$P_n^{q,n+1} > P_{n-1}^{q,n+1} > \ldots > P_0^{q,n+1} \geq P_0. \qquad (28)$$

This completes the proof of Theorem 2.   □

Theorem 2 implies the following well-posedness results, of scheme (22)–(23). Such results can be proved using the fact that the matrices involved in (22)–(23) are square and using also the *a priori* estimate (27).

Corollary 1. (Discrete well-posedness). *Under the same hypotheses of Theorem 2, there exists a unique solution* $(u_{\mathcal{D}}^n)_{n=0}^{N+1} \in \mathcal{X}_{\mathcal{D},0}^{N+2}$ *for scheme* (22)–(23) *and the following stability result holds:*

$$\max_{n=0}^{N+1} \|\Pi_{\mathcal{M}} u_{\mathcal{D}}^n\|_{L^2(\Omega)} + \left(\sum_{n=0}^{N+1} k \|\nabla_{\mathcal{D}} u_{\mathcal{D}}^n\|_{L^2(\Omega)^d}^2\right)^{\frac{1}{2}}$$

$$\leq C\left(\|f\|_{\mathcal{C}([0,T];\ L^2(\Omega))} + \|\Delta u^0\|_{L^2(\Omega)}\right). \tag{29}$$

The Lemma 2 below can be deduced from the analysis of [19]. The result (31) of such lemma is not stated in a clear manner but we highlight it for the sake of clarity and completeness.

Lemma 2. (First technical lemma, see [19]). *Let* $\mathbb{T}_3^{n+1}$ *be the error in the approximation* (14):

$$\mathbb{T}_3^{n+1}(\varphi) = \mathbb{D}_t^\omega \varphi(t_{n+1}) - \sum_{j=0}^n k P_j^{q,n+1} \partial^1 \varphi(t_{j+1}). \tag{30}$$

Then, under the assumption that $\omega \in \mathcal{C}^2([0,1])$, *the following estimate holds:*

$$\max_{n=0}^{n+1} |\mathbb{T}_3^{n+1}(\varphi)| \leq C\left(k^{1+\rho/2} + \rho^2\right), \quad \forall \varphi \in \mathcal{C}^2([0,T]). \tag{31}$$

Proof. The error $\mathbb{T}_3(\varphi)$ can be written as (recall that $\mathbb{T}_1$ and $\mathbb{T}_2^{n+1,\alpha_i}$ are given by (16) and (18))

$$\mathbb{T}_3^{n+1}(\varphi) = \mathbb{T}_1\left(\partial_t^\alpha \varphi(t_{n+1})\right) + \frac{1}{q}\sum_{i=0}^{q-1} \omega(\alpha_i) \mathbb{T}_2^{n+1,\alpha_i}(\varphi). \tag{32}$$

The second term in r.h.s. of (32) can be estimated using (19) (recall that $\alpha_i \leq \alpha_{q-1} = 1 - \frac{1}{2q}$)

$$\left|\frac{1}{q}\sum_{i=0}^{q-1} \omega(\alpha_i) \mathbb{T}_2^{n+1,\alpha_i}(\varphi)\right| \leq \frac{C}{q}\sum_{i=0}^{q-1} \omega(\alpha_i) k^{2-\alpha_i} \|\varphi''\|_{\mathcal{C}([0,T])} \leq C k^{2-\alpha_{q-1}} \|\omega\|_{\mathcal{C}([0,1])} \|\varphi''\|_{\mathcal{C}([0,T])}$$

$$\leq C k^{1+\frac{1}{2q}} \|\omega\|_{\mathcal{C}([0,1])} \|\varphi''\|_{\mathcal{C}([0,T])}. \tag{33}$$

On the other hand, using (17) and Assumption 2 implies that $|\mathbb{T}_1\left(\partial_t^\alpha \varphi(t_{n+1})\right)| \leq C\rho^2$. Gathering this with (33) yields the desired estimate (31). $\qquad\square$

The following lemma will be useful to prove Theorem 1.

Lemma 3. (New technical lemma). *Let* λ_j^{n+1,α_i} *and* $P_j^{q,n+1}$ *be defined respectively by* (13) *and* (15). *The following estimates hold, for some positive constants* γ *and* P_0 *which are independent of the parameters of the discretizations*

$$\sum_{j=0}^n k P_j^{q,n+1} \leq \gamma \quad \text{and} \quad \max_{n=0}^N \max_{j=0}^n P_j^{q,n+1} > P_0. \tag{34}$$

Proof. Using estimate (21) and definition (16) of $\mathbb{T}_1$ yields (recall that the expression in the left hand side of the first estimate in (34) is positive thanks to the Assumption 1 and to (20))

$$\sum_{j=0}^{n} k P_j^{q,n+1} = \frac{1}{q} \sum_{i=0}^{q-1} \omega(\alpha_i) \sum_{j=0}^{n} k \lambda_j^{n+1,\alpha_i} \leq \frac{1}{q} \sum_{i=0}^{q-1} \omega(\alpha_i) \frac{T^{1-\alpha_i}}{\Gamma(2-\alpha_i)}$$

$$= -\mathbb{T}_1 \left(\frac{T^{1-\alpha}}{\Gamma(2-\alpha)} \right) + \int_0^1 \omega(\alpha) \frac{T^{1-\alpha}}{\Gamma(2-\alpha)} d\alpha. \tag{35}$$

The function $\alpha \mapsto \dfrac{T^{1-\alpha}}{\Gamma(2-\alpha)}$ is $\mathcal{C}^\infty([0,1])$ and consequently, thanks to (17)

$$\left| \mathbb{T}_1 \left(\frac{T^{1-\alpha}}{\Gamma(2-\alpha)} \right) \right| \leq C \rho^2. \tag{36}$$

Gathering this with (35) and the fact that $\rho < 1$ yields the first desired estimate in (34).

The second desired estimate in (34) can be proved in a similar way by using the property (20) and the fact that $P_j^{q,n+1} > -\mathbb{T}_1 \left(\dfrac{T^{-\alpha}}{\Gamma(1-\alpha)} \right) + \displaystyle\int_0^1 \omega(\alpha) \frac{T^{-\alpha}}{\Gamma(1-\alpha)} d\alpha.$ □

Proof sketch for Theorem 1

The existence and uniqueness for scheme (22)–(23) are stated in Corollary 1. To prove the error estimates (24)–(25), we compare the finite volume scheme (22)–(23) with the following auxiliary "optimal" scheme: For any $n \in [\![0, N+1]\!]$, find $\overline{u}_{\mathcal{D}}^n \in \mathcal{X}_{\mathcal{D},0}$ such that

$$\langle \overline{u}_{\mathcal{D}}^n, v \rangle_\kappa = (-\nabla \cdot (\kappa \nabla u)(t_n), \Pi_{\mathcal{M}} v)_{L^2(\Omega)}, \quad \forall v \subset \mathcal{X}_{\mathcal{D},0}. \tag{37}$$

For any $n \in [\![0, N+1]\!]$, let $\xi_{\mathcal{D}}^n$ be the error given by $\xi_{\mathcal{D}}^n = u(t_n) - \Pi_{\mathcal{M}} \overline{u}_{\mathcal{D}}^n$. The following convergence results hold, see [6,9]:

$$\max_{n=0}^{N+1} \left(\|\xi_{\mathcal{D}}^n\|_{L^2(\Omega)} + \|\nabla u(t_n) - \nabla_{\mathcal{D}} \overline{u}_{\mathcal{D}}^n\|_{(L^2(\Omega))^d} \right) + \max_{n=1}^{N+1} \|\partial^1 \xi_{\mathcal{D}}^n\|_{L^2(\Omega)} \leq C h_{\mathcal{D}}. \tag{38}$$

We define the auxiliary error $\eta_{\mathcal{D}}^n = u_{\mathcal{D}}^n - \overline{u}_{\mathcal{D}}^n$. Comparing (37) with (22) and using the fact that $u(0) = u^0$ (subject of (3)) imply that $\eta_{\mathcal{D}}^0 = 0$. Writing scheme (37) in the level $n+1$, subtracting the result from (23), and using the fact that $\mathbb{D}_t^\omega u(t_{n+1}) = f(t_{n+1}) + \nabla \cdot (\kappa \nabla u)(t_{n+1})$ (subject of (1)) to get, for all $n \in [\![0, N]\!]$

$$\sum_{j=0}^{n} k P_j^{q,n+1} \left(\partial^1 \Pi_{\mathcal{M}} u_{\mathcal{D}}^{j+1}, \Pi_{\mathcal{M}} v \right)_{L^2(\Omega)} + \langle \eta_{\mathcal{D}}^{n+1}, v \rangle_\kappa = (\mathbb{D}_t^\omega u(t_{n+1}), \Pi_{\mathcal{M}} v)_{L^2(\Omega)}, \quad \forall v \in \mathcal{X}_{\mathcal{D},0}.$$

Subtracting now $\sum_{j=0}^{n} k P_j^{q,n+1} \left(\partial^1 \Pi_{\mathcal{M}} \overline{u}_{\mathcal{D}}^{j+1}, \Pi_{\mathcal{M}} v \right)_{L^2(\Omega)}$ from the both sides of the previous equation yields

$$\sum_{j=0}^{n} k P_j^{q,n+1} \left(\partial^1 \Pi_{\mathcal{M}} \eta_{\mathcal{D}}^{j+1}, \Pi_{\mathcal{M}} v \right)_{L^2(\Omega)} + \langle \eta_{\mathcal{D}}^{n+1}, v \rangle_\kappa = \left(\mathcal{S}^{n+1}, \Pi_{\mathcal{M}} v \right)_{L^2(\Omega)}, \quad (39)$$

where

$$\mathcal{S}^{n+1} = \mathbb{D}_t^\omega u(t_{n+1}) - \sum_{j=0}^{n} k P_j^{q,n+1} \partial^1 \Pi_{\mathcal{M}} \overline{u}_{\mathcal{D}}^{j+1}.$$

Thanks to (39), the auxiliary errors $(\eta_{\mathcal{D}}^n)_{n=0}^{N+1} = (u_{\mathcal{D}}^n - \overline{u}_{\mathcal{D}}^n)_{n=0}^{N+1}$ satisfy the hypothesis (26) of Theorem 2. This allows to apply the *a priori* estimates (27) of Theorem 2 to get

$$\max_{n=0}^{N+1} \| \Pi_{\mathcal{M}} \eta_{\mathcal{D}}^n \|_{L^2(\Omega)} + \left(\sum_{n=0}^{N+1} k \| \nabla_{\mathcal{D}} \eta_{\mathcal{D}}^n \|_{L^2(\Omega)^d}^2 \right)^{\frac{1}{2}} \leq C \left(\mathcal{S} + \| \Pi_{\mathcal{M}} \eta_{\mathcal{D}}^0 \|_{L^2(\Omega)} \right).$$
$$(40)$$

Using the triangle inequality, we have

$$\| \mathcal{S}^{n+1} \|_{L^2(\Omega)} \leq \| \mathbb{T}_3^{n+1}(u) \|_{L^2(\Omega)} + \mathbb{T}_4^{n+1}, \quad (41)$$

where $\mathbb{T}_3^{n+1}$ is given by (30) of Lemma 2 and

$$\mathbb{T}_4^{n+1} = \left\| \sum_{j=0}^{n} k P_j^{q,n+1} \partial^1 \xi_{\mathcal{D}}^{j+1} \right\|_{L^2(\Omega)}. \quad (42)$$

Using estimate (31) (resp. the first estimate in (34) and (38)) yields respectively

$$\| \mathbb{T}_3^{n+1}(u) \|_{L^2(\Omega)} \leq C \left(k^{1+\rho/2} + \rho^2 \right) \quad \text{and} \quad \mathbb{T}_4^{n+1} \leq C h_{\mathcal{D}}.$$

Gathering this with (40)–(41) and the fact that $\eta_{\mathcal{D}}^0 = 0$ leads to

$$\max_{n=0}^{N+1} \| \Pi_{\mathcal{M}} \eta_{\mathcal{D}}^n \|_{L^2(\Omega)} + \left(\sum_{n=0}^{N+1} k \| \nabla_{\mathcal{D}} \eta_{\mathcal{D}}^n \|_{L^2(\Omega)^d}^2 \right)^{\frac{1}{2}} \leq C \left(h_{\mathcal{D}} + k^{1+\rho/2} + \rho^2 \right). \quad (43)$$

Gathering (43), the triangle inequality, and (38) leads to the desired error estimates (24)–(25). □

5   Conclusion and Perspectives

We considered a distributed order diffusion equation with space-dependent conductivity. The equation is equipped with initial and homogeneous Dirichlet boundary conditions. We designed a new fully discrete finite volume scheme in which the discretization in space is performed using the finite volume method developed in [9] whereas the discretization in time is given by an approximation of the integral, defined over $(0,1)$, using the known Mid Point rule and the approximation of the Caputo derivative using the known $L1$-formula on the uniform temporal mesh.

We developed a new discrete *a priori* estimate which provides an estimate for the discrete solution in $L^\infty(L^2)$ and $L^2(H^1)$–discrete norms. This *a priori* estimate together with some additional techniques allowed to justify the well-posedness of the numerical scheme and to prove its convergence in $L^\infty(L^2)$ and $L^2(H^1)$–discrete norms.

This note is promising which led us to plan several perspectives to work on in the near future. Among these perspectives, we shall first prove error estimate in $L^\infty(H^1)$–discrete norm using a general integration rule as approximation for the integral, defined over $(0,1)$, and the so-called $L2 - 1_\sigma$ as approximation for the Caputo derivative. Another interesting task is to improve the order $k^{1+\rho/2}$ which appears in the error estimates (24)–(25). We think that such improvement is feasible, and some ideas are in progress, for a well chosen quadrature formula for the integral $\int_0^1 \omega(\alpha)\psi(\alpha)d\alpha$.

References

1. Alikhanov, A.-A.: A new difference scheme for the fractional diffusion equation. J. Comput. Phys. **280**, 424–438 (2015)
2. Benkhaldoun, F., Bradji, A.: A second order time accurate finite volume scheme for the time-fractional diffusion wave equation on general nonconforming meshes. In: Lirkov, I., Margenov, S. (eds.) LSSC 2019. LNCS, vol. 11958, pp. 95–104. Springer, Cham (2020). https://doi.org/10.1007/978-3-030-41032-2_10
3. Benkhaldoun, F., Bradji, A.: Note on the convergence of a finite volume scheme for a second order hyperbolic equation with a time delay in any space dimension. In: Klöfkorn, R., Keilegavlen, E., Radu, F.A., Fuhrmann, J. (eds.) FVCA 2020. SPMS, vol. 323, pp. 315–324. Springer, Cham (2022). https://doi.org/10.1007/978-3-030-43651-3_28
4. Benkhaldoun, F., Bradji, A., Ghoudi, T.: A finite volume scheme for a wave equation with several time independent delays. In: Lirkov, I., Margenov, S. (eds.) LSSC 2021. LNCS, vol. 13127, pp. 498–506. Springer, Cham (2022). https://doi.org/10.1007/978-3-030-97549-4_57
5. Bradji, A.: A new optimal $L^\infty(H^1)$–error estimate of a SUSHI scheme for the time fractional diffusion equation. In: Klöfkorn, R., Keilegavlen, E., Radu, F.A., Fuhrmann, J. (eds.) FVCA 2020. SPMS, vol. 323, pp. 305–314. Springer, Cham (2020). https://doi.org/10.1007/978-3-030-43651-3_27
6. Bradji, A.: A new analysis for the convergence of the gradient discretization method for multidimensional time fractional diffusion and diffusion-wave equations. Comput. Math. Appl. **79**(2), 500–520 (2020)

7. Bradji, A, Fuhrmann. J.: Convergence order of a finite volume scheme for the time-fractional diffusion equation. In: Dimov, I., Faragó, I., Vulkov, L. (eds.) NAA 2016. LNCS, vol. 10187, pp. 33–45. Springer, Cham (2017). https://doi.org/10.1007/978-3-319-57099-0_4

8. Bu, W., Xiao, A., Zeng, W.: Finite difference/finite element methods for distributed-order time fractional diffusion equations. J. Sci. Comput. **72**(1), 422–441 (2017)

9. Eymard, R., Gallouët, T., Herbin, R.: Discretization of heterogeneous and anisotropic diffusion problems on general nonconforming meshes. IMA J. Numer. Anal. **30**(4), 1009–1043 (2010)

10. Faires, J.D., Burden, R., Burden, A.M.: Numerical Methods, 10th edn. Cengage Learning, Boston (2016)

11. Gao, X., Liu, F., Li, H., Liu, Y., Turner, I., Yin, B.: A novel finite element method for the distributed-order time fractional Cable equation in two dimensions. Comput. Math. Appl. **80**(5), 923–939 (2020)

12. Gao, G.-H, Sun, Z.-Z: Two alternating direction implicit difference schemes for two-dimensional distributed-order fractional diffusion equations. J. Sci. Comput. **66**(3), 1281–1312 (2016)

13. Gao, G.H., Sun, H.W., Sun, Z.: Some high-order difference schemes for the distributed-order differential equations. J. Comput. Phys. **298**, 337–359 (2015)

14. Gao, G.-H., Sun, Z.-Z., Zhang, H.-W.: A new fractional numerical differentiation formula to approximate the Caputo fractional derivative and its applications. J. Comput. Phys. **259**, 33–50 (2014)

15. Jin, B., Lazarov, R., Liu, Y., Zhou, Z.: The Galerkin finite element method for a multi-term time-fractional diffusion equation. J. Comput. Phys. **281**, 825–843 (2015)

16. Meerschaert, M.M., Nane, E., Vellaisamy, P.: Distributed-order fractional diffusions on bounded domains. J. Math. Anal. Appl. **379**(1), 216–228 (2011)

17. Podlubny, I.: Fractional Differential Equations. An Introduction to Fractional Derivatives, Fractional Differential Equations, to Methods of their Solution and some of their Applications. Mathematics in Science and Engineering, vol. 198. Academic Press Inc, San Diego (1999)

18. Pimenov, V.G., Hendy, A.S., De Staelen, R.H.: On a class of non-linear delay distributed order fractional diffusion equations. J. Comput. Appl. Math. **318**, 433–443 (2017)

19. Ye, H., Liu, F., Anh, V., Turner, I.: Numerical analysis for the time distributed-order and Riesz space fractional diffusions on bounded domains. IMA J. Appl. Math. **80**(3), 825–838 (2015)

20. Zhang, Y.-N, Sun, Z.-Z, Liao, H.-L.: Finite difference methods for the time fractional diffusion equation on non-uniform meshes. J. Comput. Phys. **265**, 195–210 (2014)

SUSHI for a Non-linear Time Fractional Diffusion Equation with a Time Independent Delay

Fayssal Benkhaldoun[1] and Abdallah Bradji[1,2][✉]

[1] LAGA, Sorbonne Paris Nord University, Villetaneuse, Paris, France
fayssal@math.univ-paris13.fr
[2] Department of Mathematics, Faculty of Sciences, Badji Mokhtar-Annaba University, Annaba, Algeria
abdallah.bradji@gmail.com, abdallah.bradji@etu.univ-amu.fr,
abdallah.bradji@univ-annaba.dz
https://www.math.univ-paris13.fr/fayssal/,
https://www.i2m.univ-amu.fr/perso/abdallah.bradji/

Abstract. We establish a linear implicit finite volume scheme for a non-linear time fractional diffusion equation with a time independent delay in any space dimension. The fractional order derivative is given in the Caputo sense. The discretization in space is performed using the SUSHI ((Scheme Using stabilized Hybrid Interfaces) developed in [11], whereas the discretization in time is given by a constrained time step-size. The approximation of the fractional order derivative is given by $L1$-formula.

We prove rigorously new convergence results in $L^\infty(L^2)$ and $L^2(H_0^1)-$ discrete norms. The order is proved to be optimal in space and it is $k^{2-\alpha}$ in time, with k is the constant time step and α is the fractional order of the Caputo derivative.

This paper is a continuation of some of our previous works which dealt either with only the linear fractional PDEs (Partial Differential Equations) without delays, e.g. [6,7,9,10], or with only time dependent PDEs (the time derivative is given in the usual sense) with delays, e.g. [2,4,8].

Keywords: Non-linear time fractional diffusion equation · delay · SUSHI · new error estimate

MSC2010: 65M08 · 65M12

1 Introduction to the Problem to be Solved and Aim of This Note

We consider the following simple (for the sake of clarity) non-linear time fractional diffusion equation with a time independent delay

$$u_t^\alpha(\boldsymbol{x}, t) - \Delta u(\boldsymbol{x}, t) = f(u(\boldsymbol{x}, t - \tau)), \qquad (\boldsymbol{x}, t) \in \Omega \times (0, T), \qquad (1)$$

Supported by MCS team (LAGA Laboratory) of the "Université Sorbonne- Paris Nord".

where Ω is an open bounded polyhedral subset in $\mathbb{R}^d$, with $d \in \mathbb{N}^* = \mathbb{N} \setminus \{0\}$, $T > 0$, $0 < \alpha < 1$, f is a given function defined on $\mathbb{R}$, and $\tau > 0$ is the delay.

The fractional derivative φ_t^α is given in the Caputo sense

$$\partial_t^\alpha \varphi(t) = \frac{1}{\Gamma(1-\alpha)} \int_0^t (t-s)^{-\alpha} \varphi'(s) ds \qquad (2)$$

Initial condition is given by, for a given function u^0 defined on Ω

$$u(\boldsymbol{x}, t) = u^0(\boldsymbol{x}, t), \qquad \boldsymbol{x} \in \Omega, \quad -\tau \leq t \leq 0. \qquad (3)$$

Homogeneous Dirichlet boundary conditions are given by

$$u(\boldsymbol{x}, t) = 0, \qquad (\boldsymbol{x}, t) \in \partial\Omega \times (0, T). \qquad (4)$$

Delay differential equations occur in several applications such as ecology, biology, medicine, see for instance [5, 13, 20] and references therein. However, the numerical methods which are carried out with Partial (or Ordinary) Differential Equations are not enough to deal with Delay Partial Differential Equations. In fact, the implementation of schemes and some desirable accuracy and stability results which are known for PDEs (Partial Differential Equations) can be destroyed when applying these methods to Delay Partial Differential Equations, cf. [5, Pages 9–19]. Numerical methods for the delay equations are well developed for the case of Ordinary Differential Equations but the subject of numerical analysis for Delay PDEs has not attracted the attention it merits yet, see [5, 8, 20].

The time fractional PDEs have been attracted recently a considerable attention because of their applications, see for instance [14, 15, 17–19] and references therein. An interesting sub-class of time fractional PDEs is the subset of equations in which there are time delays. Such sub-class has received recently a considerable attention. Three main features distinguish time fractional equations with delays: the first is that these equations are in general nonlinear, the second one is that the time is delayed, whereas the third feature is that the order of the time derivative is fractional. These features, among others, motivate to study extensively time fractional equations with delays. We refer to [14, 18, 19] for a detailed literature about the existence and uniqueness of the solutions (since these equations are non-linear in general) and the numerical approximation of time fractional diffusion equations with delays. As examples of the numerical methods dealt with time fractional PDEs with delays, we quote Finite difference methods in [19] in one dimensional space and Finite Element methods in [14] in two dimensional space. The error estimates provided in [14, 19] are measured in $L^\infty(L^2)$ and or $L^\infty(L^\infty)$ –discrete norms (see [19, Theorem 4.7, Pages 1874] and [14, Theorem 1, Page 2]).

The aim of this contribution is twofold:

- The first aim is to establish a linear implicit finite volume scheme for the non-linear multidimensional problem (1)–(4) using SUSHI developed in [11] as discretization in space and the L1-formula on the uniform temporal mesh as approximation for the Caputo derivative.

The general class of nonconforming multidimensional meshes introduced in [11] has the following advantages:

- The scheme can be applied on any type of grid: conforming or non conforming, 2D and 3D, or more, made with control volumes which are only assumed to be polyhedral (the boundary of each control volume is a finite union of subsets of hyperplanes).
- When the family of the discrete fluxes are satisfying some suitable conditions, the matrices of the generated linear systems are sparse, symmetric, positive and definite.
- A discrete gradient for the exact solution is formulated and converges to the gradient of the exact solution.

– Prove the convergence of this scheme not only in $L^\infty(L^2)$–discrete norm but also in $L^2(H_0^1)$–discrete norm.

This paper is a continuation of some of our previous papers which dealt either with only linear time fractional PDEs (without delays) or with only time dependent PDEs with delays (the derivatives are given in the usual sense). This work is an initiation of future works on the numerical methods for time fractional PDEs with delays.

2 Space and Time Discretizations and Some Preliminaries

The space discretization is given by the non-conforming mesh, developed in [11], which we recall its definition for the sake of completeness.

Definition 1. (Space discretization, cf. [11]). *Let Ω be a polyhedral open bounded subset of $\mathbb{R}^d$, where $d \in \mathbb{N} \setminus \{0\}$, and $\partial\Omega = \overline{\Omega} \setminus \Omega$ its boundary. A discretization of Ω, denoted by $\mathscr{D}$, is defined as the triplet $\mathscr{D} = (\mathscr{M}, \mathscr{E}, \mathscr{P})$, where:*

1. *$\mathscr{M}$ is a finite family of non empty connected open disjoint subsets of Ω (the "control volumes") such that $\overline{\Omega} = \cup_{K \in \mathscr{M}} \overline{K}$. For any $K \in \mathscr{M}$, let $\partial K = \overline{K} \setminus K$ be the boundary of K; let $\mathrm{m}(K) > 0$ denote the measure of K and h_K denote the diameter of K.*

2. *$\mathscr{E}$ is a finite family of disjoint subsets of $\overline{\Omega}$ (the "edges" of the mesh), such that, for all $\sigma \in \mathscr{E}$, σ is a non empty open subset of a hyperplane of $\mathbb{R}^d$, whose $(d-1)$-dimensional measure is strictly positive. We also assume that, for all $K \in \mathscr{M}$, there exists a subset $\mathscr{E}_K$ of $\mathscr{E}$ such that $\partial K = \cup_{\sigma \in \mathscr{E}_K} \overline{\sigma}$. For any $\sigma \in \mathscr{E}$, we denote by $\mathscr{M}_\sigma = \{K, \sigma \in \mathscr{E}_K\}$. We then assume that, for any $\sigma \in \mathscr{E}$, either $\mathscr{M}_\sigma$ has exactly one element and then $\sigma \subset \partial\Omega$ (the set of these interfaces, called boundary interfaces, denoted by $\mathscr{E}_{ext}$) or $\mathscr{M}_\sigma$ has exactly two elements (the set of these interfaces, called interior interfaces, denoted by $\mathscr{E}_{int}$). For all $\sigma \in \mathscr{E}$, we denote by x_σ the barycentre of σ. For all $K \in \mathscr{M}$ and $\sigma \in \mathscr{E}$, we denote by $n_{K,\sigma}$ the unit vector normal to σ outward to K.*

3. *$\mathscr{P}$ is a family of points of Ω indexed by $\mathscr{M}$, denoted by $\mathscr{P} = (x_K)_{K \in \mathscr{M}}$, such that for all $K \in \mathscr{M}$, $x_K \in K$ and K is assumed to be x_K–star-shaped, which means that for all $x \in K$, the property $[x_K, x] \subset K$ holds. Denoting by*

$d_{K,\sigma}$ the Euclidean distance between $\boldsymbol{x}_K$ and the hyperplane including σ, one assumes that $d_{K,\sigma} > 0$. We then denote by $\mathscr{D}_{K,\sigma}$ the cone with vertex $\boldsymbol{x}_K$ and basis σ.

The time discretization is performed with a constrained time step-size k such that $\dfrac{\tau}{k} \in \mathbb{N}$. We set then $k = \dfrac{\tau}{M}$, where $M \in \mathbb{N} \setminus \{0\}$ such that $k < 1$. Denote by N the integer part of $\dfrac{T}{k}$, i.e. $N = \left[\dfrac{T}{k^\alpha} \right]$. We shall denote $t_n = nk$, for $n \in [\![-M, N]\!]$. As particular cases $t_{-M} = -\tau$, $t_0 = 0$, and $t_N \leq T$. One of the advantages of this time discretization is that the point $t = 0$ is a mesh point which is suitable since we have equation (1) defined for $t \in (0, T)$ and initial condition (3) defined for $t \in (-\tau, 0)$. We denote by ∂^1 the discrete first time derivative given by $\partial^1 v^{j+1} = \dfrac{v^{j+1} - v^j}{k}$.

We define the discrete space $\mathscr{X}_{\mathscr{D},0}$ as the set of all $v = \left((v_K)_{K \in \mathscr{M}} \,, \, (v_\sigma)_{\sigma \in \mathscr{E}} \right)$, where $v_K, v_\sigma \in \mathbb{R}$ and $v_\sigma = 0$ for all $\sigma \in \mathscr{E}_{\text{ext}}$. Let $H_{\mathscr{M}}(\Omega) \subset L^2(\Omega)$ be the space of functions which are constant on each control volume K of the mesh $\mathscr{M}$. For all $v \in \mathscr{X}_{\mathscr{D}}$, we denote by $\Pi_{\mathscr{M}} v \in H_{\mathscr{M}}(\Omega)$ the function defined by $\Pi_{\mathscr{M}} v(\boldsymbol{x}) = v_K$, for a.e. $\boldsymbol{x} \in K$, for all $K \in \mathscr{M}$.

In order to analyze the convergence, we consider the size of the discretization $\mathscr{D}$ defined by $h_{\mathscr{D}} = \sup \{ \operatorname{diam}(K), \ K \in \mathscr{M} \}$ and the regularity of the mesh given by

$$\theta_{\mathscr{D}} = \max \left(\max_{\sigma \in \mathscr{E}_{\text{int}}, K, L \in \mathscr{M}} \frac{d_{K,\sigma}}{d_{L,\sigma}}, \ \max_{K \in \mathscr{M}, \sigma \in \mathscr{E}_K} \frac{h_K}{d_{K,\sigma}} \right). \qquad (5)$$

The scheme we want to consider is based on the use of the discrete gradient given in [11]. For $u \in \mathscr{X}_{\mathscr{D}}$, we define, for all $K \in \mathscr{M}$

$$\nabla_{\mathscr{D}} u(\boldsymbol{x}) = \nabla_K u + \left(\frac{\sqrt{d}}{d_{K,\sigma}} \left(u_\sigma - u_K - \nabla_K u \cdot (\boldsymbol{x}_\sigma - \boldsymbol{x}_K) \right) \right) \mathbf{n}_{K,\sigma}, \quad \text{a.e. } \boldsymbol{x} \in \mathscr{D}_{K,\sigma}, \qquad (6)$$

where $\nabla_K u = \dfrac{1}{\operatorname{m}(K)} \displaystyle\sum_{\sigma \in \mathscr{E}_K} \operatorname{m}(\sigma) \left(u_\sigma - u_K \right) \mathbf{n}_{K,\sigma}$.

We define the following bilinear form:

$$\langle u, v \rangle_F = \int_\Omega \nabla_{\mathscr{D}} u(\boldsymbol{x}) \cdot \nabla_{\mathscr{D}} v(\boldsymbol{x}) d\boldsymbol{x}, \quad \forall (u, v) \in \mathscr{X}_{\mathscr{D},0} \times \mathscr{X}_{\mathscr{D},0}. \qquad (7)$$

Throughout this paper, the letter C stands for a positive constant independent of $h_{\mathscr{D}}$ and k.

For any $n \in [\![0, N-1]\!]$, we use the consistent approximation of $\partial_t^\alpha u(t_{n+1})$ which is defined as a linear combination of the discrete time derivatives $\{ \partial^1 u(t_{j+1}), \ j \in [\![0, n]\!] \}$ and it is given by (see [7] and references therein):

$$\partial_t^\alpha u(t_{n+1}) = \sum_{j=0}^n k \lambda_j^{n+1} \partial^1 u(t_{j+1}) + \mathbb{T}_1^{n+1}(u), \qquad (8)$$

where

$$\lambda_j^{n+1} = \frac{(n-j+1)^{1-\alpha} - (n-j)^{1-\alpha}}{k^\alpha \Gamma(2-\alpha)} \quad \text{and} \quad |\mathbb{T}_1^{n+1}(u)| \le C k^{2-\alpha}. \tag{9}$$

The following lemma summarizes some properties of the coefficients λ_j^{n+1}.

Lemma 1. (Properties of the coefficients λ_j^{n+1}, cf. [7]). *For any $n \in [\![0, N-1]\!]$ and for any $j \in [\![0,n]\!]$, let λ_j^{n+1} be defined by (9). The following properties hold:*

$$\frac{k^{-\alpha}}{\Gamma(2-\alpha)} = \lambda_n^{n+1} > \ldots > \lambda_0^{n+1} \ge \lambda_0 = \frac{T^{-\alpha}}{\Gamma(1-\alpha)} \tag{10}$$

and

$$\sum_{j=0}^{n} k\lambda_j^{n+1} \le \frac{T^{1-\alpha}}{\Gamma(2-\alpha)}. \tag{11}$$

3   Main Results of the Paper

This section is devoted to first establish a finite volume scheme for the problem (1)–(4) and then prove its convergence in several discrete norms.

3.1   Formulation of a Finite Volume Scheme for (1)–(4)

We now set a formulation of a finite volume scheme for the non-linear problem (1)–(4). The scheme we consider here is linear and implicit and its principles are inspired from our previous works [2,4,6,7,9,10]. The discrete unknowns of the scheme are the set $\{u_{\mathscr{D}}^n; n \in [\![-M, N]\!]\}$ which are expected to approximate the set of the unknowns

$$\{u(t_n); n \in [\![-M, N]\!]\}.$$

The scheme we consider is meaningful if for instance the following hypotheses are satisfied:

Assumption 1. (Assumptions on f and u). *We assume that the following hypotheses on the functions f and u^0:*

1. *The function f is satisfying $f \in \mathscr{C}(\mathbb{R})$.*
2. *The function u^0 is satisfying $u^0 \in \mathscr{C}^2(\overline{\Omega})$.*

The formulation of the scheme is given by:

1. **Approximation of initial condition** (3). The discretization of initial condition (3) can be performed as: for any $n \in [\![-M, 0]\!]$

$$\langle u_{\mathscr{D}}^n, v \rangle_F = -\left(\Delta u^0(t_n), \Pi_{\mathscr{M}} v \right)_{L^2(\Omega)}, \quad \forall v \in \mathscr{X}_{\mathscr{D},0}. \tag{12}$$

2 Approximation of (1) **and** (4). For any $n \in [\![0, N-1]\!]$, find $u_{\mathscr{D}}^n \in \mathscr{X}_{\mathscr{D},0}$ such that, for all $v \in \mathscr{X}_{\mathscr{D},0}$

$$\sum_{j=0}^{n} k\lambda_j^{n+1} \left(\partial^1 \Pi_{\mathscr{M}} u_{\mathscr{D}}^{n+1}, \Pi_{\mathscr{M}} v \right)_{L^2(\Omega)} + \langle u_{\mathscr{D}}^{n+1}, v \rangle_F$$

$$= \left(f \left(\Pi_{\mathscr{M}} u_{\mathscr{D}}^{n+1-M} \right), \Pi_{\mathscr{M}} v \right)_{L^2(\Omega)}. \tag{13}$$

To derive the error estimates we will present, for the scheme (12)–(13), we assume the following assumptions:

Assumption 2. (Assumptions on u and f). *We assume the following hypotheses on the functions u and f:*

1. *The function $f(r)$ is Lipschitz continuous with constant κ, i.e. for all $r, r' \in \mathbb{R}$:*

$$|f(r) - f(r')| \leq \kappa |r - r'|. \tag{14}$$

   Note that (14) is satisfied when for instance $f \in \mathscr{C}^1(\mathbb{R})$ and $\sup_{r \in \mathbb{R}} |f'(r)| \leq \kappa$.

2. *The solution u of function (1)–(4) is satisfying $u \in \mathscr{C}^2([-\tau, T]; \mathscr{C}^2(\overline{\Omega}))$.*

3.2   Convergence Order of Scheme (12)–(13)

The following theorem summarizes the main results of this paper:

Theorem 1. (New error estimates for the finite volume scheme (12)–(13)). *Let Ω be a polyhedral open bounded subset of $\mathbb{R}^d$, where $d \in \mathbb{N} \setminus \{0\}$, and $\partial\Omega = \overline{\Omega} \setminus \Omega$ its boundary. Let $k = \dfrac{\tau}{M}$, where $M \in \mathbb{N} \setminus \{0\}$ such that $k < 1$. Denote by N the integer part of $\dfrac{T}{k^\alpha}$, i.e. $N = \left[\dfrac{T}{k^\alpha} \right]$. We shall denote $t_n = nk$, for $n \in [\![-M, N]\!]$. As particular cases $t_{-M} = -\tau$ and $t_0 = 0$. Let $\mathscr{D} = (\mathscr{M}, \mathscr{E}, \mathscr{P})$ be a discretization in the sense of Definition 1. Assume that $\theta_{\mathscr{D}}$ (given by (5)) satisfies $\theta \geq \theta_{\mathscr{D}}$ for some given $\theta > 0$. Let $\nabla_{\mathscr{D}}$ be the discrete gradient given by (6) and denote by $\langle \cdot, \cdot \rangle_F$ the bilinear form defined by (7). Assume that Assumption 1 is satisfied.*

   Then there exists a unique solution $(u_{\mathscr{D}}^n)_{n=-M}^{N} \in \mathscr{X}_{\mathscr{D},0}^{M+N+1}$ for the discrete problem (12)–(13).

   If in addition Assumption 2 is satisfied, then the following error estimates hold:

– $L^2(H_0^1)$–*estimate.*

$$\left(\sum_{n=-M}^{N} k \|\nabla u(t_n) - \nabla_{\mathscr{D}} u_{\mathscr{D}}^n\|_{L^2(\Omega)^d}^2 \right)^{\frac{1}{2}} \leq C(k^{2-\alpha} + h_{\mathscr{D}}). \tag{15}$$

– $L^\infty(L^2)$–*estimate.*

$$\max_{n=-M}^{N} \|u(t_n) - \Pi_{\mathcal{M}} u_{\mathcal{D}}^n\|_{L^2(\Omega)} \le C(k^{2-\alpha} + h_{\mathcal{D}}). \tag{16}$$

Proof. We prove Theorem 1 item by item.

1. Existence and uniqueness for schemes (12)–(13). The schemes (12)–(13) lead to linear systems with square matrices. The uniqueness (and therefore the existence) can be justified recursively on n and using the coercivity [11, Lemma 4.2, Page 1026].

2. Proof of estimates (15)–(16). To prove (15)–(16), we compare (12)–(13) with the following auxiliary scheme: For any $n \in [\![-M, N]\!]$, find $\overline{u}_{\mathcal{D}}^n \in \mathscr{X}_{\mathcal{D},0}$ such that

$$\langle \overline{u}_{\mathcal{D}}^n, v \rangle_F = (-\Delta u(t_n), \Pi_{\mathcal{M}} v)_{L^2(\Omega)}, \quad \forall v \in \mathscr{X}_{\mathcal{D},0}. \tag{17}$$

2.1. Comparison between the solution of (17) **and the solution of problem** (1)–(4). The following convergence results hold, see [7,10,11]:
   * Discrete $L^\infty(L^2)$–error estimate. For all $n \in [\![-M, N]\!]$

$$\|u(t_n) - \Pi_{\mathcal{M}} \overline{u}_{\mathcal{D}}^n\|_{L^2(\Omega)} \le C h_{\mathcal{D}}. \tag{18}$$

   * $\mathscr{W}^{1,\infty}(L^2)$–error estimate. For all $n \in [\![-M+1, N]\!]$

$$\|\partial^1 (u(t_n) - \Pi_{\mathcal{M}} \overline{u}_{\mathcal{D}}^n)\|_{L^2(\Omega)} \le C h_{\mathcal{D}}. \tag{19}$$

   * Error estimate in the gradient approximation. For all $n \in [\![-M, N]\!]$

$$\|\nabla u(t_n) - \nabla_{\mathcal{D}} \overline{u}_{\mathcal{D}}^n\|_{(L^2(\Omega))^d} \le C h_{\mathcal{D}}. \tag{20}$$

2.2. Comparison between the solution of (12)–(13) **and the auxiliary scheme** (17). Let us define the *auxiliary* error $\eta_{\mathcal{D}}^n = u_{\mathcal{D}}^n - \overline{u}_{\mathcal{D}}^n \in \mathscr{X}_{\mathcal{D},0}$. Comparing (17) with (12) and using the fact that $u(t_n) = u^0(t_n)$ for all $n \in [\![-M, 0]\!]$ (subject of (3)) imply that

$$\eta_{\mathcal{D}}^n = 0, \quad \forall n \in [\![-M, 0]\!]. \tag{21}$$

Writing (17) in the level $n+1$ and subtracting the result from (13) to get

$$\sum_{j=0}^{n} k\lambda_j^{n+1} \left(\partial^1 \Pi_{\mathcal{M}} u_{\mathcal{D}}^{j+1}, \Pi_{\mathcal{M}} v \right)_{L^2(\Omega)} + \langle \eta_{\mathcal{D}}^{n+1}, v \rangle_F$$
$$= \left(f(\Pi_{\mathcal{M}} u_{\mathcal{D}}^{n+1-M}) + \Delta u(t_{n+1}), \Pi_{\mathcal{M}} v \right)_{L^2(\Omega)}, \forall v \in \mathscr{X}_{\mathcal{D},0}. \tag{22}$$

Subtracting $\sum_{j=0}^{n} k\lambda_j^{n+1}\left(\partial^1 \Pi_{\mathcal{M}}\overline{u}_{\mathcal{D}}^{j+1}, \Pi_{\mathcal{M}}v\right)_{L^2(\Omega)}$ from both sides of the previous equation and replacing $\Delta u(\boldsymbol{x}, t_{n+1})$ by $u_t^\alpha(\boldsymbol{x}, t_{n+1}) - f(u(\boldsymbol{x}, t_{n+1} - \tau))$ (which stems from (1)), we get, for all $v \in \mathscr{X}_{\mathcal{D},0}$ and for all $n \in [\![0, N-1]\!]$

$$\sum_{j=0}^{n} k\lambda_j^{n+1}\left(\partial^1 \Pi_{\mathcal{M}}\eta_{\mathcal{D}}^{j+1}, \Pi_{\mathcal{M}}v\right)_{L^2(\Omega)} + \langle\eta_{\mathcal{D}}^{n+1}, v\rangle_{\mathcal{D}}$$
$$= \left(\mathscr{S}^{n+1}, \Pi_{\mathcal{M}}v\right)_{L^2(\Omega)}, \tag{23}$$

where
$\mathscr{S}^{n+1}(\boldsymbol{x}) = \mathbb{T}^1(\boldsymbol{x}) + \mathbb{T}^2(\boldsymbol{x})$ with $\mathbb{T}^1$ and $\mathbb{T}^2$ are given by

$$\mathbb{T}^1(\boldsymbol{x}) = u_t^\alpha(\boldsymbol{x}, t_{n+1}) - \sum_{j=0}^{n} k\lambda_j^{n+1}\partial^1 \Pi_{\mathcal{M}}\overline{u}_{\mathcal{D}}^{j+1} \tag{24}$$

and (recall that $t_{-M} = \tau$)

$$\mathbb{T}^2(\boldsymbol{x}) = f(\Pi_{\mathcal{M}}u_{\mathcal{D}}^{n+1-M}(\boldsymbol{x})) - f(u(\boldsymbol{x}, t_{n+1-M})). \tag{25}$$

Using the triangle inequality, second estimate in (9), properties (10)–(11), and error estimate (19) yields

$$\|\mathbb{T}^1\|_{L^2(\Omega)} \le \|u_t^\alpha(\boldsymbol{x}, t_{n+1}) - \sum_{j=0}^{n} k\lambda_j^{n+1}\partial^1 u(t_{j+1})\|_{L^2(\Omega)}$$

$$+ \|\sum_{j=0}^{n} k\lambda_j^{n+1}\partial^1\left(u(t_{j+1}) - \Pi_{\mathcal{M}}\overline{u}_{\mathcal{D}}^{j+1}\right)\|_{L^2(\Omega)}$$

$$\le C\Theta, \tag{26}$$

where we have denoted $\Theta = h_{\mathcal{D}} + k^{2-\alpha}$.
Using Assumption 2 together with the triangle inequality and (18) imply that

$$\|\mathbb{T}^2\|_{L^2(\Omega)} \le C\left(\|\Pi_{\mathcal{M}}\eta_{\mathcal{D}}^{n+1-M}\|_{L^2(\Omega)} + h_{\mathcal{D}}\right). \tag{27}$$

From (26) and (27), we deduce that

$$\|\mathscr{S}^{n+1}\|_{L^2(\Omega)} \le C\left(\|\Pi_{\mathcal{M}}\eta_{\mathcal{D}}^{n+1-M}\|_{L^2(\Omega)} + \Theta\right). \tag{28}$$

Taking into account (21) and (23) and using [7, (23), Page 505] together with the Poincaré inequality [11, Lemma 5.4, Page 1038] (recall that SUSHI is a particular method of Gradient Discretization Methods which has been applied in [7]) and (28) yield, for all $n \in [\![0, N-1]\!]$

$$\lambda_n^{n+1}\|\Pi_{\mathcal{M}}\eta_{\mathcal{D}}^{n+1}\|_{L^2(\Omega)}^2 + \|\nabla_{\mathcal{D}}\eta_{\mathcal{D}}^{n+1}\|_{L^2(\Omega)^d}^2 \le C\Theta^2$$

$$+ \sum_{j=1}^{n}\left(\lambda_j^{n+1} - \lambda_{j-1}^{n+1}\right)\|\Pi_{\mathcal{D}}\eta_{\mathcal{D}}^j\|_{L^2(\Omega)}^2 + C\|\Pi_{\mathcal{M}}\eta_{\mathcal{D}}^{n+1-M}\|_{L^2(\Omega)}^2. \tag{29}$$

Let us now present a new and simple proof for some appropriate estimates for $\|\Pi_{\mathscr{M}} \eta_{\mathscr{D}}^{n+1}\|_{L^2(\Omega)}^2$ and $\|\nabla_{\mathscr{D}} \eta_{\mathscr{D}}^{n+1}\|_{L^2(\Omega)^d}^2$. It is also, maybe, possible to use the techniques of the proof [14, Theorem 6, Pages 3–5] which are based in turn on a discrete fractional Grönwall inequality mimicking the known continuous ones [17] in order to derive an estimate but for only $\|\Pi_{\mathscr{M}} \eta_{\mathscr{D}}^{n+1}\|_{L^2(\Omega)}^2$. Let us define, for any $n \in [\![0, N]\!]$ (recall that $\eta_{\mathscr{D}}^n = 0$, for all $n \in [\![-M, 0]\!]$)

$$\mu^n = \max_{j=0}^{n} \|\Pi_{\mathscr{M}} \eta_{\mathscr{D}}^j\|_{L^2(\Omega)}^2. \tag{30}$$

Dividing both sides of (29) by λ_n^{n+1} and using the property (10) imply that, for all $n \in [\![0, N-1]\!]$

$$\|\Pi_{\mathscr{M}} \eta_{\mathscr{D}}^{n+1}\|_{L^2(\Omega)}^2 \leq Ck^\alpha \Theta^2 + \left(1 - Ck^\alpha \lambda_0^{n+1} + Ck^\alpha\right) \mu^n$$
$$\leq Ck^\alpha \Theta^2 + (1 + Ck^\alpha) \mu^n. \tag{31}$$

For any $n \in [\![0, N]\!]$, let $m \in [\![0, n]\!]$ such that $\mu^n = \|\Pi_{\mathscr{M}} \eta_{\mathscr{D}}^m\|_{L^2(\Omega)}^2$. For $n \in [\![1, N]\!]$, there exists $m \in [\![1, n]\!]$ such that $\mu^n = \|\Pi_{\mathscr{M}} \eta_{\mathscr{D}}^m\|_{L^2(\Omega)}^2$ (since $\eta_{\mathscr{D}}^0 = 0$). Applying now (31) for $n \in [\![1, N]\!]$ and using the fact that $\mu^p \leq \mu^q$ for all $p \leq q$ give

$$\mu^n = \|\Pi_{\mathscr{M}} \eta_{\mathscr{D}}^m\|_{L^2(\Omega)}^2 \leq Ck^\alpha \Theta^2 + (1 + Ck^\alpha) \mu^{m-1}$$
$$\leq Ck^\alpha \Theta^2 + (1 + Ck^\alpha) \mu^{n-1}. \tag{32}$$

Applying now (32) recursively and using the fact that $\mu^0 = 0$ and the formula of the sum of the terms of a geometric sequence imply that

$$\mu^n \leq Ck^\alpha \Theta^2 \sum_{j=0}^{n-1} (1 + Ck^\alpha)^j$$
$$= Ck^\alpha \Theta^2 \frac{(1 + Ck^\alpha)^n - 1}{1 + Ck^\alpha - 1}$$
$$\leq C\Theta^2 (1 + Ck^\alpha)^N, \qquad \forall n \in [\![1, N]\!]. \tag{33}$$

This with the fact that $N = \left[\frac{T}{k^\alpha}\right]$ yield

$$\mu^n \leq C\Theta^2 (1 + Ck^\alpha)^{\frac{T}{k^\alpha}}, \qquad \forall n \in [\![1, N]\!]. \tag{34}$$

Using this estimate together with the fact that $\lim_{k \to 0} (1 + Ck^\alpha)^{\frac{T}{k^\alpha}} = \exp(CT)$, definition (30) of μ^n, and (21) yield the following $L^\infty(L^2)$–auxiliary estimate

$$\max_{n=-M}^{N} \|\Pi_{\mathscr{M}} \eta_{\mathscr{D}}^N\|_{L^2(\Omega)} \leq C\Theta. \tag{35}$$

In addition to this $L^\infty(L^2)$–auxiliary estimate, we are able also to prove an $L^2(H_0^1)$–auxiliary estimate (which is completely new even in the context of the other numerical methods applied to time fractional diffusion equation with delays, see [14,19] and references therein). Indeed, (29) can be written as

$$\sum_{j=1}^{n+1} \lambda_{j-1}^{n+1} \|\Pi_{\mathscr{M}} \eta_{\mathscr{D}}^j\|_{L^2(\Omega)}^2 + \|\nabla_{\mathscr{D}} \eta_{\mathscr{D}}^{n+1}\|_{L^2(\Omega)^d}^2 \le C\Theta^2$$

$$+ \sum_{j=1}^{n} \lambda_j^{n+1} \|\Pi_{\mathscr{M}} \eta_{\mathscr{D}}^j\|_{L^2(\Omega)}^2 + C\|\Pi_{\mathscr{M}} \eta_{\mathscr{D}}^{n+1-M}\|_{L^2(\Omega)}^2. \qquad (36)$$

Using the fact that $\lambda_{j-1}^{n+1} = \lambda_j^{n+2}$, inequality (36) implies that

$$\Lambda^{n+1} - \Lambda^n + \|\nabla_{\mathscr{D}} \eta_{\mathscr{D}}^{n+1}\|_{L^2(\Omega)^d}^2 \le C\Theta^2 + C\|\Pi_{\mathscr{M}} \eta_{\mathscr{D}}^{n+1-M}\|_{L^2(\Omega)}^2, \qquad (37)$$

where $\Lambda^n = \sum_{j=1}^{n} \lambda_j^{n+1} \|\Pi_{\mathscr{M}} \eta_{\mathscr{D}}^j\|_{L^2(\Omega)}^2$. Multiplying both sides of (37) by k,

summing the result over $n \in [\![0, N-1]\!]$, and using the fact that $\sum_{n=0}^{N-1} k\Theta^2 \le T\Theta^2$ yield

$$k\Lambda^N + \sum_{n=0}^{N-1} k\|\nabla_{\mathscr{D}} \eta_{\mathscr{D}}^{n+1}\|_{L^2(\Omega)^d}^2 \le C\Theta^2 + C \sum_{n=0}^{N-1} k\|\Pi_{\mathscr{M}} \eta_{\mathscr{D}}^{n+1-M}\|_{L^2(\Omega)}^2.$$
$$(38)$$

Gathering (35), (38), and (21) yields

$$\left(\sum_{n=-M}^{N} k\|\nabla_{\mathscr{D}} \eta_{\mathscr{D}}^n\|_{L^2(\Omega)^d}^2 \right)^{\frac{1}{2}} \le C\Theta. \qquad (39)$$

To conclude the proof, we remark

$$\nabla u(t_n) - \nabla_{\mathscr{D}} u_{\mathscr{D}}^n = \nabla u(t_n) - \nabla_{\mathscr{D}} \overline{u}_{\mathscr{D}}^n - \nabla_{\mathscr{D}} \eta_{\mathscr{D}}^n$$

and

$$u(t_n) - \Pi_{\mathscr{M}} u_{\mathscr{D}}^n = u(t_n) - \Pi_{\mathscr{M}} \overline{u}_{\mathscr{D}}^n - \Pi_{\mathscr{M}} \eta_{\mathscr{D}}^n.$$

Using these facts together with the triangle inequality, (18), (20), (35), and (39) yield the desired estimates (15)–(16).

This completes the proof of Theorem 1.   □

Remark 1. (Discrete classical Grönwall inequality and the analysis presented here) Inequality (29) implies that, thanks to (10)–(11)

$$\|\Pi_{\mathscr{M}} \eta_{\mathscr{D}}^{n+1}\|_{L^2(\Omega)}^2 \le Ck^{\alpha}\Theta^2 + Ck^{\alpha} \sum_{j=1}^{n} \left(\lambda_j^{n+1} - \lambda_{j-1}^{n+1} \right) \|\Pi_{\mathscr{D}} \eta_{\mathscr{D}}^j\|_{L^2(\Omega)}^2$$

$$+ Ck^{\alpha}\|\Pi_{\mathscr{M}} \eta_{\mathscr{D}}^{n+1-M}\|_{L^2(\Omega)}^2. \qquad (40)$$

We cannot apply the classical discrete Grönwall inequality (see [16, Lemma 1.4.2, Page 14]) to (40) for a simple reason that the coefficients $\lambda_j^{n+1} - \lambda_{j-1}^{n+1}$ of $\|\Pi_\mathscr{D} \eta_\mathscr{D}^j\|_{L^2(\Omega)}^2$ are depending on n. Such applications of the classical discrete Grönwall inequality, which fail in these situations where the coefficients are depending on n, can be found for instance in [19, Page 1875]. The use of another discrete version for Grönwall inequality which are adapted to fractional derivatives is needed in these situations, see for instance such discrete and continuous fractional Grönwall inequalities in [14, Theorem 6, Pages 3–5] and [17].

4   Conclusion and Perspectives

We established a linear implicit finite volume scheme for a non-linear time fractional diffusion equation with a time independent delay in any space dimension. The discretization in space is given by the SUSHI [11], whereas the discretization in time is given by a constrained time step-size and $L1$-formula.

We proved rigorously new convergence results in $L^\infty(L^2)$ and $L^2(H_0^1)$–discrete norms. The order is proved to be optimal in space and it is $k^{2-\alpha}$ in time. These convergence results are obtained under the assumption that the exact solution u and the data functions f and u^0 are "smooth".

There are several interesting tasks, related to this paper, which we plan to work on in the near future. We would like to extend the present results to the case when the approximation of the Caputo derivative is given by the $L2 - 1_\sigma$ developed in [1]. Such approximation will allow to get second order time accurate scheme. Other interesting possible extensions are the cases of several delays which are depending on time. Another interesting task to work on is to obtain the present results under "weak" assumptions on u, f, and u^0.

References

1. Alikhanov, A.-A.: A new difference scheme for the fractional diffusion equation. J. Comput. Phys. **280**, 424–438 (2015)
2. Benkhaldoun, F., Bradji, A., Ghoudi, T.: A finite volume scheme for a wave equation with several time independent delays. In: Lirkov, I., Margenov, S. (eds.) LSSC 2021. LNCS, vol. 13127, pp. 498–506. Springer, Cham (2022). https://doi.org/10.1007/978-3-030-97549-4_57
3. Benkhaldoun, F., Bradji, A.: A second order time accurate finite volume scheme for the time-fractional diffusion wave equation on general nonconforming meshes. In: Lirkov, I., Margenov, S. (eds.) LSSC 2019. LNCS, vol. 11958, pp. 95–104. Springer, Cham (2020). https://doi.org/10.1007/978-3-030-41032-2_10
4. Benkhaldoun, F., Bradji, A.: Note on the convergence of a finite volume scheme for a second order hyperbolic equation with a time delay in any space dimension. In: Klöfkorn, R., Keilegavlen, E., Radu, F.A., Fuhrmann, J. (eds.) FVCA 2020. SPMS, vol. 323, pp. 315–324. Springer, Cham (2020). https://doi.org/10.1007/978-3-030-43651-3_28
5. Bellen, A., Zennaro, M.: Numerical Methods for Delay Differential Equations. Numerical Mathematics and Scientific Computation, Oxford University Press, Oxford (2003)

6. Bradji, A.: A new optimal $L^\infty(H^1)$–error estimate of a SUSHI scheme for the time fractional diffusion equation. In: Klöfkorn, R., Keilegavlen, E., Radu, F.A., Fuhrmann, J. (eds.) FVCA 2020. SPMS, vol. 323, pp. 305–314. Springer, Cham (2020). https://doi.org/10.1007/978-3-030-43651-3_27

7. Bradji, A.: A new analysis for the convergence of the gradient discretization method for multidimensional time fractional diffusion and diffusion-wave equations. Comput. Math. Appl. **79**(2), 500–520 (2020)

8. Bradji, A., Ghoudi, T.: Some convergence results of a multidimensional finite volume scheme for a semilinear parabolic equation with a time delay. In: Nikolov, G., Kolkovska, N., Georgiev, K. (eds.) NMA 2018. LNCS, vol. 11189, pp. 351–359. Springer, Cham (2019). https://doi.org/10.1007/978-3-030-10692-8_39

9. Bradji, A.: Notes on the convergence order of gradient schemes for time fractional differential equations. C. R. Math. Acad. Sci. Paris **356**(4), 439–448 (2018)

10. Bradji, A, Fuhrmann. J.: Convergence order of a finite volume scheme for the time-fractional diffusion equation. In: Dimov, I., Faragó, I., Vulkov, L. (eds.) NAA 2016. LNCS, vol. 10187, pp. 33–45. Springer, Cham (2017). https://doi.org/10.1007/978-3-319-57099-0_4

11. Eymard, R., Gallouët, T., Herbin, R.: Discretization of heterogeneous and anisotropic diffusion problems on general nonconforming meshes. IMA J. Numer. Anal. **30**(4), 1009–1043 (2010)

12. Jin, B., Lazarov, R., Liu, Y., Zhou, Z.: The Galerkin finite element method for a multi-term time-fractional diffusion equation. J. Comput. Phys. **281**, 825–843 (2015)

13. Kuang, Y.: Delay Differential Equations: With Applications in Population Dynamics. Mathematics in Science and Engineering, vol. 191. Academic Press, Boston (1993)

14. Li, L., She, M., Niu, Y.: Corrigendum to "Fractional Crank-Nicolson-Galerkin finite element methods for nonlinear time fractional parabolic problems with time delay". J. Funct. Spaces, Article ID 9820258, 10 p. (2022)

15. Pimenov, V.G., Hendy, A.S., De Staelen, R.H.: On a class of non-linear delay distributed order fractional diffusion equations. J. Comput. Appl. Math. **318**, 433–443 (2017)

16. Quarteroni, A., Valli, A.: Numerical Approximation of Partial Differential Equations. Springer Series in Computational Mathematics, vol. 23. Springer, Berlin (2008). https://doi.org/10.1007/978-3-540-85268-1

17. Webb, J.R.-L.: A fractional Gronwall inequality and the asymptotic behaviour of global solutions of Caputo fractional problems. Electron. J. Differential Equations, Paper No. 80, 22 p. (2021)

18. Zhang, Y., Wang, Z.: Numerical simulation for time-fractional diffusion-wave equations with time delay. J. Appl. Math. Comput. (2022). https://doi.org/10.1007/s12190-022-01739-6

19. Zhang, Q., Ran, M., Xu, D.: Analysis of the compact difference scheme for the semilinear fractional partial differential equation with time delay. Appl. Anal. **96**(11), 1867–1884 (2017)

20. Zhang, Q., Zhang, C.: A new linearized compact multisplitting scheme for the nonlinear convection-reaction-diffusion equations with delay. Commun. Nonlinear Sci. Numer. Simul. **18**(12), 3278–3288 (2013)

On the Numerical Simulation of Exponential Decay and Outbreak Data Sets Involving Uncertainties

Milen Borisov$^{(\boxtimes)}$ (iD) and Svetoslav Markov (iD)

Institute of Mathematics and Informatics, Bulgarian Academy of Sciences,
Acad. Georgi Bonchev Str., Block 8, 1113 Sofia, Bulgaria
milen_kb@math.bas.bg, smarkov@bio.bas.bg

Abstract. Measurement data sets collected when observing epidemiological outbreaks of various diseases often have specific shapes, thereby the data may contain uncertainties. A number of epidemiological mathematical models formulated in terms of ODE's (or reaction networks) offer solutions that have the potential to simulate and fit well the observed measurement data sets. These solutions are usually smooth functions of time depending on one or more rate parameters. In this work we are especially interested in solutions whose graphs are either of "decay" shape or of a specific wave-like shape briefly denoted as "outbreak" shape. Furthermore we are concerned with the numerical simulation of measurement data sets involving uncertainties, possibly coming from one of the simplest epidemiological models, namely the two-step exponential decay process (Bateman chain). To this end we define a basic exponential outbreak function and study its properties as far as they are needed for the numerical simulations. Stepping on the properties of the basic exponential decay-outbreak functions, we propose numerical algorithms for the estimation of the rate parameters whenever the measurement data sets are available in numeric or interval-valued form.

Keywords: Least square approximation · Numerical simulation · Decay and outbreak data

1 Introduction: The Exponential Decay Chain

Consider briefly the exponential (radioactive) decay chain (Bateman chain) [2] in the special cases of one and two reaction steps. These two reaction networks find numerous applications not only in nuclear physics and nuclear medicine, but in biology, in particular in population dynamics, fishery research, pharmacodynamics and mathematical epidemiology [5, 7, 12, 15, 16].

The Exponential Decay Function. The one-step exponential (radioactive) decay (1SED) model is a first order reaction step transforming a species S into

© The Author(s), under exclusive license to Springer Nature Switzerland AG 2023
I. Georgiev et al. (Eds.): NMA 2022, LNCS 13858, pp. 85–99, 2023.
https://doi.org/10.1007/978-3-031-32412-3_8

another species P. The model is presented by the reaction network S $\xrightarrow{k_1}$ P, where k_1 is a positive rate parameter. Assuming mass action kinetics, this reaction network translates into a system of two ordinary differential equations (ODE's) for the density functions $s = s(t)$ and $p = p(t)$, corresponding to species S and P, resp. in $t \in [0, \infty]$:

$$s' = -k_1 s,$$
$$p' = k_1 s. \tag{1}$$

Under initial values $s(0) = s_0 > 0$; $p(0) = p_0 \geq 0$ the solutions to (1) can be presented in the time interval $T = [0, \infty)$ as [5]:

$$s(t) = s_0 e^{-k_1 t};$$
$$p(t) = c - s_0 e^{-k_1 t}, \quad c = s_0 + p_0. \tag{2}$$

Let us focus on solution $s = s(t)$. Assume $s_0 = 1$, then solution s from (2) is a function of variables k and t of the form:

$$\eta(k; t) = e^{-kt}. \tag{3}$$

In reality the rate parameter k in (3) takes specific distinct values in each particular biochemical process, however from mathematical perspective we cam consider parameter k as a continuous function variable. So, function $\eta(k; t)$ is continuous, differentiable and monotonically decreasing with respect to both variables k and t. Function (3) will be further referred as *basic exponential decay function*, briefly *basic decay function*. For the derivatives of (3) with respect to t we have $(e^{-kt})' = -k e^{-kt}$ and $(e^{-kt})'' = k^2 e^{-kt} > 0$. The latter inequality says that function s, resp. the basic decay function e^{-kt}, is convex in R^+.

Another characteristics of the basic decay function is:

$$t_h = \ln 2 / k, \tag{4}$$

wherein time instant t_h is known in nuclear physics as "half-life time" or just "half-life". The half-life (4) satisfies the relation $\eta(t_h) = 1/2$, resp. $s(t_h) = s_0/2$, cf. Figure 1. From (4) the rate parameter k can be expressed as:

$$k = \ln 2 / t_h \approx 0.693 / t_h. \tag{5}$$

Consider two basic decay functions: $e^{-k_1 t}$, $e^{-k_2 t}$. For a fixed t, the following monotonicity property takes place:

$$k_1 \leq k_2 \Longrightarrow e^{-k_1 t} \geq e^{-k_2 t}. \tag{6}$$

Assume that the rate parameter k in function (3) is known within an error bound $\Delta \geq 0$. As an illustration, the upper graph on Figs. 2 and 3 presents the function: $e^{-(k-\Delta)t}$, whereas the lower graph presents the function: $e^{-(k+\Delta)t}$.

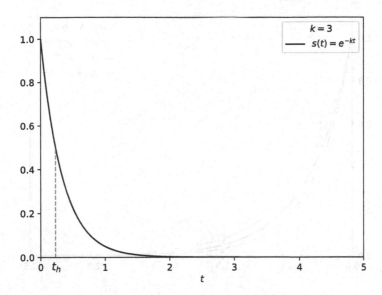

Fig. 1. Graph of function $\eta(t) = e^{-kt}$ for $k = 3$. The half-life $t_h = \ln 2/k$ is visualized.

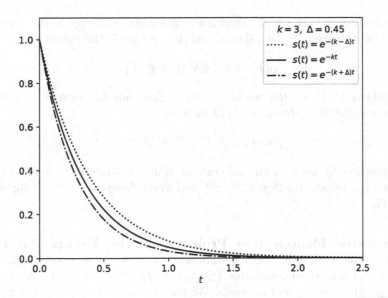

Fig. 2. Upper function $e^{-(k-\Delta)t}$ and lower function $e^{-(k+\Delta)t}$ for $k = 3, \Delta = 0.45$

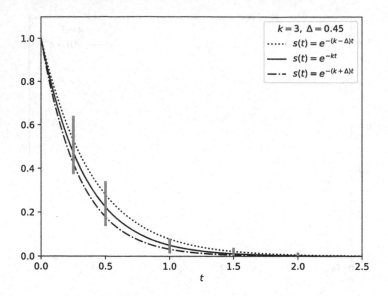

Fig. 3. A provisional interval-valued data set visualized

Denote $K = [k - \Delta, k + \Delta]$. We wish to know how the error in the parameter k influences the error in the value of function η, that is the *interval hull*:

$$\eta(K; t) = \{\eta(k; t) : k \in K\}. \tag{7}$$

Proposition 1. *Using the monotonicity of function (3) with respect to rate parameter k, for the interval hull (7) we have*

$$\eta(K; t) = [e^{-(k+\Delta)t}, e^{-(k-\Delta)t}]. \tag{8}$$

Proposition 1 allows for an easy computation of the interval hull (7). Figure 2 visualises the upper function $e^{-(k+\Delta)t}$ and lower function $e^{-(k-\Delta)t}$ for $k = 3$, $\Delta = 0.45$.

A Parameter Identification Problem for the Exponential Decay Model. Assume that at some instances t_i, $i = 1, \ldots, n$, experimental data with interval-valued uncertainties $(t_1, X_i), \ldots, (t_n, X_n)$, are available, wherein $X_i = [\underline{x}_i, \overline{x}_i]$ are some real intervals. We wish to estimate the value of the rate parameter k so that exponential decay solution η "fits well" the given experimental data. In particular, we shall additionally require that the values of function η are included in all the experimental intervals at the given time instances as shown on Fig. 3. We shall come back to the parameter identification problem for the exponential decay model in Sect. 4.

2   The Two-Step Exponential Decay-Outbreak Model: Properties

The two-step exponential (radioactive) decay-outbreak (2SED) model involves two first-order reaction steps in the transition of three species, say S, P, Q, into each other. More specifically, the 2SED model is presented by the reaction network S $\xrightarrow{k_1}$ P $\xrightarrow{k_2}$ Q, or, in canonical form [13]:

$$ S \xrightarrow{k_1} P, \quad P \xrightarrow{k_2} Q, \tag{9} $$

where k_1, k_2 are positive rate parameters. Assuming mass action kinetics, reaction network (9) translates into a system of ODE's for the density functions $s = s(t), p = p(t), q = q(t)$, corresponding to species S, P, Q, resp. in $t \in [0, \infty]$. Skipping the uncoupled equation ($q' = k_2 p$) for function q, we have:

$$ \begin{aligned} s' &= -k_1 s, \\ p' &= k_1 s - k_2 p. \end{aligned} \tag{10} $$

Assume initial value conditions to dynamical system (10) as follows:

$$ s(0) = s_0 > 0; \; p(0) = p_0 = 0. \tag{11} $$

The solutions s, p to initial value problem (10)–(11) can be explicitly presented in the time interval $T = [0, \infty)$ as follows [5]:

$$ \begin{aligned} s(k_1; t) &= s_0 e^{-k_1 t}; \\ p(k_1, k_2; t) &= s_0 \begin{cases} \frac{k_1}{k_2 - k_1}(e^{-k_1 t} - e^{-k_2 t}), & k_1 \neq k_2, \\ k\, t\, e^{-kt}, & k_1 = k_2 = k. \end{cases} \end{aligned} \tag{12} $$

In order to study analytically solution $p(k_1, k_2; t)$ from (12) we define and study a simplified function called "basic outbreak function". As seen from (12) the expression for function $p(k_1, k_2; t)$ when $k_1 \neq k_2$ makes an essential use of the difference $(e^{-k_1 t} - e^{-k_2 t})$ of two decay functions as visualized on Fig. 4. The graph of function $p(k_1, k_2; t)$ using rate parameters $k_1 = 3$ and $k_2 = 1$ is presented on Fig. 5.

2.1   A Basic Wave-Like Function

Definition 1. Define function

$$ \varepsilon(m, n; t) = \begin{cases} \frac{1}{n-m}(e^{-mt} - e^{-nt}), & m \neq n, \\ t\, e^{-kt}, & m = n = k. \end{cases} \tag{13} $$

for $m > 0, n > 0, t \in [0.\infty)$.

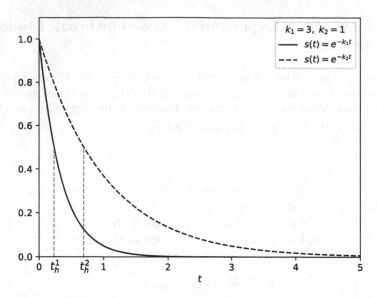

Fig. 4. Two decay functions with rate parameters $k_1 = 3$, resp. $k_2 = 1$; the half-life times are visualized.

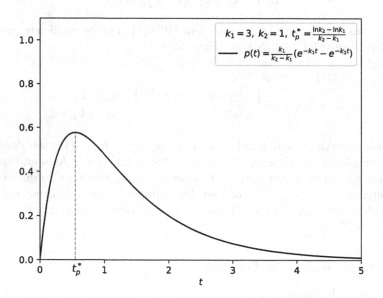

Fig. 5. A basic exponential outbreak function: $k_1 = 3, k_2 = 1, t_p^* = \frac{\ln k_2 - \ln k_1}{k_2 - k_1}$

Consider the special case $m \neq n$, that is function:

$$\varepsilon(m, n; t)|_{m \neq n} = \frac{e^{-mt} - e^{-nt}}{n - m}. \tag{14}$$

Using (14), function $\varepsilon(m, n; t)$ is defined in the case $m = n$ naturally as follows:

$$\begin{aligned} \varepsilon(m, n; t)|_{m=n} &= \lim_{k \longrightarrow n,\, k \neq n} \varepsilon(k, n; t) \\ &= -\lim_{k \longrightarrow n,\, k \neq n} \frac{e^{-kt} - e^{-nt}}{k - n} \\ &= -\frac{\partial e^{-kt}}{\partial k} = te^{-kt}. \end{aligned} \tag{15}$$

Combining cases (14) and (15) we obtain the general definition (13).

Note that $\varepsilon(m, n; t) > 0$ for $t \in R^+$. Note also that $p(k_1, k_2; t) = s_0 k_1 e(k_1, k_2; t)$.

For the derivative $\varepsilon(m, n; t)'$ of function (13) in the case $m \neq n$ we calculate:

$$\begin{aligned} \varepsilon(m, n; t)' &= (n - m)^{-1} d\left(e^{-mt} - e^{-nt}\right)/dt \\ &= \left(-me^{-mt} + ne^{-nt}\right)/(n - m), \\ &= \frac{ne^{-nt} - me^{-mt}}{n - m}. \end{aligned} \tag{16}$$

In the case $m = n$ we have:

$$\varepsilon(m, n; t)'|_{m=n=k} = (te^{-kt})' = (1 - kt)e^{-kt}. \tag{17}$$

Combining cases (16), (17), we obtain:

$$\varepsilon(m, n; t)' = \begin{cases} \frac{ne^{-nt} - me^{-mt}}{n - m} & m \neq n, \\ (1 - kt)e^{-kt}, & m = n = k. \end{cases} \tag{18}$$

Proposition 2. *Function $k_1 \varepsilon(k_1, k_2; t)$ satisfies the ODE initial value problem:*

$$p' = k_1 s - k_2 p, \; p(0) = 0, \tag{19}$$

wherein $s = s(t) = e^{-k_1 t}$ is the basic decay function (3).

Proof. Using definition (13) and Eqs. (12), (18) the proof follows. □

According to Proposition 2 function $k_1 \varepsilon(k_1, k_2)$ satisfies the 2SED reaction network in the case $s_0 = 1, p_0 = 0$. This property allows us to name function $\varepsilon(m, n)$ a basic outbreak function (using "epidemiological" terminology). We next give some more properties of function $\varepsilon(m, n)$.

2.2 The Basic Outbreak Function: Maximum Value and Inflection Point

Let us compute the time instant t_ε^* when the outbreak function $\varepsilon(m, n)$ attains its maximum. Using $(e^{-kt})' = -ke^{-kt}$, and (18), we have for case $m \neq n$:

$$\varepsilon(m, n; t)' = \frac{-me^{-mt} + ne^{-nt}}{n - m}. \tag{20}$$

To compute t_ε^* we need to solve equation $\varepsilon(m, n; t)' = 0$ with respect to t. We obtain that $t = t_\varepsilon^*$ satisfies equation $-me^{-mt} + ne^{-nt} = 0$, or

$$\frac{e^{-mt}}{e^{-nt}} = e^{-(m-n)t}. \tag{21}$$

From (21) we obtain:

$$e^{-(m-n)t_\varepsilon^*} = \frac{n}{m},$$

hence (for $n \neq m$):

$$t_\varepsilon^* = \ln\left(\frac{n}{m}\right)/(n - m) = \frac{\ln n - \ln m}{n - m} = \ln\left(\frac{n}{m}\right)^{\frac{1}{n-m}}.$$

To compute inflection point t_ε^* for the case $k_1 = k_2 = k$ we solve equation:

$$\varepsilon(k_1, k_2; t)'|_{k_1 = k_2 = k} = (te^{-kt})' = (1 - kt)e^{-kt} = 0,$$

for $t = t_\varepsilon^*$ to obtain $t_\varepsilon^* = 1/k$. In the general case $m > 0, n > 0$ we obtain the following

Proposition 3. *The basic outbreak function attains its maximum at time instant*

$$t_\varepsilon^* = \begin{cases} \frac{\ln n - \ln m}{n - m} = \ln\left(\frac{n}{m}\right)^{\frac{1}{n-m}}, & n \neq m, \\ 1/k. & n = m = k. \end{cases} \tag{22}$$

We next compute the (maximum) value of the outbreak function $\varepsilon(k_1, k_2; t)$ at point $t = t_\varepsilon^*$. Using (13) we compute

$$\varepsilon(k_1, k_2; t_\varepsilon^*) = \begin{cases} \frac{e^{-k_1 t_\varepsilon^*} - e^{-k_2 t_\varepsilon^*}}{k_2 - k_1}, & k_1 \neq k_2, \\ t_\varepsilon^* e^{-kt_\varepsilon^*}, & k_1 = k_2 = k. \end{cases} \tag{23}$$

We have for $k_1 \neq k_2$ and $i = 1, 2$:

$$e^{-k_i t_\varepsilon^*} = e^{-k_i \ln\left(\frac{k_2}{k_1}\right)^{\frac{1}{k_2 - k_1}}} = e^{\ln\left(\frac{k_2}{k_1}\right)^{\frac{-k_i}{k_2 - k_1}}} = \left(\frac{k_2}{k_1}\right)^{\frac{-k_i}{k_2 - k_1}}.$$

Thus formula (23) becomes

$$\varepsilon(k_1, k_2; t_\varepsilon^*) = \begin{cases} [\left(\frac{k_2}{k_1}\right)^{\frac{-k_1}{k_2 - k_1}} - \left(\frac{k_2}{k_1}\right)^{\frac{-k_2}{k_2 - k_1}}]/(k_2 - k_1), & k_1 \neq k_2, \\ \frac{1}{ke}, & k_1 = k_2 = k. \end{cases} \tag{24}$$

Let us now compute the time instant t_ε^{**}, when the outbreak function $\varepsilon(m, n)$ attains its inflection point. To this end we have to solve equation $\varepsilon(m, n; t)'' = 0$ for $t = t_\varepsilon^{**}$, or

$$(-m\eta(m; t) + n\eta(m; t))' = m^2\eta(m; t) - n^2\eta(n; t) = 0,$$

or

$$\frac{\eta(m;t)}{\eta(n;t)} = \frac{n^2}{m^2} = \left(\frac{n}{m}\right)^2.$$

We obtain

$$\eta(m - n; t_\varepsilon^{**}) = \left(\frac{n}{m}\right)^2,$$

or

$$-(m - n)t_\varepsilon^{**} = 2\ln\frac{n}{m},$$

hence

Proposition 4. *The outbreak function* $\varepsilon(m, n)$ *attains its inflection point at*

$$t_\varepsilon^{**} = 2\frac{\ln n - \ln m}{n - m} = 2t_\varepsilon^*. \tag{25}$$

The outbreak function plays an important role in a number of epidemiological and pharmacokinetic models.

3 Least-Square Approximations: Fitting the Exponential Decay Function

3.1 Fitting the Exponential Decay Function to a Numeric Data Set

Consider the exponential function

$$x(t) = ae^{bt}, \tag{26}$$

wherein $a > 0$ and $b < 0$ are unknown parameters. Let $(t_i, x_i), i = 1, 2, ..., n$, be a given (numeric) measurement data set observed from some decay process, to be briefly denoted as a "decay data set". To determine the optimum parameter values of function (26), a and b, using least-square approximation, we minimize the functional

$$\Phi(a, b) = \sum_{i=1}^{n} \left(x_i - ae^{bt_i}\right)^2, \tag{27}$$

with respect to a and b.

Fitting the One-Parameter Exponential Decay function. Let the parameter a be known as $a = 1$, so that the function to be fitted is of the form

$$x(t) = e^{bt}, \tag{28}$$

then functional (27) obtains the form:

$$\Phi_1(b) = \sum_{i=1}^{n} \left(x_i - e^{bt_i}\right)^2. \tag{29}$$

For the derivative of (29) with respect to b we obtain:

$$\frac{d\Phi_1(b)}{db} = -2\sum_{i=1}^{n}\left(x_i - e^{bt_i}\right)t_i e^{bt_i} = 0. \tag{30}$$

Equation (30) can be written as:

$$\sum_{i=1}^{n} t_i x_i e^{bt_i} - \sum_{i=1}^{n} t_i e^{2bt_i} = 0. \tag{31}$$

To solve equation (31) with respect to b we shall need to use a numeric procedure, such as the Newton-Raphson method. An important part of such an iteration procedure is to determine an initial approximation for the parameter b. This can be done by a visual inspection of the shape of the decay data set $(t_i, x_i), i = 1, 2, ..., n$, in order to establish approximately the half-life time instant t_h. Using expression (4) we can determine an initial approximation for the rate parameter b, namely $b \approx \ln 2/t_h$.

The Two-Parameter Exponential Decay Function. Consider now the case when both parameters a, b in (27) are unknown. From equation:

$$\frac{\partial\Phi(a,b)}{\partial a} = -2\sum_{i=1}^{n} e^{bt_i}\left(x_i - ae^{bt_i}\right) = 0, \tag{32}$$

we calculate:

$$a = \frac{\sum_{i=1}^{n} x_i e^{bt_i}}{\sum_{i=1}^{n} e^{2bt_i}} = \frac{S_1}{S_2}. \tag{33}$$

The condition for minimization of functional (27) with respect to b obtains the form:

$$\frac{\partial\Phi(a,b)}{\partial b} = -2\sum_{i=1}^{n}\left(x_i - ae^{bt_i}\right)\left(\frac{da}{db}e^{bt_i} + at_i e^{bt_i}\right) = 0. \tag{34}$$

We now need to solve equation

$$\sum_{i=1}^{n}\left(x_i - ae^{bt_i}\right)\left(\frac{da}{db}e^{bt_i} + at_i e^{bt_i}\right) = 0, \tag{35}$$

wherein

$$\frac{da}{db} = \frac{1}{S_2^2}\left(S_2\frac{dS_1}{db} - S_1\frac{dS_2}{db}\right), \quad \frac{dS_1}{db} = \sum_{i=1}^{n} t_i x_i e^{bt_i}, \quad \frac{dS_2}{db} = 2\sum_{i=1}^{n} x_i e^{bt_i}. \tag{36}$$

We can use a numerical iterative procedure for solving problem (35)–(36). In practice, the two-parameter setting can be avoided via a suitable normalization of the decay data set, so that the parameter a can be set to 1 (possibly by visual inspection of the data set).

3.2 Fitting the Exponential Decay Function to an Interval-Valued Data Set

In the literature there are a number of research papers dealing with numerical simulation of the exponential decay models. Many authors focus on the case when imprecise/uncertain input data sets are available. Such data sets can be considered under various statistical assumptions, see e.g. [8,9,11]. A number of researchers assume that the input data are available in the form of intervals, that is the data are "interval-valued", see e.g. [1,3,4,6,10,14].

For the exponential decay model let us assume that an interval-valued decay data set is available, (t_i, X_i), $i = 1, 2, ..., k$, $X_i = [\underline{x}_i, \overline{x}_i]$, as mentioned at the end of Sect. 1 and depicted in Fig. 3.

A Numerical Algorithm. For the solution of the interval-valued optimization problems we propose the following algorithm:

Step 1. Find an initial numeric value for the parameter k, such that the solution e^{-kt} is included in all interval-valued data (t_i, X_i), $i = 1, 2, ..., k$. To this end a visual inspection of the decay data set can be performed aiming at finding an approximation of the half-life time t_h, resp. the rate parameter $k = \ln 2 / t_h$.

Step 2. Choose a sufficiently small value for $\Delta > 0$. Check if the interval enclosure (7): e^{-Kt} for $K = [k - \Delta, k + \Delta]$, $\Delta > 0$, for the chosen Δ is included in the interval-valued decay data set, if so, then increase Δ and repeat the procedure. If the interval enclosure is not included in the interval-valued data set, then STOP. The final result of this procedure is the last Δ, resp, last interval $K = [k - \Delta, k + \Delta]$.

4 Numerical Simulations Involving the Basic Exponential Decay-Outbreak Model

Consider next a biological process based on reaction network (9) involving two species S and P, resp. the two rate parameters k_1 and k_2 in the induced dynamical system (10). In reality such a situation appears in a number of areas, e.g. in epidemiology, where researchers collect information simultaneously for the decay of the susceptible and for the outbreak of the infected/infectious individuals. Since the decay part does not depend on the outbreak part of the process, we shall assume that the decay part has been solved using the methods presented in the previous Sect. 3 and thus the decay rate parameter (k_1) has been computed. Hence we can concentrate on the numerical computation of the outbreak rate parameter (k_2) using an available outbreak data set for the decay-outbreak process.

The Numeric Problem for the Decay-Outbreak Process. Mathematically, consider the basic outbreak function (13). Assume that the decay rate parameter m is already available, as computed from a given decay data set (t_i, x_i), $i = 1, 2, ..., k$, according to instructions given in Sect. 3. Assume that an "outbreak data set" (t_j, y_j), $j = 1, 2, ..., l$, is available from experimental observations on the decay-outbreak process. Then, we have to minimize the functional

$$\Psi(m, n) = \sum_{j=1}^{l} (y_i - \varepsilon(m, n; t_i))^2, \tag{37}$$

with respect to rate parameter n. Expression (37) leads to the following

Proposition 5. *To minimize functional (37) with respect to rate parameter n we need to solve equation*

$$\frac{\partial \Psi(m, n)}{\partial n} = -2 \sum_{j=1}^{l} (y_i - \varepsilon(m, n; t_i)) \cdot \frac{\partial \varepsilon(m, n)}{\partial n}|_{t_i} = 0, \tag{38}$$

wherein

$$\frac{\partial \varepsilon(m, n)}{\partial n}|_{t_i} = \frac{m}{n - m} \left[(n - m)te^{-nt} - (e^{-mt} - e^{-nt}) \right]|_{t=t_i}. \tag{39}$$

Problem (38)–(39) can be solved numerically using appropriate numerical methods.

Initial conditions for the rate parameter n. By visual inspection on the outbreak data set one can find approximately an initial condition for the outbreak maximum time t_ε^*. Using expression (13) we have

$$\ln k_2 - k_2 t_\varepsilon^* = \ln k_1 - k_1 t_\varepsilon^*. \tag{40}$$

Equation (40) can be solved for k_2 using appropriate numerical method. If the available data set does not provide a good (visual) approximation for the value t_ε^*, then one may try to inspect visually the inflection point $2t_\varepsilon^*$, cf. expression (25) for t_ε^{**}.

The interval-valued problem for the decay-outbreak process can be formulated and treated in similar lines as done for the decay function, see previous Sect. 3. For the relevant numerical procedure the following condition may be applied.

Consistency Condition. Using (13) we have for $K_1 = [k_1 - \Delta, k_1 + \Delta]$, $\Delta > 0$, resp. for interval enclosure

$$\varepsilon(K_1, k_2; t) = \{\varepsilon(k_1, k_2; t) : k_1 \in K_1\}, \tag{41}$$

the following expression:

$$\varepsilon(K_1, k_2; t) = \begin{cases} \frac{e^{-K_1 t} - e^{-k_2 t}}{k_2 - K_1}, & k_2 \notin K_1, \\ t\, e^{-K_1 t}, & k_2 \in K_1. \end{cases} \tag{42}$$

The consistency condition now reads: $\varepsilon(K_1, k_2; t_j) \in (t_j, Y_j)$, $j = 1, 2, ..., l$.

Two basic decay functions with interval-valued rates $k_1 = 3, \Delta_1 = 0.45, k_2 = 1, \Delta_2 = 0.15$ are visualised on Fig. 6. A basic exponential outbreak function with interval valued rates: $k_1 = 3, \Delta_1 = 0.45, k_2 = 1, \Delta_2 = 0.15$ is presented on Fig. 7. A basic exponential outbreak function fitted to interval-valued data set; $k_1 = 3, \Delta_1 = 0.45, k_2 = 1, \Delta_2 = 0.15$ visualized graphically on Fig. 8.

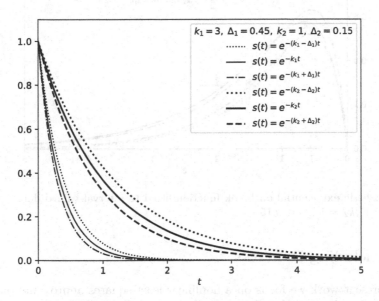

Fig. 6. Two basic decay functions with interval-valued rates: $k_1 = 3, \Delta_1 = 0.45, k_2 = 1, \Delta_2 = 0.15$

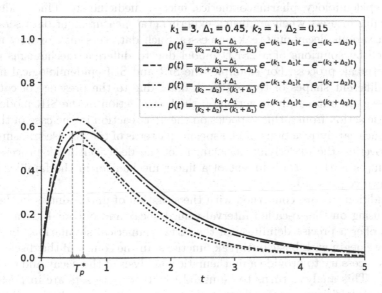

Fig. 7. A basic exponential outbreak function with interval valued rates: $k_1 = 3, \Delta_1 = 0.45, k_2 = 1, \Delta_2 = 0.15$

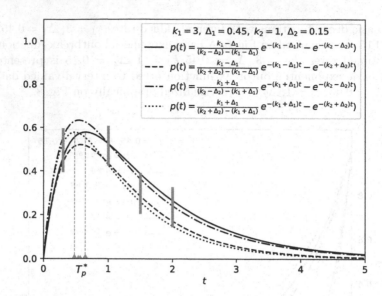

Fig. 8. A basic exponential outbreak function fitted to interval-valued data set; $k_1 = 3, \Delta_1 = 0.45, k_2 = 1, \Delta_2 = 0.15$

5   Concluding Remarks

In the present work we focus on a nonlinear least-squares approximation problem related to decay-outbreak biological processes. In particular we consider the two-step exponential decay chain (2SED) model, that has numerous applications in epidemiology, pharmacokinetics, nuclear medicine etc. This model is of considerable interest when analysing experimental measurement data sets coming from various decay-outbreak processes. Such data sets may or may not be fitted well by solutions of a 2SED model due to different mechanisms of the decay-outbreak process. For example, the SIR and SEI epidemiological models possess different shapes of the outbreak data, due to the presence of catalytic actions, e.g. an autocatalytic action on the first reaction in the SIR model, and a catalytic action from a third species on the first reaction in the case of the SEI model. Such catalytic actions model specific patterns of the disease transmission. So, discovering the underlying mechanism of the decay-outbreak process is as important as is the establishment of a linear mechanism in the familiar linear case (linear regression).

In addition, we are concerned with the situation of uncertainties in the data sets, focusing on the so-called interval-valued decay and outbreak data sets. In order to offer a precise definition of the term "numerical simulation" in such a setting, we begin with basic outbreak functions. In the course of the discussion it becomes necessary to provide a mathematical analysis of the decay and outbreak functions. This analysis turns to be useful when the data sets are inspected for specific peculiarities in their shapes, which can produce initial approximations for the induced numerical problems.

Acknowledgements. The second author (SM) gratefully acknowledges financial support for project KP-06-N52/1 of 2021 at Bulgarian NSF entitled "New mathematical models for analysis of big data, with applications to medicine and epidemiology".

References

1. Ahn, M., Peng, C., Park, Y., Jeon, Y.: A resampling approach for interval-valued data regression. Stat. Anal. Data Min. **5**(4), 336–348 (2012)
2. Bateman, H.: The solution of a system of differential equations occurring in the theory of radio-active transformations. Proc. Cambridge Phil. Soc. **15**, 423–427 (1910)
3. Angela, B.-F., et al.: Extensions of linear regression models based on set arithmetic for interval data. arXiv: Statistics Theory (2012): n. pag
4. Billard, L., Diday, E.: Regression analysis for interval-valued data. In: Proceedings of the Seventh Conference of the International Federation of Classification Societies on Data Analysis, Classification and Related Methods, pp. 369–374 (2000)
5. Borisov, M., Markov, S.: The two-step exponential decay reaction network: analysis of the solutions and relation to epidemiological SIR models with logistic and Gompertz type infection contact patterns. J. Math. Chem. **59**(5), 1283–1315 (2021). https://doi.org/10.1007/s10910-021-01240-8
6. Brito, P., Silva, A.P.D.: Modeling interval data with normal and skew-normal distributions. J. Appl. Stat. **39**, 157–170 (2012)
7. Chellaboina V., Bhat, S.P., Haddat, W.M., Bernstein, D.S.: Modeling and Analysis of Mass-Action Kinetics. IEEE Control Syst. Mag. 60–78 (2009)
8. Dette, H., Neugebauer, H.-M.: Bayesian optimal one point designs for one parameter nonlinear models. J. Statist. Plann. Infer. **52**(1), 17–31 (1996)
9. Dette, H., Martinez Lopez, I., Ortiz Rodriguez, I.M., Pepelyshev, A.: Efficient design of experiment for exponential regression models. Technical Report, Universität Dortmund, SFB 475, No. 08 (2004).http://hdl.handle.net/10419/49337
10. Gil, M., Lubiano, M., Montenegro, M., López, M.: Least squares fitting of an affine function and strength of association for interval-valued data. Metrika **56**(2), 97–111 (2002)
11. Han, C., Chaloner, K.: D- and c-optimal designs for exponential regression models used in pharmacokinetics and viral dynamics. J. Statist. Plann. Infer. **115**, 585–601 (2003)
12. Hethcote, H.W.: The mathematics of infectuous diseases. SIAM Rev. **42**(4), 599–653 (2000)
13. Lente, G.: Deterministic Kinetics in Chemistry and Systems Biology. Briefs in Molecular Science. Springer (2016)
14. Lima Neto, E., Cordeiro, G., De Carvalho, F.: Bivariate symbolic regression models for interval-valued variables. J. Statist. Comput. Simul. **81**(11), 1727–1744 (2011)
15. Muench, H.: Catalytic Models in Epidemiology. Harvard University Press, Cambridge (MA) (1959)
16. Murray, J.D.: Mathematical Biology: I. An Introduction, 3rd edn. Springer, New York, NY (2002). https://doi.org/10.1007/b98868

Rational Approximation Preconditioners
for Multiphysics Problems

Ana Budiša[1]([✉]) [iD], Xiaozhe Hu[2] [iD], Miroslav Kuchta[1] [iD],
Kent-André Mardal[1,3] [iD], and Ludmil Zikatanov[4] [iD]

[1] Simula Research Laboratory, 0164 Oslo, Norway
{ana,miroslav,kent-and}@simula.no
[2] Tufts University, Medford, MA 02155, USA
Xiaozhe.Hu@tufts.edu
[3] University of Oslo, 0316 Oslo, Norway
[4] Penn State University, University Park, PA 16802, USA
ludmil@psu.edu

Abstract. We consider a class of mathematical models describing multiphysics phenomena interacting through interfaces. On such interfaces, the traces of the fields lie (approximately) in the range of a weighted sum of two fractional differential operators. We use a rational function approximation to precondition such operators. We first demonstrate the robustness of the approximation for ordinary functions given by weighted sums of fractional exponents. Additionally, we present more realistic examples utilizing the proposed preconditioning techniques in interface coupling between Darcy and Stokes equations.

Keywords: Rational approximation · Preconditioning · Multiphysics

1 Introduction

Fractional operators arise in the context of preconditioning of coupled multiphysics systems and, in particular, in the problem formulations where the coupling constraint is enforced by a Lagrange multiplier defined on the interface. Examples include the so-called EMI equations in modeling of excitable tissue [30], reduced order models of microcirculation in $2d$-$1d$ [22,23] and $3d$-$1d$ [20,21] setting or Darcy/Biot–Stokes models [2,24]. We remark that fractional operators have been recently utilized also in monolithic solvers for formulations of Darcy/Biot–Stokes models without the Lagrange multipliers, see [5,6].

The coupling in the multiplier formulations is naturally posed in Sobolev spaces of fractional order. However, for parameter robustness of iterative methods, a more precise setting must be considered, where the interface problem is posed in the intersection (space) of parameter-weighted fractional order Sobolev spaces [17]. Here the sums of fractional operators induce the natural inner product. In the examples mentioned earlier, however, the interface preconditioners were realized by eigenvalue decomposition, and thus, while being parameter robust, the resulting solvers do not scale with mesh size. Using Darcy–Stokes

© The Author(s), under exclusive license to Springer Nature Switzerland AG 2023
I. Georgiev et al. (Eds.): NMA 2022, LNCS 13858, pp. 100–113, 2023.
https://doi.org/10.1007/978-3-031-32412-3_9

system as the canonical example, we aim to show that rational approximations are crucial for designing efficient preconditioners for multiphysics systems.

There have been numerous works on approximating/preconditioning problems such as $\mathcal{D}^s u = f$, e.g. [4,9,27]. The rational approximation (RA) approach has been advocated, and several techniques based on it were proposed by Hofreither in [15,16] (where also a Python implementation of Best Uniform Rational Approximation (BURA) for x^λ is found). Further rational approximations used in preconditioning can be found in [12,13]. An interesting approach, which always leads to real-valued poles and uses Padé approximations of suitably constructed time-dependent problems, is found in [11]. In our numerical tests, we use Adaptive Antoulas-Anderson (AAA) algorithm [26] which is a greedy strategy for locating the interpolation points and then using the barycentric representation of the rational interpolation to define a function that is close to the target. In short, for a given continuous function f, the AAA algorithm returns a rational function that approximates the best uniform rational approximation to f.

The rest of the paper is organized as follows. In Sect. 2, we introduce a model problem that leads to the sum of fractional powers of a differential operator on an interface acting on weighted Sobolev spaces. In Sect. 3, we introduce the finite element discretizations that we employ for the numerical solution of such problems. Next, Sect. 4 presents some details on the rational approximation and the scaling of the discrete operators. In Sect. 5, we test several relevant scenarios and show the robustness of the rational approximation as well as the efficacy of the preconditioners. Conclusions are drawn in Sect. 6.

2   Darcy–Stokes Model

We study interaction between a porous medium occupying $\Omega_D \subset \mathbb{R}^d$, $d = 2, 3$ surrounded by a free flow domain $\Omega_S \supset \Omega_D$ given by a Darcy–Stokes model [24] as: For given volumetric source terms $f_S : \Omega_S \to \mathbb{R}^d$ and $f_D : \Omega_D \to \mathbb{R}$ find the Stokes velocity and pressure $u_S : \Omega_S \to \mathbb{R}^d$, $p_S : \Omega_S \to \mathbb{R}$ and the Darcy flux and pressure $u_D : \Omega_D \to \mathbb{R}^d$, $p_D : \Omega_D \to \mathbb{R}$ such that

$$-\nabla \cdot \sigma(u_S, p_S) = f_S \text{ and } \nabla \cdot u_S = 0 \qquad \text{in } \Omega_S,$$
$$u_D + K\mu^{-1}\nabla p_D = 0 \text{ and } \nabla \cdot u_D = f_D \qquad \text{in } \Omega_D,$$
$$u_S \cdot \nu_s + u_D \cdot \nu_D = 0 \qquad \text{on } \Gamma, \qquad (1)$$
$$-\nu_S \cdot \sigma(u_S, p_s) \cdot \nu_S - p_D = 0 \qquad \text{on } \Gamma,$$
$$-P_{\nu_S}(\sigma(u_S, p_S) \cdot \nu_S) - \alpha\mu K^{-1/2} P_{\nu_S} u_S = 0 \qquad \text{on } \Gamma.$$

Here, $\sigma(u, p) := 2\mu\epsilon(u) - pI$ with $\epsilon(u) := \frac{1}{2}(\nabla u + \nabla u^T)$. Moreover, $\Gamma := \partial\Omega_D \cap \partial\Omega_S$ is the common interface, ν_S and $\nu_D = -\nu_S$ represent the outward unit normal vectors on $\partial\Omega_S$ and $\partial\Omega_D$. Given a surface with normal vector ν, $P_\nu := I - \nu \otimes \nu$ denotes a projection to the tangential plane with normal ν. The final three equations in (1) represent the coupling conditions on Γ. Parameters of the model (which we shall assume to be constant) are viscosity $\mu > 0$, permeability $K > 0$ and Beavers-Joseph-Saffman parameter $\alpha > 0$.

Finally, let us decompose the outer boundary $\partial \Omega_S \setminus \Gamma = \Gamma_u \cup \Gamma_\sigma$, $|\Gamma_i| > 0$, $i = u, \sigma$ and introduce the boundary conditions to close the system (1)

$$u_S \cdot \nu_S = 0 \text{ and } P_{\nu_S}(\sigma(u_S, p_S) \cdot \nu_S) = 0 \quad \text{on } \Gamma_u,$$
$$\sigma(u_S, p_S) \cdot \nu_S = 0 \quad \text{on } \Gamma_\sigma.$$

Thus, Γ_u is an impermeable free-slip boundary. We remark that other conditions, in particular no-slip on Γ_u, could be considered without introducing additional challenges. However, unlike the tangential component, constraints for the normal component of velocity are easy[1] to implement in $H(\text{div})$-conforming discretization schemes considered below.

Letting $V_S \subset H_1(\Omega_S)$, $V_D \subset H(\text{div}, \Omega_D)$, $Q_S \subset L^2(\Omega_S)$, $Q_D \subset L^2(\Omega_D)$ and $Q \subset H^{1/2}(\Gamma)$ the variational formulation of (1) seeks to find $w := (u, p, \lambda) \in W$, $W := V \times Q \times \Lambda$, $V := V_S \times V_D$, $Q := Q_S \times Q_D$ and $u := (u_S, u_D)$, $p := (p_S, p_D)$ such that $\mathcal{A}w = L$ in W', the dual space of W, where L is the linear functional of the right-hand sides in (1) and the problem operator $\mathcal{A}$ satisfies

$$\langle \mathcal{A}w, \delta w \rangle = a_S(u_S, v_S) + a_D(u_D, v_D) + b(u, q) + b_\Gamma(v, \lambda) + b(v, p) + b_\Gamma(u, \delta\lambda), \quad (2)$$

where $\delta w := (v, q, \delta\lambda)$, $v := (v_S, v_D)$, $q := (q_S, q_D)$ and $\langle \cdot, \cdot \rangle$ denotes a duality pairing between W and W'. The bilinear forms in (2) are defined as

$$a_S(u_S, v_S) := \int_{\Omega_S} 2\mu\epsilon(u_S) : \epsilon(v_S) \, dx + \int_\Gamma \alpha\mu K^{-1/2} P_{\nu_S} u_S \cdot P_{\nu_S} v_S \, dx,$$
$$a_D(u_D, v_D) := \int_{\Omega_D} \mu K^{-1} u_D \cdot v_D \, dx,$$
$$b(v, p) := -\int_{\Omega_S} p_S \nabla \cdot v_S \, dx - \int_{\Omega_D} p_D \nabla \cdot v_D \, dx,$$
$$b_\Gamma(v, \lambda) := \int_\Gamma (v_S \cdot \nu_S + v_D \cdot \nu_D)\lambda \, ds.$$

(3)

Here, λ represents a Lagrange multiplier whose physical meaning is related to the normal component of the traction vector on Γ, $\lambda := -\nu_S \cdot \sigma(u_S, p_S) \cdot \nu_S$, see [24] where also well-posedness of the problem in the space W above is established.

Following [17], parameter-robust preconditioners for Darcy–Stokes operator $\mathcal{A}$ utilize weighted sums of fractional operators on the interface. Specifically, we shall consider the following operator

$$S := \mu^{-1}(-\Delta_\Gamma + I_\Gamma)^{-1/2} + K\mu^{-1}(-\Delta_\Gamma + I_\Gamma)^{1/2}, \quad (4)$$

where we have used the subscript Γ to emphasize that the operators are considered on the interface. We note that $-\Delta_\Gamma$ is singular in our setting as Γ is a closed surface and adding lower order term I_Γ thus ensures positivity.

---

[1] Conditions on the normal component can be implemented as Dirichlet boundary conditions and enforced by the constructions of the finite element trial and test spaces. The tangential component can be controlled e.g., by the Nitsche method [29] which modifies the discrete problem operator.

Letting A_S be the operator induced by the bilinear form a_S in (3) we define the Darcy–Stokes preconditioner as follows,

$$\mathcal{B} := \text{diag} \left(A_S, \mu K^{-1}(I - \nabla\nabla \cdot), \mu^{-1}I, K\mu^{-1}I, S \right)^{-1}. \tag{5}$$

Note that the operators I in the pressure blocks of the preconditioner act on different spaces/spatial domains, i.e., Q_S and Q_D. We remark that in the context of Darcy–Stokes preconditioning, [14] consider BURA approximation for a simpler interfacial operator, namely, $K\mu^{-1}(-\Delta_\Gamma + I_\Gamma)^{1/2}$. However, the preconditioner cannot yield parameter robustness, cf. [17].

3   Mixed Finite Element Discretization

In order to assess numerically the performance of rational approximation of S^{-1} in (5), stable discretization of the Darcy–Stokes system is needed. In addition to parameter variations, here we also wish to show the algorithm's robustness to discretization and, in particular, the construction of the discrete multiplier space. To this end, we require a family of stable finite element discretizations.

Let $k \geq 1$ denote the polynomial degree. For simplicity, to make sure that the Lagrange multiplier fits well with the discretization of both the Stokes and Darcy domain, we employ the same $H(\text{div})$ based discretization in both domains. That is, we discretize the Stokes velocity space V_S by Brezzi-Douglas-Marini $\mathbb{BDM}_k$ elements [8] over simplicial triangulations Ω_S^h of Ω_S and likewise, approximations to V_D are constructed with $\mathbb{BDM}_k$ elements on Ω_D^h. Here, h denotes the characteristic mesh size. Note that by construction, the velocities and fluxes have continuous normal components across the *interior* facets of the respective triangulations. However, on the interface Γ, we do not impose any continuity between the vector fields. The pressure spaces Q_S, Q_D shall be approximated in terms of discontinuous piecewise polynomials of degree $k - 1$, $\mathbb{P}_{k-1}^{\text{disc}}$. Finally, the Lagrange multiplier space is constructed by $\mathbb{P}_k^{\text{disc}}$ elements on the triangulation $\Gamma^h := \Gamma_S^h$, with Γ_S^h being the trace mesh of Ω_S^h on Γ. For simplicity, we assume $\Gamma_D^h = \Gamma_S^h$.

Approximation properties of the proposed discretization are demonstrated in Fig. 1. It can be seen that all the quantities converge with order k (or better) in their respective norms. This is particularly the case for the Stokes velocity, where the error is measured in the H^1 norm.

Let us make a few remarks about our discretization. First, observe that by using $\mathbb{BDM}$ elements on a global mesh $\Omega_S^h \cup \Omega_D^h$ the Darcy–Stokes problem (1) can also be discretized such that the mass conservation condition $u_S \cdot \nu_S + u_D \cdot \nu_D = 0$ on Γ is enforced by construction, i.e. no Lagrange multiplier is required. Here, u is the global $H(\text{div})$-conforming vector field, with $u_i := u|_{\Omega_i}$, $i = S, D$. Second, we note that the chosen discretization of V_S is only $H(\text{div}, \Omega_S)$-conforming. In turn, stabilization of the tangential component of the Stokes velocity is needed, see, e.g., [3], which translates to modification of the bilinear form a_S in (3) as

$$a_S^h(u,v) := a_S(u,v) - \sum_{e \in F_S^h} \int_e 2\mu \{\!\!\{\epsilon(u)\}\!\!\} \cdot [\![P_{\nu_e} v]\!] \, ds$$

$$- \sum_{e \in F_S^h} \int_e 2\mu \{\!\!\{\epsilon(v)\}\!\!\} \cdot [\![P_{\nu_e} u]\!] \, ds + \sum_{e \in F_S^h} \int_e \frac{2\mu\gamma}{h_e} [\![P_{\nu_e} u]\!] \cdot [\![P_{\nu_e} v]\!] \, ds, \tag{6}$$

see [18]. Here, $F_S^h = F(\Omega_S^h)$ is the collection of interior facets e of triangulation Ω_S^h, while ν_e, h_e denote respectively the facet normal and facet diameter. The stabilization parameter $\gamma > 0$ has to be chosen large enough to ensure the coercivity of a_S^h. However, the value depends on the polynomial degree. Finally, for interior facet e shared by elements T^+, T^- we define the (facet) jump of a vector v as $[\![v]\!] := v|_{T^+ \cap e} - v|_{T^- \cap e}$ and the (facet) average of a tensor ϵ as $\{\!\!\{\epsilon\}\!\!\} := \frac{1}{2}(v|_{T^+ \cap e} \cdot \nu_e + v|_{T^- \cap e} \cdot \nu_e)$.

We conclude this section by discussing discretization of the operator $(-\Delta+I)$ needed for realizing the interface preconditioner (4). Since, in our case, the multiplier space Λ_h is only L^2-conforming we adopt the symmetric interior penalty approach [10] so that in turn for $u, v \in \Lambda_h$,

$$\langle(-\Delta+I)u,v\rangle := \int_\Gamma (\nabla u \cdot \nabla v + uv)\, dx - \sum_{e \in F(\Gamma_h)} \int_e \{\!\!\{\nabla u\}\!\!\} [\![v]\!]\, ds$$

$$- \sum_{e \in F(\Gamma_h)} \int_e \{\!\!\{\nabla v\}\!\!\} [\![u]\!]\, ds + \sum_{e \in F(\Gamma_h)} \int_e \frac{\gamma}{h_e} [\![u]\!][\![v]\!]\, ds. \tag{7}$$

Here, the jump of a scalar f is computed as $[\![f]\!] := f|_{T^+ \cap e} - f|_{T^- \cap e}$ and the average of a vector v reads $\{\!\!\{v\}\!\!\} := \frac{1}{2}(v|_{T^+ \cap e} \cdot \nu_e + v|_{T^- \cap e} \cdot \nu_e)$. As before, $\gamma > 0$ is a suitable stabilization parameter.

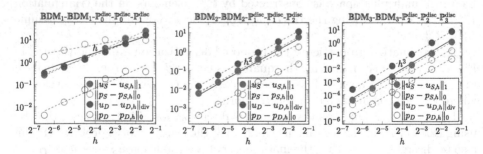

Fig. 1. Approximation properties of BDM_k-BDM_k-$\mathbb{P}_{k-1}^{\mathrm{disc}}$-$\mathbb{P}_{k-1}^{\mathrm{disc}}$-$\mathbb{P}_k^{\mathrm{disc}}$ discretization of the Darcy–Stokes problem. Two-dimensional setting is considered with $\Omega_S = (0,1)^2$ and $\Omega_D = (\frac{1}{4}, \frac{3}{4})^2$. Parameter values are set as $K = 2$, $\mu = 3$ and $\alpha = \frac{1}{2}$. Left figure is for $k = 1$ while $k = 2, 3$ is shown in the middle and right figures, respectively.

4   Rational Approximation for the General Problems of Sums of Fractional Operators

Let $s, t \in [-1, 1]$ and $\alpha, \beta \geq 0$ where at least one of α, β is not zero. For interval $I \subset \mathbb{R}^+$, consider a function $f(x) = (\alpha x^s + \beta x^t)^{-1}$, $x \in I$.. The basic idea is to find a rational function $R(x)$ approximating f on I, that is, $R(x) = \frac{P_{k'}(x)}{Q_k(x)} \approx f(x)$, where $P_{k'}$ and Q_k are polynomials of degree k' and k, respectively. Assuming $k' \leq k$, the rational function can be given in the following partial fraction form

$$R(x) = c_0 + \sum_{i=1}^{N} \frac{c_i}{x - p_i},$$

for $c_0 \in \mathbb{R}$, $c_i, p_i \in \mathbb{C}$, $i = 1, 2, \ldots, N$. The coefficients p_i and c_i are called *poles* and *residues* of the rational approximation, respectively.

We note that the rational approximation has been predominantly explored to approximate functions with only one fractional power, that is $x^{-\bar{s}}$ for $\bar{s} \in (0, 1)$ and $x > 0$. Additionally, the choice of the rational approximation method that computes the poles and residues is not unique. One possibility is the BURA method which first computes the best uniform rational approximation $\bar{r}_\beta(x)$ of $x^{\beta - \bar{s}}$ for a positive integer $\beta > \bar{s}$ and then uses $\bar{r}(x) = \frac{\bar{r}_\beta(x)}{x}$ to approximate $x^{-\bar{s}}$. Another possible choice is to use the rational interpolation of $z^{-\bar{s}}$ to obtain $\bar{r}(x)$. The AAA algorithm proposed in [26] is a good candidate. The AAA method is based on the representation of the rational approximation in barycentric form and greedy selection of the interpolation points. Both approaches lead to the poles $p_i \in \mathbb{R}$, $p_i \leq 0$ for the case of one fractional power. An overview of rational approximation methods can be found in [15].

The location of the poles is crucial in rational approximation preconditioning. For $\bar{f}(x) = x^{\bar{s}}$, $\bar{s} \in (0, 1)$ the poles of the rational approximation for $\bar{f}$ are all real and negative. Hence, in the case of a positive definite operator $\mathcal{D}$, the approximation of $\mathcal{D}^{\bar{s}}$ requires only inversion of positive definite operators of the form $\mathcal{D} + |p_i|I$, for $i = 1, 2, \ldots, N$, $p_i \neq 0$. Such a result for rational approximation of $\bar{f}$ with $\bar{s} \in (-1, 1)$ is found in a paper by H. Stahl [28]. In the following, we present an extensive set of numerical tests for the class of *sum of fractional operators*, which gives a wide class of efficient preconditioners for the multiphysics problems coupled through an interface. The numerical tests show that the poles remain real and nonpositive in most combinations of fractional exponents s and t.

Let V be a Hilbert space and V' be its dual. Consider a symmetric positive definite (SPD) operator $A : V \to V'$. Then, the rational function $R(\cdot)$ can be used to approximate $f(A)$ as follows,

$$z = f(A)r \approx c_0 r + \sum_{i=1}^{N} c_i (A - p_i I)^{-1} r$$

with $z \in V$ and $r \in V'$. The overall algorithm is shown in Algorithm 1.

Without loss of generality we let the operator A be a discretization of the Laplacian operator $-\Delta$, and I is the identity defined using the standard L^2

Algorithm 1. Compute $z = f(A)r$ using rational approximation.

1: Solve for w_i: $(A - p_i I) w_i = r, \quad i = 1, 2, \dots, N.$

2: Compute: $z = c_0 r + \sum\limits_{i=1}^{N} c_i w_i$

---

inner product. In particular, unlike in (4) we assume that $-\Delta$ is SPD (e.g. by imposing boundary conditions eliminating the constant nullspace). Therefore, the equations in Step 1 of Algorithm 1 can be viewed as discretizations of the shifted Laplacian problems $-\Delta w_i - p_i w_i = r, \ p_i < 0$, and we can use efficient numerical methods, such as Algebraic MultiGrid (AMG) methods [7,32], for their solution.

We would like to point out that in the implementation, the operators involved in Algorithm 1 are replaced by their matrix representations on a concrete basis and are properly scaled. We address this in more detail in the following section.

4.1   Preconditioning

Let A be the stiffness matrix associated with $-\Delta$ and M a corresponding mass matrix of the L^2 inner product. Also, denote with n_c the number of columns of A. The problem we are interested in is constructing an efficient preconditioner for the solution of the linear system $F(A)x = b$. Thus, we would like to approximate $f(A) = F(A)^{-1}$ using the rational approximation $R(x)$ of $f(x) = \frac{1}{F(x)}$.

Let I be a $n_c \times n_c$ identity matrix and let U be an M-orthogonal matrix of the eigenvectors of the generalized eigenvalue problem $Au_j = \lambda_j Mu_j, j = 1, 2, \dots, n_c$, namely,

$$AU = MU\Lambda, \quad U^T MU = I \quad \Longrightarrow \quad U^T AU = \Lambda. \tag{8}$$

For any continuous function $G : [0, \rho] \to \mathbb{R}$ we define

$$G(A) := MUG(\Lambda)U^T M, \quad \text{i.e.,} \quad f(A) = MUf(\Lambda)U^T M. \tag{9}$$

where $\rho := \rho\left(M^{-1}A\right)$ is the spectral radius of the matrix $M^{-1}A$.

A simple consequence from the Chebyshev Alternation Theorem is that the residues/poles for the rational approximation of $f(x)$ for $x \in [0, \rho]$ are obtained by scaling the residues/poles of the rational approximation of $g(y) = f(\rho y)$ for $y \in [0, 1]$. Indeed, we have, $c_i(f) = \rho c_i(g)$, and $p_i(f) = \rho p_i(g)$. Therefore, in the implementation, we need an upper bound on $\rho\left(M^{-1}A\right)$. For $\mathbb{P}_1$ finite elements, such a bound can be obtained following the arguments from [31],

$$\rho\left(M^{-1}A\right) \leq \frac{1}{\lambda_{\min}(M)} \|A\|_\infty = d(d+1) \left\|\text{diag}(M)^{-1}\right\|_\infty \|A\|_\infty \tag{10}$$

with d the spatial dimension[2].

---

[2] Such estimates can be carried out for Lagrange finite elements of any polynomial degree because the local mass matrices are of a special type: a constant matrix, which depends only on the dimension and the polynomial degree, times the volume of the element.

Proposition 1. *Let $R_f(\cdot)$ be the rational approximation for the function $f(\cdot)$ on $[0, \rho]$. Then for the stiffness matrix A and mass matrix M satisfying (8), we have*

$$f(A) \approx R_f(A) = c_0 M^{-1} + \sum_{i=1}^{N} c_i \left(A - p_i M\right)^{-1}. \tag{11}$$

Proof. The relations (8) imply that $UU^T = M^{-1}$, and therefore,

$$AU - p_i MU = MU(\Lambda - p_i I) \iff (A - p_i M)U = MU(\Lambda - p_i I).$$

This is equivalent to $(\Lambda - p_i I)^{-1} \underbrace{U^{-1} M^{-1}}_{U^T} = U^{-1}(A - p_i M)^{-1}$. Hence,

$$(\Lambda - p_i I)^{-1} = U^{-1}(A - p_i M)^{-1} U^{-T}.$$

A straightforward substitution in (9) then shows (11). □

In addition, to apply the rational approximation preconditioner, we need to compute the actions of M^{-1} and each $(A - \rho p_i M)^{-1}$. If $p_i \in \mathbb{R}$, $p_i \leq 0$, this leads to solving a series of elliptic problems where the AMG methods are very efficient.

5   Numerical Results

In this section, we present two sets of experiments: (1) on the robustness of the rational approximation with respect to the scaling parameters and the fractional exponents; and (2) on the efficacy of the preconditioned minimal residual (MinRes) method as a solver for Darcy–Stokes coupled model. We use the AAA algorithm [26] to construct a rational approximation. The discretization and solver tools are Python modules provided by FEniCS_ii [19], cbc.block [25], and interfaced with the HAZmath library [1].

5.1   Approximating the Sum of Two Fractional Exponents

In this example, we test the approximation power of the rational approximation computed by AAA algorithm regarding different fractional exponents s, t and parameters α, β. That is, we study the number of poles N required to achieve

$$\|f - R\|_\infty = \max_{x \in [0,1]} \left| (\alpha x^s + \beta x^t)^{-1} - \left(c_0 + \sum_{i=1}^{N} \frac{c_i}{x - p_i} \right) \right| \leq \epsilon_{RA},$$

for a fixed tolerance $\epsilon_{RA} = 10^{-12}$. In this case, we consider the fractional function f to be defined on the unit interval $I = (0, 1]$. As we noted earlier, however, the approximation can be straightforwardly extended to any interval. We also consider the scaling regarding the magnitude of parameters α and β. Specifically, in case when $\alpha > \beta$, we rescale the problem with $\gamma_\alpha = \frac{\beta}{\alpha} < 1$ and approximate

$$\widetilde{f}(x) = (x^s + \gamma_\alpha x^t)^{-1} \approx R_{\widetilde{f}}(x).$$

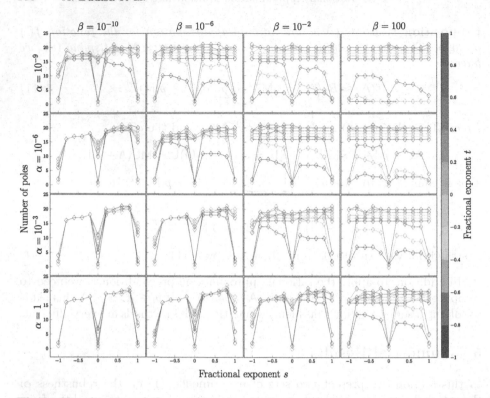

Fig. 2. Visualization of the number of poles in the rational approximations of the function $f(x) = (\alpha x^s + \beta x^t)^{-1}$ for $x \in (0,1]$ with regards to varying fractional exponents s, t and coefficients α, β.

Then the rational approximation for the original function f is given as

$$f(x) \approx R(x) = \frac{1}{\alpha} R_{\widetilde{f}}(x).$$

Similar can be done in case when $\beta > \alpha$.

The results are summarized in Fig. 2. To obtain different parameter ratios γ_α, we take $\alpha \in \{10^{-9}, 10^{-6}, 10^{-3}, 1\}$ and $\beta \in \{10^{-10}, 10^{-6}, 10^{-2}, 10^2\}$. Furthermore, we vary the fractional exponents $s, t \in [-1, 1]$ with the step 0.2. We observe that the number of poles N remains relatively uniform with varying the exponents, except in generic cases when $s, t = \{-1, 0, 1\}$. For example, for the combination $(s, t) = (1, 1)$, the function we are approximating is $f(x) = \frac{1}{2x}$, thus the rational approximation should return only one pole $p_1 = 0$ and residues $c_0 = 0, c_1 = \frac{1}{2}$. We also observe that for the fixed tolerance of $\epsilon_{\text{RA}} = 10^{-12}$, we obtain a maximum of 22 poles in all cases.

Additionally, we remark that in most test cases, we retain real and negative poles, which is a desirable property to apply the rational approximation as a positive definite preconditioner. However, depending on the choice of fractional exponents s, t and tolerance ϵ_{RA}, the algorithm can produce positive or a pair of complex conjugate poles. Nevertheless, these cases are rare, and the number and the values of those poles are small. Therefore numerically, we do not observe any significant influence on the rational approximation preconditioner. More concrete analytical results on the location of poles for sums of fractionalities are part of our future research.

5.2 The Darcy–Stokes Problem

In this section, we discuss the performance of RA approximation of S^{-1} in the Darcy–Stokes preconditioner (5). To this end, we consider $\Omega_S = (0,1)^2$, $\Omega_D = (\frac{1}{4}, \frac{3}{4})^2$ and fix the value of Beavers-Joseph-Saffman parameter $\alpha = 1$ while permeability and viscosity are varied[3] $10^{-6} \leq K \leq 1$ and $10^{-6} \leq \mu \leq 10^2$. The system is discretized by BDM_k-BDM_k-$\mathbb{P}_{k-1}^{\mathrm{disc}}$-$\mathbb{P}_{k-1}^{\mathrm{disc}}$-$\mathbb{P}_k^{\mathrm{disc}}$ elements, $k = 1, 2, 3$, which were shown to provide convergent approximations in Fig. 1. A hierarchy of meshes Ω_S^h, Ω_D^h is obtained by uniform refinement. We remark that the stabilization constants γ in (6) and (7) are chosen as $\gamma = 20k$ and $\gamma = 10k$ respectively. The resulting linear systems are then solved by preconditioned MinRes method. To focus on the RA algorithm in the preconditioner, for all but the multiplier block, an exact LU decomposition is used. Finally, the iterative solver is always started from a zero initial guess and terminates when the preconditioned residual norm drops below 10^{-10}.

We present two sets of experiments. First, we fix the tolerance in the RA algorithm at $\epsilon_{\mathrm{RA}} = 2^{-40} \approx 10^{-12}$ and demonstrate that using RA in the preconditioner (5) leads to stable MinRes iterations for any practical values of the material parameters, the mesh resolution h and the polynomial degree k in the finite element discretization (see Fig. 3). Here it can be seen that the number of iterations required for convergence is bounded in the above-listed quantities. Results for different polynomial degrees are largely similar and appear to be mostly controlled by material parameters. However, despite the values of K, μ spanning several orders of magnitude, the iterations vary only between 30 to 100.

Next, we assess the effects of the accuracy in RA on the performance of preconditioned MinRes solver. Let us fix $k = 1$ and vary the material parameters as well as the RA tolerance ϵ_{RA}. In Fig. 4, we observe that the effect of ϵ_{RA} varies with material properties (which enter the RA algorithm through scaling). In particular, it can be seen that for $K = 1$, the number of MinRes iterations is practically constant for any $\epsilon_{\mathrm{RA}} \leq 10^{-1}$. On the other hand, when $K = 10^{-6}$, the counts vary with ϵ_{RA} and to a lesser extent with μ. Here, lower accuracy typically leads to a larger number of MinRes iterations. However, for $\epsilon_{\mathrm{RA}} \leq 10^{-4}$ the iterations behave similarly. We remark that with $K = 10^{-6}$ (and any $10^{-6} \leq \mu \leq 10^2$), the tolerance $\epsilon_{\mathrm{RA}} = 10^{-1}$ leads to 2 poles, cf. Figure 5, while for $K = 1$ there are at least 5 poles needed in the RA approximation.

Our results demonstrate that RA approximates S^{-1} in (4), which leads to a robust, mesh, and parameter-independent Darcy–Stokes solver. We remark that though the algorithm complexity is expected to scale with the number of degrees of freedom on the interface, $n_h = \dim \Lambda_h$, which is often considerably

---

[3] These ranges are identified as relevant for many applications in biomechanics [6].

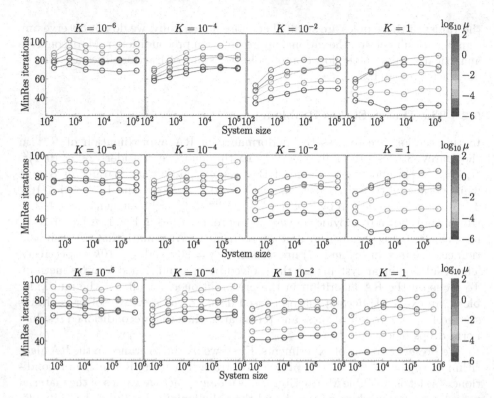

Fig. 3. Number of MinRes iterations required for convergence with preconditioner (5) using the RA with tolerance $\epsilon_{\mathrm{RA}} = 2^{-40}$. Setup from Fig. 1 is considered with the system discretized by $\mathbb{BDM}_k$-$\mathbb{BDM}_k$-$\mathbb{P}_{k-1}^{\mathrm{disc}}$-$\mathbb{P}_{k-1}^{\mathrm{disc}}$-$\mathbb{P}_k^{\mathrm{disc}}$ elements. (Top) $k = 1$, (middle) $k = 2$, (bottom) $k = 3$.

smaller than the total problem size, the setup cost may become prohibitive. This is particularly true for spectral realization, which often results in $\mathcal{O}(n_h^3)$ complexity. To address such issues, we consider how the setup time of RA and the solution time of the MinRes solver depend on the problem size.

In Fig. 5 we show the setup time of RA (for fixed material parameters) as function of mesh size and ϵ_{RA}. It can be seen that the times are $< 0.1\,\mathrm{s}$ and practically constant with h (and n_h). As with the number of poles, the small variations in the timings with h are likely due to different scaling of the matrices A and M. We note that in our experiments $32 \le n_h \le 1024$. Moreover, since Λ_h is in our experiments constructed from $\mathbb{P}_1^{\mathrm{disc}}$ we apply the estimate (10).

In Fig. 5, we finally plot the dependence of the solution time of the preconditioned MinRes solver on the problem size. Indeed we observe that the solver is of linear complexity. In particular, application of rational approximation of S^{-1} in our implementation requires $\mathcal{O}(n_h)$ operations. We remark that here the solvers for the shifted Laplacian problems are realized by the conjugate gradient method with AMG as a preconditioner. Let us also recall that the remaining blocks of the preconditioner are realized by LU, where the setup cost is not

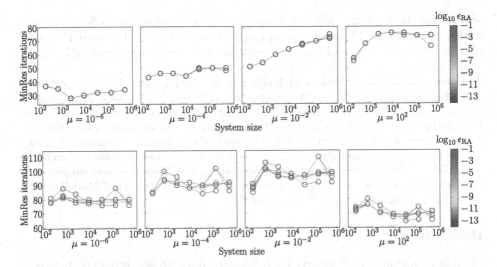

Fig. 4. Number of MinRes iterations required for convergence with preconditioner (5) using the RA with varying tolerance ϵ_{RA}. (Top) $K = 1$. (Bottom) $K = 10^{-6}$. Setup as in Fig. 3 is used with discretization by $\mathbb{BDM}_1$-$\mathbb{BDM}_1$-$\mathbb{P}_0^{disc}$-$\mathbb{P}_0^{disc}$-$\mathbb{P}_1^{disc}$ elements.

included in our timings. However, LU does not define efficient preconditioners for the respective blocks. Here instead, multilevel methods could provide order optimality, and this is a topic of current and future research.

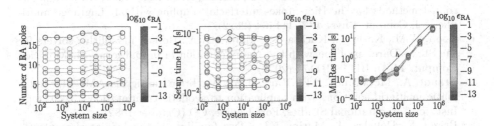

Fig. 5. Dependence of the number of poles (left), the setup time of RA (center) and runtime of the MinRes solver (right) on mesh size h and RA tolerance ϵ_{RA}. Parameters in the Darcy–Stokes problem are fixed at $K = 10^{-6}$, $\mu = 10^{-2}$ and $\alpha = 1$. Setup as in Fig. 3 is used with discretization by $\mathbb{BDM}_1$-$\mathbb{BDM}_1$-$\mathbb{P}_0^{disc}$-$\mathbb{P}_0^{disc}$-$\mathbb{P}_1^{disc}$ elements.

6 Conclusions

We have demonstrated that RA provides order optimal preconditioners for sums of fractional powers of SPD operators and can thus be utilized to construct parameter robust and order optimal preconditioners for multiphysics problems. The results are of practical interest for constructing efficient preconditioning on interfaces in models for which fractional weighted Sobolev spaces are the natural setting for the resulting differential operators, for example, when flow interacts with porous media. The techniques presented here could aid the numerical simulations in a wide range of biology, medicine, and engineering applications.

Acknowledgments. A. Budiša, M. Kuchta, K. A. Mardal, and L. Zikatanov acknowledge the financial support from the SciML project funded by the Norwegian Research Council grant 102155. The work of X. Hu is partially supported by the National Science Foundation under grant DMS-2208267. M. Kuchta acknowledges support from the Norwegian Research Council grant 303362. The work of L. Zikatanov is supported in part by the U. S.-Norway Fulbright Foundation and the U. S. National Science Foundation grant DMS-2208249. The collaborative efforts of Hu and Zikatanov were supported in part by the NSF DMS-2132710 through the Workshop on Numerical Modeling with Neural Networks, Learning, and Multilevel FE.

References

1. Adler, J., Hu, X., Zikatanov, L.: HAZmath solver library. https://github.com/ HAZmathTeam/hazmath
2. Ambartsumyan, I., Khattatov, E., Yotov, I., Zunino, P.: A Lagrange multiplier method for a Stokes-Biot Fluid-Poroelastic structure interaction model. Numer. Math. **140**(2), 513–553 (2018)
3. Ayuso, B., Brezzi, F., Marini, L.D., Xu, J., Zikatanov, L.: A simple preconditioner for a discontinuous Galerkin method for the Stokes problem. J. Sci. Comput. **58**(3), 517–547 (2014)
4. Bonito, A., Pasciak, J.E.: Numerical approximation of fractional powers of elliptic operators. Math. Comp. **84**(295), 2083–2110 (2015)
5. Boon, W.M., Hornkjøl, M., Kuchta, M., Mardal, K.A., Ruiz-Baier, R.: Parameter-robust methods for the Biot-Stokes interfacial coupling without Lagrange multipliers. J. Comput. Phys. **467**, 111464 (2022)
6. Boon, W.M., Koch, T., Kuchta, M., Mardal, K.A.: Robust monolithic solvers for the Stokes-Darcy problem with the Darcy equation in primal form. SIAM J. Sci. Comput. **44**(4), B1148–B1174 (2022)
7. Brandt, A., McCormick, S.F., Ruge, J.W.: Algebraic multigrid (AMG) for automatic multigrid solutions with application to geodetic computations. Tech. rep., Inst. for Computational Studies, Fort Collins, CO (October 1982)
8. Brezzi, F., Douglas, J., Marini, L.D.: Two families of mixed finite elements for second order elliptic problems. Numer. Math. **47**(2), 217–235 (1985)
9. Chen, L., Nochetto, R.H., Otárola, E., Salgado, A.J.: Multilevel methods for nonuniformly elliptic operators and fractional diffusion. Math. Comp. **85**(302), 2583–2607 (2016)
10. Di Pietro, D.A., Ern, A.: Mathematical aspects of discontinuous Galerkin methods, vol. 69. Springer Science & Business Media (2011)
11. Duan, B., Lazarov, R.D., Pasciak, J.E.: Numerical approximation of fractional powers of elliptic operators. IMA J. Numer. Anal. **40**(3), 1746–1771 (2020)
12. Harizanov, S., Lazarov, R., Margenov, S., Marinov, P.: Numerical solution of fractional diffusion-reaction problems based on BURA. Comput. Math. Appl. **80**(2), 316–331 (2020)
13. Harizanov, S., Lazarov, R., Margenov, S., Marinov, P., Pasciak, J.: Analysis of numerical methods for spectral fractional elliptic equations based on the best uniform rational approximation. J. Comput. Phys. 408, 109285, 21 (2020)

14. Harizanov, S., Lirkov, I., Margenov, S.: Rational approximations in robust precon-
    ditioning of multiphysics problems. Mathematics **10**(5), 780 (2022)
15. Hofreither, C.: A unified view of some numerical methods for fractional diffusion.
    Comput. Math. Appl. **80**(2), 332–350 (2020)
16. Hofreither, C.: An algorithm for best rational approximation based on barycentric
    rational interpolation. Numer. Algorithms **88**(1), 365–388 (2021)
17. Holter, K.E., Kuchta, M., Mardal, K.A.: Robust preconditioning of monolithically
    coupled multiphysics problems. arXiv preprint arXiv:2001.05527 (2020)
18. Hong, Q., Kraus, J., Xu, J., Zikatanov, L.: A robust multigrid method for discon-
    tinuous Galerkin discretizations of Stokes and linear elasticity equations. Numer.
    Math. **132**(1), 23–49 (2016)
19. Kuchta, M.: Assembly of multiscale linear PDE operators. In: Vermolen, F.J., Vuik,
    C. (eds.) Numerical Mathematics and Advanced Applications ENUMATH 2019.
    LNCSE, vol. 139, pp. 641–650. Springer, Cham (2021). https://doi.org/10.1007/
    978-3-030-55874-1_63
20. Kuchta, M., Laurino, F., Mardal, K.A., Zunino, P.: Analysis and approximation of
    mixed-dimensional PDEs on 3D–1D domains coupled with Lagrange multipliers.
    SIAM J. Numer. Anal. **59**(1), 558–582 (2021)
21. Kuchta, M., Mardal, K.A., Mortensen, M.: Preconditioning trace coupled 3D–1D
    systems using fractional Laplacian. Numer. Methods Partial Diff. Eqn. **35**(1), 375–
    393 (2019)
22. Kuchta, M., Nordaas, M., Verschaeve, J.C., Mortensen, M., Mardal, K.A.: Pre-
    conditioners for saddle point systems with trace constraints coupling $2d$ and $1d$
    domains. SIAM J. Sci. Comput. **38**(6), B962–B987 (2016)
23. Lamichhane, B.P., Wohlmuth, B.I.: Mortar finite elements for interface problems.
    Computing **72**(3), 333–348 (2004)
24. Layton, W.J., Schieweck, F., Yotov, I.: Coupling fluid flow with porous media flow.
    SIAM J. Numer. Anal. **40**(6), 2195–2218 (2002)
25. Mardal, K.A., Haga, J.B.: Block preconditioning of systems of PDEs. In: Auto-
    mated Solution of Differential Equations by the Finite Element Method, pp.
    643–655. Springer, Berlin, Heidelberg (2012). https://doi.org/10.1007/978-3-642-
    23099-8_35
26. Nakatsukasa, Y., Sète, O., Trefethen, L.N.: The AAA algorithm for rational approx-
    imation. SIAM J. Sci. Comput. **40**(3), A1494–A1522 (2018)
27. Nochetto, R.H., Otárola, E., Salgado, A.J.: A PDE approach to space-time frac-
    tional parabolic problems. SIAM J. Numer. Anal. **54**(2), 848–873 (2016)
28. Stahl, H.R.: Best uniform rational approximation of x^α on $[0,1]$. Acta Math.
    190(2), 241–306 (2003)
29. Stenberg, R.: On some techniques for approximating boundary conditions in the
    finite element method. J. Comput. Appl. Math. **63**(1–3), 139–148 (1995)
30. Tveito, A., Jæger, K.H., Kuchta, M., Mardal, K.A., Rognes, M.E.: A cell-based
    framework for numerical modeling of electrical conduction in cardiac tissue. Front.
    Phys. 5 (2017)
31. Wathen, A.J.: Realistic eigenvalue bounds for the Galerkin mass matrix. IMA J.
    Numer. Anal. **7**(4), 449–457 (1987)
32. Xu, J., Zikatanov, L.: Algebraic multigrid methods. Acta Numer **26**, 591–721
    (2017)

Analysis of Effective Properties of Poroelastic Composites with Surface Effects Depending on Boundary Conditions in Homogenization Problems

Mikhail Chebakov[1], Maria Datcheva[2]([✉]), Andrey Nasedkin[1],
Anna Nasedkina[1], and Roumen Iankov[2]

[1] Institute of Mathematics, Mechanics and Computer Sciences,
Southern Federal University, 344090 Rostov-on-Don, Russia
{mchebakov,avnasedkin,aanasedkina}@sfedu.ru
[2] Institute of Mechanics, Bulgarian Academy of Sciences, 1113 Sofia, Bulgaria
datcheva@imbm.bas.bg, iankovr@abv.bg

Abstract. The paper considers the homogenization problems for two-component poroelastic composites with a random structure of nanosized inclusions. The nanoscale nature of the inclusions was taken into account according to the generalized Gurtin–Murdoch theory by specifying surface elastic and porous stresses at the interface boundaries, the scale factor of which was related to the size of the inclusions. The formulation of homogenization problems was based on the theory of effective moduli, considering Hill's energy relations. The problems of static poroelasticity were solved in accordance with the Biot and filtration models. A feature of this investigation was the comparison of solutions of four types of homogenization problems with different boundary conditions. Modeling of representative volumes and solving problems of determining the effective material moduli were carried out in the ANSYS finite element package. Representative volumes were built in the form of a cubic grid of hexahedral finite elements with poroelastic properties of materials of one of the two phases and with a random arrangement of elements of the second phase. To consider interface effects, the interfaces were covered with shell elements with options for membrane stresses. The results of computational experiments made it possible to study the effective moduli depending on the boundary conditions, on the percentage of inclusions, their characteristic nanosizes, and areas of interface boundaries.

Keywords: Poroelastic composite · Nanosized inclusion · Poroelastic interface · Gurtin–Murdoch model · Effective modulus · Finite element method

1 Introduction

Recently, the Gurtin–Murdoch model of surface stresses has been widely used to describe size effects at the nanolevel. The popularity of this model can be judged

I. Georgiev et al. (Eds.): NMA 2022, LNCS 13858, pp. 114–126, 2023.
https://doi.org/10.1007/978-3-031-32412-3_10

by the reviews [4–6,8,25,26]. The Gurtin–Murdoch model was also used to analyze thermoelastic composites with nanosized inclusions or pores. For example, in [2,3,13], the thermomechanical properties of materials with spherical nanoinclusions or nanopores and fibrous nanocomposites were studied in the framework of the theory of thermal stresses with surface or interface effects.

To model the effective thermal conductivity of composites with imperfect interface boundaries, a strong thermal conductivity model similar to the Gurtin–Murdoch surface model has been used. The effective thermal conductivity coefficients for micro- and nanocomposites with imperfect interfaces have been studied, for example, in [12,14].

Nanostructured poroelastic composites have been investigated to a lesser extent [23], and this also applies to conventional poroelastic macroscale composites [1,7,15–17]. However, the well known thermoelastic-poroelastic analogy [22, 27] allows poroelastic composites to be studied using the same approach as that applied to thermoelastic composites. Therefore, anisotropic poroelastic materials with nano-sized inclusions are generally considered in this paper by applying the approach used in [18] for thermoelastic composites, but with modifications regarding the calculation of the effective Biot's coefficients. To account for the nanoscale interfaces between constituents with different properties, an interface surface tension model and an interface filtration model are used. Furthermore, different types of boundary conditions are used when applying the method to determine the effective properties of the considered poroelastic composites.

2   Determination of Effective Moduli of Poroelastic Nanocomposites Using Homogenization Approach and Finite Element Method

Let V denotes the volume of a representative volume element (RVE) of a two-component composite poroelastic material with nanosized inclusions. Therefore, $V = V^{(1)} \cup V^{(2)}$ where $V^{(1)}$ is the volume, occupied by the main material (matrix), and $V^{(2)}$ is the total volume occupied by the inclusions. Furthermore, $S = \partial V$ is the external boundary of the RVE, S^i represents the interface boundaries between the two materials ($S^i = \partial V^{(1)} \cap \partial V^{(2)}$), $\mathbf{n}$ is the vector of the outer unit normal, outer with respect to $V^{(1)}$, i.e. to the volume occupied by the matrix material, $\mathbf{x} = \{x_1, x_2, x_3\}$ is the radius vector of a point in the Cartesian coordinate system. We assume that the volumes $V^{(1)}$ and $V^{(2)}$ are filled with different anisotropic poroelastic materials. Then, in the framework of the classical static linear Biot theory of poroelasticity, we have the following system of differential equations in V:

$$\nabla \cdot \boldsymbol{\sigma} = 0, \quad \boldsymbol{\sigma} = \mathbf{c} : \boldsymbol{\varepsilon} - \mathbf{b}\,p, \quad \boldsymbol{\varepsilon} = (\nabla \mathbf{u} + (\nabla \mathbf{u})^{\mathrm{T}})/2\,, \tag{1}$$

$$\nabla \cdot \mathbf{v}_f = 0, \quad \mathbf{v}_f = -\frac{\mathbf{k}_f}{\mu_f} \cdot \nabla p\,, \tag{2}$$

where $\boldsymbol{\sigma}$ and $\boldsymbol{\varepsilon}$ are the second-order tensors of stresses and strains with components σ_{ij} and ε_{ij}, respectively; $\mathbf{u}$ is the displacement vector, p is the pore pressure, $\mathbf{c}$ is the fourth-order tensor of elastic stiffness with components c_{ijkl}, $\mathbf{b}$ is the second-order tensor of Biot poroelastic coefficients with components b_{ij}, $\mathbf{v}_f$ is the filtration velocity vector, $\mathbf{k}_f$ is the second-order permeability tensor, $\mathbf{k}_f = (\mathbf{K}_f \mu_f)/(\rho_f g)$, $\mathbf{K}_f$ is the second-order tensor of filtration with components K_{fij}, μ_f is the fluid viscosity, ρ_f is the fluid density, g is magnitude of the gravitational acceleration vector. Note that for the material properties we have $\mathbf{c} = \mathbf{c}^{(m)}$, $\mathbf{b} = \mathbf{b}^{(m)}$, $\mathbf{k}_f = \mathbf{k}_f^{(m)}$, $\mathbf{x} \in V^{(m)}$, $m = 1, 2$.

In accordance with the Gurtin–Murdoch model, we assume that the following condition is satisfied on the nanoscale interface boundaries S^i:

$$\mathbf{n} \cdot [\boldsymbol{\sigma}] = \nabla^i \cdot \boldsymbol{\sigma}^i, \quad \mathbf{x} \in S^i, \tag{3}$$

where $[\boldsymbol{\sigma}] = \boldsymbol{\sigma}^{(1)} - \boldsymbol{\sigma}^{(2)}$; ∇^i is the surface nabla operator, $\nabla^i = \nabla - \mathbf{n}(\mathbf{n} \cdot \nabla)$, $\boldsymbol{\sigma}^i$ is the second-order tensor of surface (interface) stresses.

It is assumed that the tensor $\boldsymbol{\sigma}^i$ is related to the displacements $\mathbf{u}$ and the pore pressure p according to the following set of equations:

$$\boldsymbol{\sigma}^i = \mathbf{c}^i : \boldsymbol{\varepsilon}^i - \mathbf{b}^i p, \quad \boldsymbol{\varepsilon}^i = (\nabla^i \mathbf{u}^i + (\nabla^i \mathbf{u}^i)^{\mathrm{T}})/2,, \quad \mathbf{u}^i = \mathbf{A} \cdot \mathbf{u}, \quad \mathbf{A} = \mathbf{I} - \mathbf{n}\mathbf{n}^*, \tag{4}$$

where $\boldsymbol{\varepsilon}^i$ is the second-order tensor of surface (interface) strains, $\mathbf{c}^i$ is the fourth-order tensor of surface (interface) elastic moduli, $\mathbf{b}^i$ is the second-order tensor of surface's (interface's) Biot coefficients.

Similarly to (3), for nanoscale interface boundaries S^i we introduce the following relations:

$$\mathbf{n} \cdot [\mathbf{v}_f] = \nabla^i \cdot \mathbf{v}_f^i, \quad \mathbf{v}_f^i = -\frac{\mathbf{k}_f^i}{\mu_f^i} \cdot \nabla^i p, \quad \mathbf{x} \in S^i, \tag{5}$$

where $\mathbf{v}_f^i$ is the interface filtering rate vector, $\mathbf{k}_f^i$ is the second-order tensor of surface (interface) permeability coefficients, μ_f^i is the surface (interface) fluid viscosity.

In the following, we will also use the poroelastic material constitutive relations given with the second equation in (1) and with the second equation (Darcy's law) in (2), but expressing strains and pore pressure in terms of stresses and fluid velocity:

$$\boldsymbol{\varepsilon} = \mathbf{s} : \boldsymbol{\sigma} + \boldsymbol{\alpha} p, \tag{6}$$

$$\nabla p = -\mu_f \mathbf{r}_f \cdot \mathbf{v}_f, \tag{7}$$

where $\mathbf{s} = \mathbf{c}^{-1}$ is the elastic compliance tensor with components s_{ijkl}, $\boldsymbol{\alpha} = \mathbf{s} : \mathbf{b}$ is the second-order tensor of Biot filtration expansion coefficients α_{ij}, $\mathbf{r}_f = \mathbf{k}_f^{-1}$.

According to the method of obtaining the effective moduli in mechanics of composites, to determine the stiffness and filtration properties of a poroelastic composite material, one has to solve the system of Eqs. (1)–(5) with special boundary conditions on the outer surface S of the RVE, allowing for a homogeneous comparison medium constant fields of stresses, strains, pore pressure for Eqs. (1), (3), (4) or filtration velocity and pore pressure gradient for

Eqs. (2), (5). The theoretical basis of the considered homogenization method is the Hill–Mandel condition, which reflects the equality of the energy of the composite and the energy of a homogeneous reference medium under the same external loading conditions.

It can be shown that the effective elastic stiffness moduli c_{ijkl}^{eff} can be determined by solving the Eqs. (1), (3), (4) at zero pore pressure $p = 0$ along V and with displacements imposed on the outer boundary S of the RVE that depend linearly on the spatial coordinates. Therefore, for each pair $\{kl\}$, $k, l = 1, 2, 3$, we have:

$$\mathbf{u} = \varepsilon_0(x_k \mathbf{e}_l + x_l \mathbf{e}_k)/2, \mathbf{x} \in S; \quad p = 0, \mathbf{x} \in V \Rightarrow \quad c_{ijkl}^{\text{eff}} = \langle \sigma_{ij} \rangle / \varepsilon_0, \quad (8)$$

where $\varepsilon_0 = \text{const}$; $\mathbf{e}_k$, $\mathbf{e}_l$ are the orts of the Cartesian coordinate system; and the average values in the presence of interface surface effects are calculated both over the volume V and over the interface boundaries S^i:

$$\langle (...) \rangle = \frac{1}{|V|} \left(\int_V (...) \, dV + \int_{S^i} (...)^i \, dS \right). \quad (9)$$

Once we find the effective stiffness moduli, the effective Biot coefficients b_{ij}^{eff}, can be determined by solving the equations (1), (3), (4) with the same boundary conditions (8) for the displacements, but at constant non-zero pore pressure $(p = p_0 = \text{const} \neq 0)$. For each pair $\{kl\}$, $k, l = 1, 2, 3$, we have:

$$\mathbf{u} = \varepsilon_0(x_k \mathbf{e}_l + x_l \mathbf{e}_k)/2, \mathbf{x} \in S; \quad p = p_0 \neq 0, \mathbf{x} \in V \quad \Rightarrow \quad (10)$$

$$b_{ij}^{\text{eff}} = (c_{ijkl}^{\text{eff}} \varepsilon_0 - \langle \sigma_{ij} \rangle)/p_0. \quad (11)$$

An alternative option is to use boundary conditions in terms of stresses for the system of Eqs. (1), (3), (4). In accordance with Eq. (6), the effective compliance moduli s_{ijkl}^{eff} can be immediately determined. Thus for each pair $\{kl\}$, $k, l = 1, 2, 3$ instead of (8) we get:

$$\mathbf{n} \cdot \boldsymbol{\sigma} = \sigma_0(n_k \mathbf{e}_l + n_l \mathbf{e}_k)/2, \mathbf{x} \in S; \quad p = 0, \mathbf{x} \in V \quad \Rightarrow \quad s_{ijkl}^{\text{eff}} = \langle \varepsilon_{ij} \rangle / \sigma_0, \quad (12)$$

where $\sigma_0 = \text{const}$.

Further, after determining the effective elastic moduli, in accordance with (6), the effective moduli of the filtration extension can be obtained by solving the system of Eqs. (1), (3), (4) with the same stress boundary conditions, but at nonzero pore pressure $(p = p_0 = \text{const} \neq 0)$. For each pair $\{kl\}$, $k, l = 1, 2, 3$, we have:

$$\mathbf{n} \cdot \boldsymbol{\sigma} = \sigma_0(n_k \mathbf{e}_l + n_l \mathbf{e}_k)/2, \mathbf{x} \in S; \quad p = p_0 \neq 0, \mathbf{x} \in V \quad \Rightarrow \quad (13)$$

$$\alpha_{ij}^{\text{eff}} = (\langle \varepsilon_{ij} \rangle - s_{ijkl}^{\text{eff}} \sigma_0)/p_0. \quad (14)$$

Note that the effective Biot coefficients can be determined from Eqs. (10) and (11) or from Eqs. (13) and (14) independently of how the effective elastic moduli were calculated, e.g. by using Eq. (8) or Eq. (12).

We also emphasize that in the problems for the system of Eqs. (1), (3), (4) with boundary conditions in displacements or in stresses, the pore pressure is known and is constant: $p = 0$ for the cases (8) and (12) ; $p = p_0$ for the cases (10) and (13).

To determine the effective permeability coefficients k_{fil}^{eff}, one can find the average filtration velocities over the volume and over the interface boundaries from the solution of the system of Eqs. (2) and (5) with the boundary condition of pore pressure that depends linearly on the position on the external surface. Thus for each $l = 1, 2, 3$ we get:

$$p = x_l G_0, \quad \mathbf{x} \in S \quad \Rightarrow \quad k_{fil}^{\text{eff}} = -\mu_f \langle v_{fi} \rangle / G_0, \tag{15}$$

where $G_0 = \text{const}$.

An alternative option is to use the boundary conditions for the normal component of the filtration velocity, which, in accordance with (7). This allows us to find the effective inverse permeability coefficients r_{fil}^{eff} from the solution of the system of Eqs. (2) and (5) and under the following conditions for each $l = 1, 2, 3$:

$$n_i v_{fi} = n_l v_{f0}, \quad \mathbf{x} \in S \quad \Rightarrow \quad r_{fil}^{\text{eff}} = -\langle p_{,i} \rangle / (\mu_f v_{f0}), \tag{16}$$

where $v_{f0} = \text{const}$.

As can be seen, the problems formulated above to determine the effective moduli of nanostructured poroelastic composites differ from those formulated to obtain the effective characteristics of poroelastic composites with micro- and macro-sized inclusions. The difference consists in the presence of interface boundary conditions (3), (4) and (5) introduced to account for the nanoscale dimensions in the Gurtin–Murdoch model for interface surface stresses and in the model for the interface filtration.

For numerical solution using the finite element method of the considered above boundary value problems, as usual, one should pass to weak formulation by applying standard transformations associated with the multiplication of field equations by projection functions, integration of the resulting equalities over the volume, and the use of divergence theorems taking into account interface effects. To discretize the obtained weak form of the corresponding relations, one can use classical Lagrangian or serendipal finite elements with degrees of freedom nodal displacements and pore pressure. Note that, due to the structure of the interface mechanical and filtration fields (3)–(5), shell elements with membrane stress options should be used as surface elements, for which the degrees of freedom are only nodal displacements, in combination with filtration shell elements. For these elements, one can take a fictitious unit thickness so that the surface moduli on the interface S^i are determined through the products of specially given bulk moduli and the shell thickness. The RVE can be of dimensionless size, and the minimum side of an individual finite element can be taken equal to 1. Thus, the dimensionless parameter in spatial coordinates can be taken equal to the minimum inclusion size.

When using standard comercial finite element software, in the absence of shell poroelastic finite elements, one can use the well-known analogy between poroelasticity and thermoelasticity [21,22,27] and solve poroelasticity problems as a thermoelasticity problem with the corresponding variation in moduli and variables. It is this approach that was used by the authors in solving the considered homogenization problems in the ANSYS environment.

3   RVE Finite Element Model

As noted, the described above homogenization strategy was implemented in the general purpose software ANSYS considering the analogy between the problems of poroelasticity and thermoelasticity. The RVE is constructed in the form of a cube, regularly subdivided into smaller geometrically identical cubes, which are SOLID226 thermoelastic hexahedral twenty-node finite elements with the option of coupling the temperature filed and the stress–strain fields. Initially, the material properties of the base thermoelastic (i.e. poroelastic) material are assigned to all bulk finite elements.

Furthermore, consistent with the assumption that the composite has a nanostructure, elements with the properties of the second material were isolated to represent the inclusions. In the case of a composite with an irregular stochastic structure, based on a given volume fraction of the inclusions, for a randomly selected fraction of finite elements their moduli were changed to the moduli of the second type material. Note that this algorithm is the simplest, but it does not support the connectivity of the first type material (matrix) and does not reflect the connectivity structure of the second type material (closed or open inclusions). Other methods supporting the connectivity of the matrix elements and the isolation of inclusions, and supporting the connectivity of the both types of elements are described in [9–11,20].

The following algorithm was used for automated placement of interface finite elements [18]. First, finite elements with material properties of the second material type were isolated. The outer boundaries of this set of elements were covered by contact eight-node TARGE170 elements of the QUA8 type using the ANSYS TSHAP, QUA8 command. Then, the contact elements located on the outer boundary S of the RVE were removed, and the remaining contact elements were replaced by SHELL281 eight-node shell elements with the option of membrane stresses. As a result of this procedure, all interfaces of bulk elements with different material properties were covered with membrane finite element.

Two examples of RVE having 20 elements along each of the axes, with 5 and 15 percent inclusions are shown in Figs. 1 and 2, respectively. In these figures, on the left side all bulk solid elements are shown, in the center only bulk solid elements of the second material type are shown, and on the right side only the shell elements are shown.

As can be seen from Figs. 1 and 2, with an increase in the proportion of inclusions (of the second material type), the area of the interfaces also increases. So, with a fraction of inclusions of 5 %, there are 400 cubic elements of the second

Fig. 1. Example of RVE with 5% inclusions, 2170 surface elements.

Fig. 2. Example of RVE with 15% inclusions, 5815 surface elements.

type and 2170 surface elements (Fig. 1), and with a fraction of inclusions of 15 %, there are already 1200 cubic elements of the second phase and 5815 surface elements (Fig. 2). Here, with a three-fold increase in the number of elements of the second type, the number of surface elements increases approximately 2.68 times. With a further increase in the fraction of inclusions, the interface area reaches a maximum at 50 % of inclusions, and then decreases [18], which is quite understandable, since then the materials of the matrix and inclusions seem to change places.

After constructing the RVE, the stationary thermoelasticity problems corresponding to the poroelasticity problems consisting of the system of Eqs. (1), (3), (4) with the appropriate boundary conditions (8), (10), (12) or (13) were solved in the next step. Then, in the ANSYS postprocessor, the average stresses $\langle \sigma_{ij} \rangle$ or average strains $\langle \varepsilon_{ij} \rangle$ were calculated, and this was done according to formula (9). i.e. the averaging is done with respect to both volumetric finite elements and surface elements. The average stresses found using the formulas (8), (11) made it possible to determine the effective stiffness moduli and the effective Biot coefficients. With the alternative approach with boundary conditions in stresses, the average strains obtained using the formulas (12), (14) can be used to determine the effective compliance moduli and the effective coefficients of filtration extension.

Using a similar approach but with volume and shell temperature finite elements, heat conduction problems were solved in ANSYS, thus solving the corresponding filter homogenization problems (2), (5), (15) or (2), (5), (16).

4   Numerical Results

In the numerical calculations, for simplicity, we consider a composite consisting of two isotropic materials. Then the stiffness moduli, can be represented as $c_{ijkl} = \lambda\delta_{ij}\delta_{kl} + \mu(\delta_{ik}\delta_{jl} + \delta_{il}\delta_{jk})$, where δ_{ij} is the Kronecker symbol, and λ and μ are the Lame coefficients. Instead of the Lame coefficients, other pairs of coefficients are commonly used: Young's modulus E and Poisson's ratio ν, or bulk modulus K and shear modulus G. These moduli are interconnected as follows: $\lambda = K - (2/3)G$, $\mu = G$; $E = 9KG/(3K + G)$, $\nu = (3K - 2G)/2/(3K + G)$. Similarly, compliance moduli can be represented in the following form $s_{ijkl} = \lambda_s\delta_{ij}\delta_{kl} + \mu_s(\delta_{ik}\delta_{jl} + \delta_{il}\delta_{jk})$, where $\lambda_s = -\lambda/(2\mu(3\lambda + 2\mu)) = (2G - 3K)/(18KG)$, $\mu_s = 1/(4\mu) = 1/(4G)$.

The Biot coefficients and the permeability tensors are diagonal: $b_{ij} = b\delta_{ij}$, $k_{fij} = k_f\delta_{ij}$. The same applies to the inverse tensor $\alpha_{ij} = \alpha\delta_{ij}$, $r_{fij} = r_f\delta_{ij}$.

Thus, each material constituting the composite is characterized by two elastic moduli, one Biot coefficient, and one permeability coefficient. The interface modules have similar representations and are also four, but are defined for two-dimensional surfaces.

Two types of composites are considered hereafter. In both cases, for the fluid-saturated matrix, the elastic moduli $K^{(1)}$ and $G^{(1)}$ are 17.6 GPa and 16.8 GPa, respectively, and the Biot coefficient $b^{(1)}$ is 0.6. For the case I, we consider stiff inclusions with the following poroelastic moduli: $K^{(2)} = 24.2$ GPa, $G^{(2)} = 28.8$ GPa, $b^{(2)} = 0.33$. For the case II we consider soft inclusions with moduli: $K^{(2)} = 15.16$ GPa, $G^{(2)} = 7.2$ GPa, $b^{(2)} = 0.82$. These values are taken from [16] and correspond to the moduli of saturated sandstones at porosity 20 % (matrix), 10 % (inclusions, case I), and 30 % (inclusions, case II). For the permeability coefficients we take the following values ($\mu_f = 1\cdot 10^{-3}$ Pa·s): $k_f^{(1)} = 61\cdot 10^{-9}$ m^2, $k_f^{(2)} = 88.45\cdot 10^{-9}$ m^2 (case I), and $k_f^{(2)} = 36.6\cdot 10^{-9}$ m^2 (case II).

To model the interface effects for the elastic shell finite elements in ANSYSe, we indicated the membrane stress option and a fictitious shell thickness h. We assume the shell stiffness moduli c_{ijkl}^s to be proportional to the stiffness moduli of the matrix $c_{ijkl}^s = \kappa^i c_{ijkl}^{(1)}$, where κ^i is a dimensionless multiplier. For the interface stiffness moduli c_{ijkl}^i defined in (4) we assume that $c_{ijkl}^i = l_c c_{ijkl}^{(1)}$, $l_c = 0.1$ nm, and consequently, $c_{ijkl}^s = \kappa^i c_{ijkl}^{(1)} = (\kappa^i/l_c)c_{ijkl}^i$. As the dimension analysis shows, $l_0 = l_c/\kappa^i$, and so the factor κ^i is inversely proportional to the minimum inclusion size l_0. Next, the dimensionless fictitious thickness of the shell elements was taken to be $h = 1$. In the performed numerical simulations the multiplier κ^i and the fraction of inclusions $\phi = |V^{(2)}|/|V|$ were varied. The interface Biot coefficients $b_{jj}^i = b^i$ and the interface permeability moduli $k_{fjj}^i = k_f^i$ have been introduced in the same way.

The results of the numerical calculations of the effective moduli are presented here in terms of relative values to those of the matrix, e.g. $\tilde{K} = K^{\text{eff}}/K^{(1)}$, etc. Figures 3 and 4 show the variation of the relative effective moduli with respect to the percentage of inclusions, $P = 100\phi$ % and for RVE with 20 elements along each axis. In these figures curves I are for composites with hard inclusions, while curves II represent results for composites with soft inclusions. Furthermore, the relative effective moduli are obtained for two values of the κ^i coefficient, namely at $\kappa^i = 10^{-5}$, which gives a composite without interface effects, i.e. with macrosized inclusions (the curves are denoted by "m") and at $\kappa^i = 1$, yielding a composite with significant interface effects, i.e. with nanosized inclusions (the curves are denoted by "n").

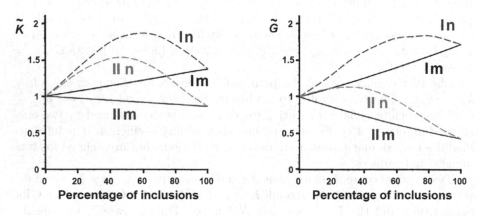

Fig. 3. Relative effective elastic moduli $\tilde{K}$ and $\tilde{G}$ versus the inclusions proportion.

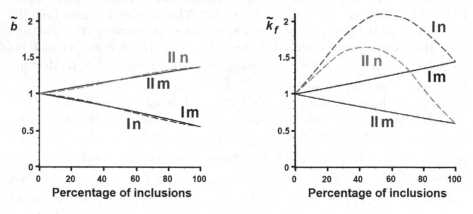

Fig. 4. Relative effective Biot coefficient $\tilde{b}$ and permeability coefficient $\tilde{k}_f$ versus the inclusions proportion.

As it can be seen from Figs. 3 and 4, for small values of the factor κ^i the processes at the interfaces do not affect the obtained poroelastic composite effective moduli. The obtained values in this case are well correlated with approximate analytical formulas for the elastic moduli and for the Biot coefficients [16], as well as for the permeability coefficient [17,19,24]

$$\tilde{e} = 1 + \phi\left(\frac{1}{\tilde{e}^{(2)} - 1} + \frac{1 - \phi}{\tilde{e}_A}\right)^{-1}, \tilde{e} = K, G; \quad \tilde{K}_A = 1 + \frac{4\tilde{g}^{(1)}}{3}, \quad \tilde{G}_A = \frac{5(3 + 4\tilde{g}^{(1)})}{6(1 + 2\tilde{g}^{(1)})},$$

$$\tilde{b} = 1 + \frac{\phi(\tilde{b}^{(2)} - 1)}{\tilde{K}^{(2)} - 1}\left(\frac{1}{\tilde{K}^{(2)} - 1} + \frac{1 - \phi}{\tilde{K}_A}\right)^{-1}, \quad \tilde{k}_f = \frac{2(1 - \phi) + (1 + 2\phi)\tilde{k}_f^{(2)}}{2 + \phi + (1 - \phi)\tilde{k}_f^{(2)}},$$

where $\tilde{g}^{(1)} = G^{(1)}/K^{(1)}$, $\tilde{K}^{(2)} = K^{(2)}/K^{(1)}$, $\tilde{G}^{(2)} = G^{(2)}/G^{(1)}$, $\tilde{k}_f^{(2)} = k_f^{(2)}/k_f^{(1)}$.

It can be concluded that for all values of the volume content of the inclusions, accounting for interface effects leads to larger values of the effective characteristics K^{eff}, G^{eff}, and k_f^{eff}. Moreover, there are cases where the nanoporous material have greater elastic stiffness and permeability coefficient than the corresponding dense material (e.g. curves 'II n' with values of the relative effective property greater than 1).

We also emphasize that inclusion fraction and interfacial effects have opposite effects on the effective stiffness and permeability coefficient for the composite with soft inclusions. As the fraction of inclusions increases, the effective stiffness and the permeability coefficient decrease, while changing the interfacial effects can lead to an increase in the values of these effective material parameters, for example in the case of increasing the corresponding interface modules or decreasing the size of the inclusions.

As stated above, increasing the inclusion fraction to a certain limit increases the total interface area. Therefore, as the inclusion fraction increases in the nanostructured material, the values of the effective elastic modulus and the permeability coefficient can both decrease and increase. Then, for large values of κ^i, an increase in inclusion content starting from small ϕ leads to an increase in the corresponding effective material parameters, and with further increase in inclusion volume content the values of these effective characteristics decrease. Note that similar effects are known for nanoporous elastic and thermoelastic materials [5,18].

Furthermore, as can be seen from Fig. 4 (left), the Biot coefficient b^{eff} is little dependent on the interface effects, while it is different for the coefficient of the filtration expansion α^{eff} which depends on the interface fields in the same way as the compliance moduli.

Finally, we note that the results of the performed numerical homogenization by solving the corresponding RVE–problems with boundary conditions in terms of stresses and filtration velocity are practically the same as those, which are obtained by solving the corresponding problems with boundary conditions in terms of displacements and pore pressure. However, the boundary conditions for displacements and pore pressure are more convenient in the homogenization procedure for composites with very soft inclusions.

5 Conclusion

Thus, the paper describes a technique for solving homogenization problems for poroelastic composites with interface effects caused by nanosized inclusions according to the Gurtin–Murdoch theory. We used the effective moduli method to formulate the homogenization problem and the finite element method to solve it numerically. The results of the homogenization procedure obtained for a representative volume with a random distribution of inclusions without taking into account surface effects are in good agreement with the outcome of known approximate analytical formulas. The main advantage of the homogenization approach proposed here, however, is that it enables a unified approach for the study of poroelastic composites with different internal nanostructures.

Acknowledgements. The financial support of the Russian Foundation for Basic Research according to the research project No 19-58-18011 Bulg_a (M.C., A.N., A.N.), the National Science Fund of Bulgaria (project KP-06-Russia-1/27.09.2019) and by the Science and Education for Smart Growth Operational Program (2014-2020) and the ESIF through grant BG05M2OP001-1.001-0003 (M.D.) is gratefully acknowledged.

References

1. Berryman, J.G.: Effective medium theories for multicomponent poroelastic composites. J. Eng. Mech. **132**, 519–531 (2006). https://doi.org/10.1061/(ASCE)0733-9399(2006)132:5(519)
2. Chen, T., Dvorak, G.J., Yu, C.C.: Solids containing spherical nano-inclusions with interface stresses: effective properties and thermal-mechanical connections. Int. J. Solids Struct. **44**(3–4), 941–955 (2007). https://doi.org/10.1016/j.ijsolstr.2006.05.030
3. Duan, H.L., Karihaloo, B.L.: Thermo-elastic properties of heterogeneous materials with imperfect interfaces: generalized Levin's formula and Hill's connections. J. Mech. Phys. Solids **55**(5), 1036–1052 (2007). https://doi.org/10.1016/j.jmps.2006.10.006
4. Duan, H.L., Wang, J., Karihaloo, B.L.: Theory of elasticity at the nanoscale. Adv. Appl. Mech. **42**, 1–68 (2009). https://doi.org/10.1016/S0065-2156(08)00001-X
5. Eremeyev, V.A.: On effective properties of materials at the nano- and microscales considering surface effects. Acta Mech. **227**, 29–42 (2016). https://doi.org/10.1007/s00707-015-1427-y
6. Firooz, S., Steinmann, P., Javili, A.: Homogenization of composites with extended general interfaces: comprehensive review and unified modeling. Appl. Mech. Rev. **73**(4), 040802 (2021). https://doi.org/10.1115/1.4051481
7. Giraud, A., Huynh, Q.V., Hoxha, D., Kondo, D.: Effective poroelastic properties of transversely iso-tropic rock-like composites with arbitrarily oriented ellipsoidal inclusions. Mech. Mater. **39**(11), 1006–1024 (2007). https://doi.org/10.1016/j.mechmat.2007.05.005
8. Javili, A., McBride, A., Steinmann, P.: Thermomechanics of solids with lower-dimensional energetics: On the importance of surface, interface and curve structures at the nanoscale. A unifying review. Appl. Mech. Rev. **65**(1), 010802 (2013). https://doi.org/10.1115/1.4023012

9. Kudimova, A.B., Nadolin, D.K., Nasedkin, A.V., Nasedkina, A.A., Oganesyan, P.A., Soloviev, A.N.: Models of porous piezocomposites with 3–3 connectivity type in ACELAN finite element package. Mater. Phys. Mech. **37**(1), 16–24 (2018). https://doi.org/10.18720/MPM.3712018_3

10. Kudimova, A.B., Nadolin, D.K., Nasedkin, A.V., Oganesyan, P.A., Soloviev, A.N.: Finite element homogenization models of bulk mixed piezocomposites with granular elastic inclusions in ACELAN package. Mater. Phys. Mech. **37**(1), 25–33 (2018). https://doi.org/10.18720/MPM.3712018_4

11. Kudimova, A.B., Nadolin, D.K., Nasedkin, A.V., Nasedkina, A.A., Oganesyan, P.A., Soloviev, A.N.: Finite element homogenization of piezocomposites with isolated inclusions using improved 3–0 algorithm for generating representative volumes in ACELAN-COMPOS package. Mater. Phys. Mech. **44**(3), 392–403 (2020). https://doi.org/10.18720/MPM.4432020_10

12. Kushch, V.I., Sevostianov, I., Chernobai, V.S.: Effective conductivity of composite with imperfect contact between elliptic fibers and matrix: Maxwell's homogenization scheme. Int. J. Eng. Sci. **83**, 146–161 (2014). https://doi.org/10.1016/j.ijengsci.2014.03.006

13. Le Quang, H., He, Q.-C.: Estimation of the effective thermoelastic moduli of fibrous nanocomposites with cylindrically anisotropic phases. Arch. Appl. Mech. **79**, 225–248 (2009). https://doi.org/10.1007/s00419-008-0223-8

14. Le Quang, H., Pham, D.S., Bonnet, G., He, Q.-C.: Estimations of the effective conductivity of anisotropic multiphase composites with imperfect interfaces. Int. J. Heat Mass Transfer. **58**, 175–187 (2013). https://doi.org/10.1016/j.ijheatmasstransfer.2012.11.028

15. Levin, V.M., Alvarez-Tostado, J.M.: On the effective constants of inhomogeneous poroelastic medium. Sci. Eng. Compos. Mater. **11**(1), 35–46 (2004). https://doi.org/10.1515/SECM.2004.11.1.35

16. Levin, V., Kanaun, S., Markov, M.: Generalized Maxwell's scheme for homogenization of poroelastic composites. Int. J. Eng. Sci. **61**, 75–86 (2012). https://doi.org/10.1016/j.ijengsci.2012.06.011

17. Milton, G.W.: Mechanics of Composites. Cambridge University Press (2002). https://doi.org/10.1017/CBO9780511613357

18. Nasedkin, A., Nasedkina, A., Rajagopal, A.: Homogenization of dispersion-strengthened thermoelastic composites with imperfect interfaces by using finite element technique. In: Parinov, I.A., Chang, S.-H., Kim, Y.-H. (eds.) Advanced Materials. SPP, vol. 224, pp. 399–411. Springer, Cham (2019). https://doi.org/10.1007/978-3-030-19894-7_30

19. Nasedkin, A., Nassar, M.E.: About anomalous properties of porous piezoceramic materials with metalized or rigid surfaces of pores. Mech. Mat. **162**, 104040 (2021). https://doi.org/10.1016/j.mechmat.2021.104040

20. Nasedkin, A.V., Shevtsova, M.S.: Improved finite element approaches for modeling of porous piezocomposite materials with different connectivity. In: Parinov, I.A. (ed.) Ferroelectrics and superconductors: Properties and applications, pp. 231–254. Nova Science Publishers, NY (2011)

21. Nasedkina, A.A., Nasedkin, A.V., Iovane, G.: A model for hydrodynamic influence on a multi-layer deformable coal seam. Comput. Mech. **41**(3), 379–389 (2008). https://doi.org/10.1007/s00466-007-0194-6

22. Norris, A.: On the correspondence between poroelasticity and thermoelasticity. J. Appl. Phys. **71**(3), 1138–1141 (1992). https://doi.org/10.1063/1.351278

23. Ren, S.-C., Liu, J.-T., Gu, S.-T., He, Q.-C.: An XFEM-based numerical procedure for the analysis of poroelastic composites with coherent imperfect interface. Comput. Mater. Sci. **94**, 173–181 (2014). https://doi.org/10.1016/j.commatsci.2014.03.047

24. Tuncer, E.: Dielectric mixtures-importance and theoretical approaches. IEEE Electr. Insul. Mag. **29**(6), 49–58 (2013). https://doi.org/10.1109/MEI.2013.6648753

25. Wang, J., Huang, Z., Duan, H., Yu, S., Feng, X., Wang, G., Zhang, W., Wang, T.: Surface stress effect in mechanics of nanostructured materials. Acta Mech. Solida Sin. **24**(1), 52–82 (2011). https://doi.org/10.1016/S0894-9166(11)60009-8

26. Wang, K.F., Wang, B.L., Kitamura, T.: A review on the application of modified continuum models in modeling and simulation of nanostructures. Acta. Mech. Sin. **32**(1), 83–100 (2016). https://doi.org/10.1007/s10409-015-0508-4

27. Zimmerman, R.W.: Coupling in poroelasticity and thermoelasticity. Int. J. Rock Mech. Min. Sci. **37**(1–2), 79–87 (2000). https://doi.org/10.1016/S1365-1609(99)00094-5

Ant Algorithm for GPS Surveying Problem with Fuzzy Pheromone Updating

Stefka Fidanova[1]([envelope]) [iD] and Krassimir Atanassov[2] [iD]

[1] Institute of Information and Communication Technologies – Bulgarian Academy of Sciences, Acad. G. Bonchev str. bl.25A, 1113 Sofia, Bulgaria
stefka.fidanova@iict.bas.bg
[2] Institute of Biophysics and Biomedical Engineering – Bulgarian Academy of Sciences, Sofia, Bulgaria

Abstract. GPS system has important part of our life. We use it everyday for navigation, but it has other important use too. The GPS navigation is used by vessels, aircrafts and vehicles for safety passage. Other use is geodetic survey. GPS gives more precise coordinates comparing with traditional methods. Scientific use of GPS includes measurement of crystal deformation, monitoring of earthquake processes, volcano eruption process, ruffing, mountain building. These applications needs very precise measurement. We propose Ant Colony Optimization (ACO) algorithm with two variants intuitionistic fuzzy pheromone updating. The proposed algorithm is tested on 8 test problems with dimension from 38 to 443 sessions. The variants of our algorithm is compared with traditional ACO algorithm.

1 Introduction

Many engineering applications lead to complex optimization problems. As the size of the problem increases, so do the resources to solve it. Exponential growth makes it impossible to solve the problem using exact methods or traditional numerical methods. This type of problems are a great challenge for scientists and developer of new methodologies. The proposed methods must be easy to apply, fast, use a reasonable amount of computer resources and be able to apply to a variety of problems.

Artificial intelligence is a powerful tool for solving difficult, computational problems. Nature inspired methods are part of artificial intelligence methods and are very suitable for solving hard Non Polynomial (NP-hard) problems. Ideas coming from a natural phenomenon are used as their basis: biological, physical, chemical etc. The main categories of these methods are: evolutionary [11]; swarm intelligence based [4]; physics based [16]. Other categorization is: population based [20]; single solution based [14]. The most prominent representative of evolutionary algorithms is genetic algorithm [11,22]. It is based on Darwinian evolutionary theory. As a representatives of physical phenomena based algorithms, we will mention simulated annealing [14] and gravitation

I. Georgiev et al. (Eds.): NMA 2022, LNCS 13858, pp. 127–133, 2023.
https://doi.org/10.1007/978-3-031-32412-3_11

search algorithm [16]. Human behavior based algorithms are interior search [18] and tabu search [17]. The nature inspired swarm algorithms are based on animal and insect behavior. We can mention ant colony optimization [4], artificial bee colony [12], bat algorithm [20], firefly algorithm [21], particle swarm optimization [13], gray wolf algorithm, emperor penguin colony [15] and several others.

GPS surveying problem is NP-hard combinatorial optimization problem and Ant Colony Optimization (ACO) is very appropriate to be applied on it. Ants can find the shortest way to the source of food, even when there is an obstacle. Prof. Dorigo proposed a method for solving NP-hard optimization problems, based on ants behavior [5]. The problem is described with a graph and the solution is a path in a graph. The optimal solution for minimization problem is a shorter path, respecting problem constraints. An ant starts solution construction from random (ore semi-random) node of the graph. After the next nodes are included according probabilistic rule called transition probability which includes the pheromone concentration and heuristic information. The pheromone is a numerical information related with a nodes or arcs of the graph and its value corresponds to a quality of the solution. Thus elements of better solutions accumulate more pheromone and attracts more ants.

ACO is appropriate for solving combinatorial optimization problems, especially with strict constraints. ACO can escape local minima, because it is constructive method. It can be adapted to the dynamic changes of the problem [7,8,19]. It is one of the best methods for combinatorial optimization. Last decades the method has been applied on large number of hard combinatorial optimization problems with real application. There are several variants of the method: ant system [6]; elitist ants [6]; ant colony system [5]; max-min ant system [19]; rank-based ant system [6]; ant algorithm with additional reinforcement [9]. The difference between them is pheromone updating.

In this paper we propose intuitionistic fuzzy pheromone updating, which is applied on GPS surveying problem. GPS surveying problem is an ordered problem, the order of the elements of the solution are with high importance. We compare algorithm performance with application on subset problem. As a representative of subset problems is used Multidimensional Knapsack Problem (MKP).

The paper is organized as follows: in Sect. 2 intuitionistic fuzzy logic is briefly introduced; The description of GPS surveying problem is in Sect. 3; in Sect. 4 is described ACO with intuitionistic fuzzy pheromone and its application on GPS surveying problem; Sect. 5 is devoted to numerical experiments and analysis of the results; in Sect. 6 consist of concluding remarks and direction for future work.

2 Short Remark on Intuitionistic Fuzzy Logic

Intuitionistic fuzzy logic is proposed by Krassimir Atanassov [1]. In the classical logic, to every proposition, it can be assigned value 1 or 0: 1 in the sense of true and 0 in the sense of false. In fuzzy logic [23], there is a truth degree $m \in [0, 1]$. In intuitionistic fuzzy (propositional) logic [2,3], each proposition is evaluated by two degrees: degree of validity (or of truth) $m \in [0, 1]$ and degree of non-validity (or of false) $n \in [0, 1]$, so that $m + n \leq 1$. These degrees generate a third degree $1 - m - n$, called degree of uncertainty (or indeterminacy) and, obvious, the sum of all three components is 1. The pair $\langle m, n \rangle$ is called an Intuitionistic Fuzzy Pair (IFP). For two IFPs a lot of operations, relations and operators are defined (see Reference [3]). For example, the simplest modal operators, defined over IFPs are: $\square \langle m, n \rangle = \langle m, 1 - m \rangle, \diamond \langle m, n \rangle = \langle 1 - n, n \rangle$.

3 GPS Surveying Problem Formulation

During orbiting around the Earth the GPS satellites transmit signals. A receiver with an unknown position receives the signals from satellites and calculates useful measurement. On the basis of the measurement is calculated 3D location of the receiver.

The GPS network is defined as a set of stations $(x_1, x_2, \ldots x_n)$ and placing receivers on them to determine sessions $(x_1 x_2, x_1 x_3, x_2 x_3, \ldots)$. The problem is to search for the best order the sessions to be observed consecutively. The solution is represented by weighted linear graph. The nodes represent the stations and the edges represent the moving cost. The objective function of the problem is the cost of the solution which is the sum of the costs (time) to move from one point to another one. The problem can be redefined for searching the best order the sessions to be observed, with respect the time. The problem is NP-hard combinatorial optimization problem and to solve it on reasonable time it is appropriate to apply some metaheuristic method. In this paper we apply ACO with two variants of intuitinistic fuzzy pheromone updating.

4 ACO Algorithm with Intuitionistic Fuzzy Pheromone

When the ACO is applied first step is representation of the problem by graph. Then the ants walk on the graph and construct solutions, which are paths in a graph. In our case the graph is fully connected and the arcs are weighted. The graph is asymmetric, because to go from A to B can take different time than to go from B to A. The ants constructs feasible solutions beginning from random node. Every step ants compute a set of possible moves and select the best one, according probabilistic rule called transition probability. The transition probability p_{ij}, to chose the node j when the current node is i, is based on the heuristic information η_{ij} and pheromone trail level τ_{ij} of the move, where $i, j = 1, \ldots, n$.

$$p_{ij} = \frac{\tau_{ij}^{\alpha} \eta_{ij}^{\beta}}{\sum_{k \in Unused} \tau_{ik}^{\alpha} \eta_{ik}^{\beta}}. \tag{1}$$

If the value of the heuristic information or pheromone is higher, the move is more attractive. the initial pheromone level is set to a small positive constant value τ_0 and then ants update this value after completing the construction stage. Let ρ be the evaporation. The pheromone updating rule is:

$$\tau_{lj} \leftarrow (1 - \rho)\tau_{lj} + \Delta\tau_{lj}. \tag{2}$$

In traditional ACO algorithm $\Delta\tau_{lj} = \rho F$, where F is the value of the objective function and ρ is a constant [6].

We propose using an *intuitionistic fuzzy pheromone updating* and the pheromone updating rule becomes:

$$\tau_{lj} \leftarrow (1 - \rho)\tau_{lj} + \alpha F, \tag{3}$$

where $(1 - \rho) + \alpha \leq 1$, $\alpha \in (0,1)$. The parameter ρ is generated as a random number from the interval $(0,1)$. The parameter α is generated in a random way in the interval $(0, \rho)$. Thus the sum of the two parameters is less than 1, and the updating rule is intuitionistic fuzzy.

Two variants of intuitionistic fuzzy pheromone updating are proposed. The parameters ρ and α are generated before the first iteration of the algorithm and stay unchanged during the end of the run. Thus, in the same run, the values of the two parameters are the same, but, in a different runs of the algorithm, they are different. The second idea is the parameters ρ and α to be generated every iteration. Thus, their values are different at each iteration in the same run. Thus the search process will be more diversified. As a consequence we assume achieving better solutions.

5   Numerical Experiments and Discussion

We have tested our idea for intuicionistic fuzzy pheromone on 8 test problems with various size, from 38 to 443 sessions. The smallest tests, 38 and 78 sessions, are real data and the others are from the Operational Research Library "OR-Library" available within WWW, accessed at http://people.brunel.ac.uk/mastjjb/jeb/orlib/periodtspinfo.html (21 Jun 2022). The two versions of the algorithm are software implemented on C programming language and run on Pentium desktop computer at 2.8 GHz with 4 GB of memory. The ACO parameters are the same for all tests and variants of the algorithm and are reported on Table 1. We fixed the parameters experimentally after several runs of the algorithm. With every of the test problems we performed 30 independent runs for calculation of average result. ANOVA test is applied with significance 95% thus the difference between the average results are guaranteed.

On Table 2 are reported average results achieved with the variants of the ACO algorithm. On the first column are the size of the problem. The average results achieved with traditional ACO algorithm are reported on second column. The third column is ACO with intuitionistic fuzzy pheromone updating, when the parameters are calculated at the beginning of the algorithm. The average results,

Table 1. ACO parameter settings.

Parameters	Value
Number of iterations	Number of sessions
Number of ants	20
ρ	0.9
τ_0	0.5
a	1
b	2

calculated with intuitionistic fuzzy pheromone updating in every iteration are reported on forth column. The last row shows the average result over all tests. We observe that the intuitionistic fuzzy pheromone updating in every iterations outperforms the traditional pheromone updating. It achieves better solutions for all test problems and has better average over all problems. The intuitionistic fuzzy pheromone updating at the beginning of the algorithm achieves similar or worst results comparing with traditional ACO and its average result over all tests is also worst according the traditional ACO. Intuitionistic fuzzy pheromone updating increase the level of diversification of the search of good solutions and the chance to improve achieved solutions are higher. The variant with generation of the pheromone parameters at the beginning of the algorithm less diversifies the search according the variant with generation of the pheromone parameters every iteration. It means that this problem needs more diversification of the search, to improve achieved solutions. It is our explanation of better performance of the algorithm in this case.

Table 2. ACO performance with different pheromone updating.

Instance	Traditional ACO	Int Fuzzy Beginning	Int. Fuzzy Every
GPS 38	899.50	898.66	**897.33**
GPS 78	922.06	926.83	**905.26**
GPS 100	40910.60	41644.26	**40522.30**
GPS 171	3341.93	3464.00	**3268.93**
GPS 323	1665.93	1685.43	**1639.20**
GPS 358	1692.66	1710.40	**1665.20**
GPS 403	3428.56	3466.16	**3403.06**
GPS 443	3765.8	3794.43	**3723.93**
average	7078.38	7198.77	**7003.05**

We tested a variant where only the parameter α is randomly generated keeping the condition to be intuitionistic fuzzy, $(1 - \rho) + \alpha \leq 1$. The achieved results are very similar to the variant with generating both parameters. We prefer the second because we do not need tuning the parameter ρ. We applied intuitionistic fuzzy pheromone updating on Multidimensional Knapsack Problem (MKP) in our previous work [10]. For this problem less diversification is better, and the variant with parameter generation at the beginning of the algorithm gives better algorithm performance. Comparing both applications we can conclude that ordered problems need more diversification to achieve good solutions than subset problems.

6   Conclusions

ACO algorithm with intuitionistic fuzzy pheromone updating for solving GPS surveying problem is proposed in this paper. We introduced two pheromone parameters and their sum is less or equal to 1. The coefficients are randomly generated. Two variants of parameters generations are proposed. In the first variant the pheromone parameters are generated before the first iteration of the algorithm and are unchanged till the end of the algorithm. Thus they are different in different runs, but the same during the same run. In the second variant the parameters are generated every iteration. We test our idea on 8 test problems with different size from 38 to 443. The second variant with parameters generation on every iteration outperforms the traditional ACO. The other variant gives very similar results to the traditional ACO. Thus more diversification is better for ordering problems, part of which is GPS. Comparing with MKP which is a subset problem we observe that for MKP is better to have more balanced diversification.

In the future work we will try with other variants of ACO and will research which variant is better for different classes of problems.

Acknowledgment. The presented work is partially supported by the grant No BG05M2OP011-1.001-0003, financed by the Science and Education for Smart Growth Operational Program and co-financed by European Union through the European structural and Investment funds. The work is supported too by National Scientific Fund of Bulgaria under the grant DFNI KP-06-N52/5.

References

1. Atanassov, K.: Two variants of intuitionistic fuzzy propositional calculus. Bioautomation **20**, 17–26 (2016)
2. Atanassov, K.T.: Intuitionistic Fuzzy Sets. In: Intuitionistic Fuzzy Sets. Studies in Fuzziness and Soft Computing 123, vol. 35. Springer, Heidelberg (1999). https://doi.org/10.1007/978-3-7908-1870-3_1
3. Atanassov, K.: Intuitionistic Fuzzy Logics. Studies in Fuzziness and Soft Computing 351. Springer, Heidelberg (2017). https://doi.org/10.1007/978-3-319-48953-7
4. Bonabeau, E., Dorigo, M., Theraulaz, G.: Swarm Intelligence: From Natural to Artificial Systems. Oxford University Press, Oxford, UK (1999)
5. Dorigo, M., Gambardella, L.: Ant colony system: a cooperative learning approach to the traveling salesman problem. IEEE Trans. Evol. Comput. **1**, 53–66 (1996)

6. Dorigo, M., Stutzle, T.: Ant Colony Optimization. MIT Press, Cambridge, MA, USA (2004)
7. Fidanova, S., Lirkov, I.: 3D protein structure prediction. Analele Univ. Vest Timis. **XLVII**, 33–46 (2009)
8. Fidanova, S.: An improvement of the grid-based hydrophobic-hydrophilic model. Bioautomation **14**, 147–156 (2010)
9. Fidanova, S.: ACO algorithm with additional reinforcement. In: Dorigo, M. (ed.) Proceedings of From Ant Colonies to Artificial Ants Conference, pp. 292–293. Brussels, Belgium (2002)
10. Fidanova S., Atanassov K.: ACO with intuitionistic fuzzy pheromone updating applied on multiple knapsack problem. Mathematics **9**(13), MDPI 1–7 (2021). https://doi.org/10.3390/math9131456
11. Goldberg, D.E., Korb, B., Deb, K.: Messi genetic algorithms: motivation analysis and first results. Complex Syst. **5**, 493–530 (1989)
12. Karaboga, D., Basturk, B.: Artificial bee colony (abc) optimization algorithm for solving constrained optimization problems. In: Melin, P., Castillo, O., Aguilar, L.T., Kacprzyk, J., Pedrycz, W. (eds.) IFSA 2007. LNCS (LNAI), vol. 4529, pp. 789–798. Springer, Heidelberg (2007). https://doi.org/10.1007/978-3-540-72950-1_77
13. Kennedy, J., Eberhart, R.: Particle swarm optimization. In: Proceedings of the IEEE International Conference on Neural Networks IV, Perth, Australia, pp. 1942–1948 (1995)
14. Kirkpatrick, S., Gelatt, C.D., Vecchi, M.P.: Optimization by simulated annealing. Science **13**, 671–680 (1983)
15. Mirjalili, S., Mirjalili, S.M., Lewis, A.: Grey wolf optimizer. Adv. Eng. Softw. **69**, 46–61 (2014)
16. Mosavi, M.R., Khishe, M., Parvizi, G.R., Naseri, M.J., Ayat, M.: Training multi-layer perceptron utilizing adaptive best-mass gravitational search algorithm to classify sonar dataset. Arch. Acoust **44**, 137–151 (2019)
17. Osman, I.H.: Metastrategy simulated annealing and Tabu search algorithms for the vehicle routing problem. Ann. Oper. Res. **41**, 421–451 (1993)
18. Ravakhah, S., Khishe, M., Aghababaee, M., Hashemzadeh, E.: Sonar false alarm rate suppression using classification methods based on interior search algorithm. Comput. Sci. Netw. Secur. **17**, 58–65 (2017)
19. Stutzle, T., Hoos, H.: Max min ant system. Futur. Gener. Comput. Syst. **16**, 889–914 (2000)
20. Vikhar, P.A.: Evolutionary algorithms: a critical review and its future prospects. In: Proceedings of the 2016 International Conference on Global Trends in Signal Processing, Information Computing and Communication (ICGTSPICC), Jalgaon, India, pp. 261–265 (2016)
21. Yang, X.S.: A new metaheuristic bat-inspired algorithm. In: Pelta, D.A., Cruz, C., Terrazas, G., Krasnogor, N. (eds.) Nature Inspired Cooperative Strategies for Optimization (NICSO 2010). Studies in Computational Intelligence, vol. 284, pp. 65–74. Springer, Heidelberg (2010). https://doi.org/10.1007/978-3-642-12538-6_6
22. Yang, X.S.: Nature-Inspired Metaheuristic Algorithms. Luniver Press, Beckington, UK (2008)
23. Zadeh, L.: Fuzzy Sets. Inf. Control **8**(3), 338–353 (1965)

Covering a Set of Points with a Minimum Number of Equal Disks via Simulated Annealing

Stefan M. Filipov[✉] and Fani N. Tomova

Department of Computer Science, University of Chemical Technology and
Metallurgy, blvd. Kl. Ohridski 8, 1756 Sofia, Bulgaria
sfilipov@uctm.edu

Abstract. This paper considers the following problem. Given n points
in the plane, what is the minimum number of disks of radius r needed
to cover all n points? A point is covered if it lies inside at least one
disk. The problem is equivalent to the unit disc cover problem. It is
known to be NP-hard. To solve the problem, we propose a stochastic
optimization algorithm of estimated time complexity $O(n^2)$. First, the
original problem is converted into an unconstrained optimization problem by introducing an objective function, called energy, in such a way
that if a configuration of disks minimizes the energy, then it necessarily
covers all points and does this with a minimum number of disks. Thus,
the original problem is reduced to finding a configuration of disks that
minimizes the energy. To solve this optimization problem, we propose
a Monte Carlo simulation algorithm based on the simulated annealing
technique. Computer experiments are performed demonstrating the ability of the algorithm to find configurations with minimum energy, hence
solving the original problem of covering the points with minimum number
of disks. By adjusting the parameters of the simulation, we can increase
the probability that the found configuration is a solution.

Keywords: UDC · unit disk cover · optimization · stochastic
algorithm · Monte Carlo · simulated annealing

1 Introduction

This work considers the problem of covering n given points in the plane by a
minimum number of disks of equal radius. The input data for the problem is
the radius r and n pairs of real numbers $(x_i, y_i), i = 1, 2, ..., n$ which are the
Cartesian coordinates of the points. We have to answer the following question:
"What is the minimum number of disks of radius r needed to cover all n points?".
A point is covered if it lies inside one or more than one disk. Let a disk of radius
r have a center $(\tilde{x}, \tilde{y})$. By definition, the point i lies inside the disk if the distance
between the point and the center of the disk is less than the radius of the disk:

$$(x_i - \tilde{x})^2 + (y_i - \tilde{y})^2 < r^2 \tag{1}$$

© The Author(s), under exclusive license to Springer Nature Switzerland AG 2023
I. Georgiev et al. (Eds.): NMA 2022, LNCS 13858, pp. 134–145, 2023.
https://doi.org/10.1007/978-3-031-32412-3_12

We need to find a configuration of disks that covers all points and the number of disks is minimum. A configuration of m disks, called also *a state*, is uniquely determined by the number m and the centers of the disks $(\tilde{x}_j, \tilde{y}_j)$, $j = 1, 2, ..., m$. As an example, in Fig. 1, we have shown a system of nine points (the red dots) and six different configurations of disks (states). State A consists of four disks. The configuration covers one point. States B and C both consist of two disks. State B has three covered points, while the number of covered points in C is six. Configurations D, E, and F cover all nine points. State D has six disks, while states E and F have three disks. Three is, in fact, the minimum number of disks (of that radius) needed to cover all nine points. Any state (configuration of disks) that covers all points with a minimum number of disks is *a solution* to our problem. State E and state F are both solutions. Obviously, there are infinitely many solutions. Indeed, given a solution state, we can always displace any of its disks by an arbitrary but small enough displacement so that the points covered by the disk remain covered, hence the new state is also a solution. Finding any solution answers the original question of "what is the minimum number of disks needed to cover all points". We just need to count the number of disks in that solution state.

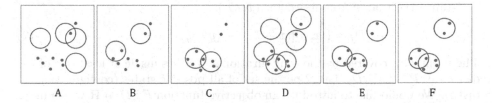

A   B   C   D   E   F

Fig. 1. Six different states for a system of nine points.

The described problem is equivalent to the unit disk cover problem [1 10]. The problem is known to be NP-hard [11]. Various approximate algorithms have been proposed [1–9]. Comprehensive study comparing the performance of the algorithms can be found in [10]. In this paper, to solve the problem, we propose a stochastic optimization algorithm based on the simulated annealing technique. We start from the observation that the problem can be viewed as a constrained optimization problem. Indeed, the number of disks needs to be minimized under the constraint that all points are covered. In this form the problem is hard to tackle since we cannot use the great number of powerful unconstrained optimization techniques available [12–14]. This work proposes a way of converting the original problem into an unconstrained optimization problem. We introduce an objective function (fitness function), called *energy*, in such a way that if a configuration of disks minimizes the energy, then it covers all points with a minimum number of disks. Thus, the original problem is reduced to finding a configuration of disks that minimizes the energy. To solve this optimization problem, we propose a Monte Carlo importance sampling simulation algorithm which is a variant of the simulated annealing technique [15–17]. Since the algorithm is probabilistic, it guarantees that the final state found in a simulation is a solution only with certain, usually high, probability. There is some probability that the found state is not a solution but just an approximate solution. However, adjusting the simulation parameters, we can always increase the probability that the final state found is a solution. The algorithm is tested on several systems demonstrating its ability to find states with minimum energy, i.e. configurations of disks that cover all points with a minimum number of disks. The advantage of the proposed algorithm over the above mentioned approximate algorithms is that it finds, with high probability, exact solutions of the problem, and not just approximate solutions.

2 Converting the Problem into Unconstrained Optimization

In this section, we propose a way of converting the considered problem into an unconstrained optimization problem. We are given a system of n points $(x_i, y_i), i = 1, 2, ..., n$ denoted by $P_i, i = 1, 2, ..., n$. Let us consider a configuration (a state) ω of m disks $D_j, j = 1, 2, ..., m$ with radius r and centers $(\tilde{x}_j, \tilde{y}_j), j = 1, 2, ..., m$. The disk D_j is defined as the set of all points that are at a distance less than r from its center $(\tilde{x}_j, \tilde{y}_j)$:

$$D_j = \{(x, y) | (x - \tilde{x}_j)^2 + (y - \tilde{y}_j)^2 < r^2\}. \tag{2}$$

The point P_i is covered by the configuration ω if it lies inside at least one of the disks, i.e. $P_i \in \cup_{j=1}^m D_j$. Let Ω be the set of all possible states (configurations of disks). We would like to introduce an objective function $E : \Omega \to \mathbf{R}$ which maps the states into the real numbers in such a way that (i) E has a minimum and (ii) if $\omega_0 \in \Omega$ minimizes E, i.e.

$$E(\omega_0) = \min_{\omega \in \Omega} E(\omega), \tag{3}$$

then the configuration ω_0 covers all points and does this with a minimum number of disks, i.e. ω_0 is a solution. The function E will be called *energy*. Defining E is not a trivial task as demonstrated by the following example. A most natural way to define E is to attribute an energy 1 to each uncovered point and energy 1 to each disk. The covered points have energy 0. The energy of a state is the sum of the energy of all elements (points and disks). Then, it is not hard to show that if all points are covered by a minimum number of disks, the energy is minimum. Unfortunately, it is not true that if the energy is minimum, then all points are covered with a minimum number of disks. This can be seen very clearly form the one-point system shown in Fig. 2. In the figure, you can see two states. In state A, all points are covered with a minimum number of disks, i.e. the state is a solution. The energy of the state is minimum, namely $E = 1$. In state B, however, the energy is also 1, i.e. minimum, but the state is not a solution. We can draw the conclusion that defining the energy in the suggested way ensures that "a solution $\Longrightarrow$ minimum energy"but not "minimum energy $\Longrightarrow$ a solution". The latter implication, however, is exactly what we need in order to convert the problem into an unconstrained optimization problem. We would like to minimize the energy, i.e. find a state with minimum energy, and then, from the implication "minimum energy $\implies$ a solution", we can claim that the found state is a solution.

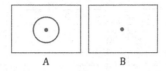

Fig. 2. Two states, A and B, of a one-point system. The state A is a solution, but the state B is not.

Now we proceed to define the energy in such a way that minimum energy implies a solution. Let the energy of each uncovered point be 2, the energy of each covered point 0, and the energy of each disk 1, i.e.

$$e(P_i) = \begin{cases} 2, P_i \notin \cup_{j=1}^m D_j \\ 0, P_i \in \cup_{j=1}^m D_j \end{cases} \tag{4}$$

and

$$e(D_j) = 1. \tag{5}$$

The energy of the state ω is the sum of the energies of all elements, i.e.

$$E(\omega) = \Sigma_{i=1}^n e(P_i) + \Sigma_{j=1}^m e(D_j) = \Sigma_{i=1}^n e(P_i) + m. \tag{6}$$

This definition resolves the problem described above. Now, state A in Fig. 2, which is a solution, has energy 1, while state B, which is not a solution, has energy 2. Other possible states for a one-point system are shown in Fig. 3. As can be seen, the state with the minimum energy ($E = 1$) is a solution. The other states are not solutions and their energy is greater than 1. The same holds for a two-point system (Fig. 3 - right). The state with the minimum energy ($E = 1$) is a solution, while the other states, which are not solutions, have energy greater than 1.

Now we prove rigorously that the energy definition (4)-(6) implies "minimum energy $\implies$ a solution".

Theorem 1. *Let $\omega_0 \in \Omega$ be a configuration of disks such that (3) holds, i.e. $E(\omega_0)$ is minimum. Then, all points are covered by a minimum number of disks.*

Proof. First, we prove that all points are covered. Suppose that there is a point that is not covered. Then, adding a disk to cover the point will reduce the energy by 1, which contradicts the fact that ω_0 is a configuration with minimum energy. Now we prove that ω_0 is a configuration with minimum number of disks. If there is a configuration of disks that covers all points and has at least one disk less

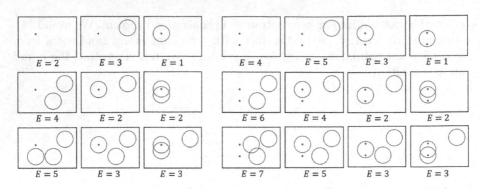

Fig. 3. Possible states and corresponding energies for a one-point system (left) and a two-point system (right).

than the number of disks in ω_0, then its energy will be at least 1 less than the energy of ω_0, which again contradicts the fact that ω_0 is a configuration with a minimum energy.

This result reduces the original problem to an unconstrained optimization problem. If we find a configuration ω_0 that minimizes the energy, then, by theorem 1, this configuration is a solution to the original problem of covering the points with a minimum number of disks. What remains to be done is to implement an optimization method that is well suited for finding states with minimum energy. As shown in the next section, due to possible existence of local minima of the energy, not all optimization methods are appropriate. In this paper, we design a method, based on the simulated annealing technique, which is able to overcome local minima and find the global minimum.

3   Simulated Annealing Algorithm

In this section we design a stochastic algorithm (a Monte Carlo simulation) that can solve the optimization problem discussed in the previous section, namely, find a configuration of disks (a state) with minimum energy. We propose the following approach. Suppose we are given a state ω with energy E. From this state, using a certain rule that involves a probabilistic element, we generate a new state ω_{new}. Let the energy of the new state be E_{new}. We calculate the change of energy ΔE, where $\Delta E = E_{new} - E$, and depending on the value of ΔE we decide whether to accept the move or not. If we accept the move, we change the configuration from ω to ω_{new}. Then, we repeat the procedure. If the rules of generating new states and accepting moves are chosen properly, repeating the procedure a certain number of times should lead, with certain probability, to a state with minimum energy.

The rule by which we generate new states should be such that starting from any state $\omega \in \Omega$, any other state $\omega' \in \Omega$ should, in principle, be reachable in a finite number of steps. All states that can be reached directly, i.e. in one step, from the state ω are called neighboring states (neighbors) of ω. We propose the following way of generating neighboring states. Given a state ω we do, with certain probability, one of the following:

(i) Choose at random one of the disks from ω and displace it (move it) by randomly chosen displacement vector.
(ii) Choose at random a position within the allowed domain and place a new disk there.
(iii) Choose at random one of the disks from ω and remove it.

The generated in this fashion new state is denoted ω_{new}. Technical details concerning the allowed domain, the generation of the random displacement vector, and the relative frequencies of moves (i), (ii), and (iii) are discussed later on. For the time being, it suffices to say that these elements/parameters are chosen in such a way that any sate within the allowed domain is reachable from any other state in a finite number of steps.

Once we have generated a new state ω_{new}, the rule by which we decide whether to accept the move, i.e. replace ω by ω_{new}, or reject it, i.e. keep ω, should be such that the tendency of the process should be towards replacing states with higher energy by states with lower energy. Thus, the hope is that after a certain number of steps, a state with minimum energy will be reached. Seemingly, the most natural rule that can be adopted is to accept moves that decrease the energy ($\Delta E < 0$) and reject other moves. This acceptance rule, however, is not good enough to ensure convergence of the process towards a state with minimum energy, i.e. towards a *global minimum* of the system. The problem is that the described procedure of generating neighboring states does not exclude the existence of local minima. Let us define a *local minimum* as a state that does not have neighbors with lower energy. A local minimum may not be a global minimum. Consider, for example, the two states shown in Fig. 4.

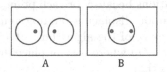

A   B

Fig. 4. Example of a local minimum (state A) and a global minimum (state B).

State A has energy 2, while state B has energy 1. State B is a global minimum, while state A is a not. In order to go from A to B, we can do one of the following. We can remove any of the two disks in A and then move the other disk to cover the two points. We can place a new disk to cover the two points in A and then remove, one by one, the remaining two disks. Finally, we can move any of the disks in A to cover the two points, and then remove the other disk. In each case, the first step leads to a state with energy greater than or equal to the energy of state A. Obviously, if we accept only moves that decrease the energy, we cannot go from state A to state B. We say that the state A is a local minimum. If a simulation has reached a local minimum, it will get stuck in it and won't be able to reach the global minimum. The example shows clearly that algorithms which accept only moves that decrease the energy are not applicable.

A way to overcome this problem is to adopt the acceptance rule of the importance sampling algorithms used in Mote Carlo molecular simulations [18]. According to this rule, if the energy decreases ($\Delta E < 0$), the move is accepted, but if the energy increases ($\Delta E > 0$), the move is not rejected, but accepted with probability

$$P_{accept} = e^{-\frac{\Delta E}{T}}, \qquad (7)$$

where $T > 0$ is a parameter called *temperature*. The probability of accepting a move for which $\Delta E > 0$ decreases exponentially with ΔE. Thus, although moves that increase the energy are also accepted, the overall tendency is towards moves that decrease the energy. This tendency increases as T decreases. Now we are ready to state the proposed algorithm. Starting from a randomly chosen initial state and some temperature T, at each step, we generate a new state and accept or reject it according to the described acceptance rule. The new states are generated according to rules (i), (ii), and (iii). In the proposed algorithm, at each 5 attempts to displace a randomly chosen disk (i), we make, with equal probability, either an attempt to add a new disk at a random position (ii), or remove a randomly selected disk (iii). After $100n$ attempts are performed, the temperature is lowered, according to some specified schedule, and the procedure is repeated. The process is ended when the temperature has become sufficiently low, i.e. the system has been cooled down enough. The described algorithm is a variance of the simulated annealing algorithm [15–17].

Note that the number of attempts at each temperature is proportional to the number of points n. Since the average number of disks involved in a simulation is proportional to n, this ensures that the number of attempts per disk is independent of n. At each attempt, we need to calculate the change of energy ΔE. This requires n operations because for each point we need to check whether its energy (4) is changed upon displacing (adding/removing) the selected disk. Therefore, the time complexity of the algorithm is $O(n^2)$.

The region in the plane xOy where we can place disks, called domain, is the set of all points (x, y) such that

$$\min_i x_i - r < x < \max_i x_i + r, \qquad (8)$$

$$\min_i y_i - r < y < \max_i y_i + r, \qquad (9)$$

where $i \in \{1, 2, ..., n\}$. This is a rectangular region. A disk with center outside this region cannot cover any of the given points, hence, we restrict ourselves to placing disks inside this region only.

To construct the random displacement vector needed in point (i), we generate two uniformly distributed random real numbers

$$\Delta x \in \left[-\frac{r}{2}, \frac{r}{2}\right], \Delta y \in \left[-\frac{r}{2}, \frac{r}{2}\right]. \tag{10}$$

The vector $(\Delta x, \Delta y)$ is the random displacement vector. Let the disk D_k with center $(\tilde{x}_k, \tilde{y}_k)$ be the disk selected to be displaced (moved). Then, the disk's new center is $(\tilde{x}_k + \Delta x, \tilde{y}_k + \Delta y)$.

```
1    Input data: n points and radius; Determine domain;
2    Choose initial configuration of m disks, 1<=m<=n;
3    Calculate energy E; T=1;
4    while(T>0.01)
5        for attempt=1:100*n // attempt to move a disk
6            Choose disk at random; Choose displacement at random;
7            Calculate new position of disk;
8            if(new position within domain)
9                Calculate change of energy dE if disk is moved to new position;
10               if(dE<0||rand<exp(-dE/T))
11                   Move disk; E=E+dE;
12               end
13           end
14           if(mod(attempt,5)==0)
15               choice=randi(2);
16               if(choice==1 && m<n) // attempt to add a disk
17                   Choose random position within domain;
18                   Calculate change of energy dE if new disk is added there;
19                   if(dE<0||rand<exp(-dE/T))
20                       Add disk; E=E+dE; m=m+1;
21                   end
22               end
23               if(choice==2 && m>1) // attempt to remove a disk
24                   Choose disk at random;
25                   Calculate change of energy dE if disk is removed;
26                   if(dE<0||rand<exp(-dE/T))
27                       Remove disk; E=E+dE; m=m-1;
28                   end
29               end
30           end
31       end
32       T=T/1.5;
33   end
34   Print out final configuration;
```

Fig. 5. MATLAB code structure of the proposed stochastic algorithm.

In the algorithm, the initial temperature is chosen to be 1. At each temperature level, $100n + 20n$ attempts to replace the current state with a new state are performed and then the temperature is reduced by $1/3$ of its current value, i.e. $T \longleftarrow T - T/3$. The process is ended when the temperature has become less than 0.01. In general, the higher the initial temperature, the lower the final temperature, and the slower the cooling is, the greater the probability is that the final state found is a solution. In the performed computer experiments, we have established that the stated values are good enough to ensure high probability of success in finding a solution.

The proposed algorithm is implemented on MATLAB. The structure of the code is presented in Fig. 5. In the code, rand generates random "real"number between 0 and 1. The number is uniformly distributed on the interval. The function randi(2) generates at random 1 or 2 with equal probability. The function mod(a, 5) returns the remainder after division of a by 5.

4   Computer Experiments

The proposed stochastic algorithm was tested on a system of 24 points. The points are (3,8.5), (1.5,5), (−3,2.5), (−4,4), (2.5,3.5), (12,3), (−3.5,6.5), (−2.5,−8.5), (−1.5,−10), (−2.5,−8), (3.5,1.5), (1.5,1), (0.5,1.5), (−5,−4.5), (6,−2.5), (−4,1), (0,−1), (−5.5,0.5), (−1,3), (−2,−0.5), (4,−1), (−1,1), (−5,1.5), (−4.5,2). To cover the points, we used disks with radius $r = 2$, then $r = 3$, $r = 4$, and $r = 5$. For $r = 2$ the minimum number of disks needed to cover the given 24 points is 10. A particular solution found by the algorithm is shown in the upper left corner of Fig. 6. In the figure, particular solutions for $r = 2$, $r = 3$, $r = 4$, and $r = 5$ are shown.

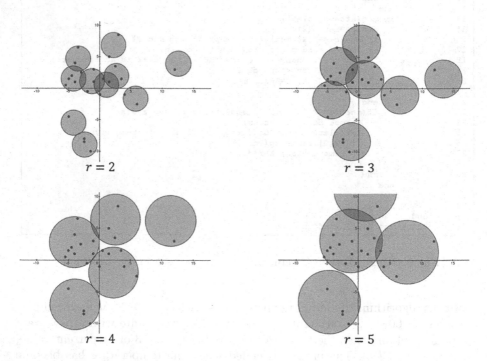

Fig. 6. Final states reached in a simulation for four different radii.

As you can see, the number of disks needed to cover the points is 10, 7, 5, and 4, respectively. The cooling schedule used in the simulations is

$$T = \left(\frac{2}{3}\right)^{k-1}, k = 1, 2, \dots \tag{11}$$

where k is the temperature level. As we established experimentally, using initial temperature $T = 1$ and ending the process when T has become less than 0.01 is enough to ensure that the final state found in a simulation is a solution with high degree of certainty. More than 99% of the performed simulations ended in states with minimum energy, i.e. found exact solutions. The rest ended in states that had energy just one unit more than the minimum energy. The energy of the system during simulation as a function of the temperature level k is shown in Fig. 7. As an initial state, we always used 5 disks with centers (1.5,0), (−1,−6), (3,2), (−5,6), (11,3). In the figure, the energy at temperature level 0 is the energy of the initial state. As the results indicate, in all the cases shown in the figure, the system reached its minimum energy about temperature level 6 and then stayed at the minimum till the final temperature level 12.

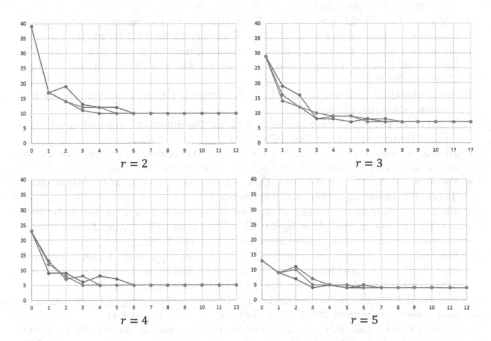

Fig. 7. Energy E as a function of temperature level k during simulation for four different radii. For each radius three simulations are shown.

5   Conclusion

This paper considered the problem of covering a set of points by a minimum number of equal disks. The problem is equivalent to the unit disk cover problem. To solve the problem, we proposed a reduction to unconstrained optimization by introducing an objective function called energy. To minimize the energy, a Monte Carlo simulation algorithm was developed. The algorithm is a variant of the simulated annealing method. It was demonstrated that the algorithm is efficient in finding states with minimum energy, that is, configurations that cover all points with minimum number of disks. The advantage of the algorithm over the many approximate algorithms available is that it finds, with high degree of certainty, exact solutions.

References

1. Hochbaum, D.S., Maass, W.: Approximation schemes for covering and packing problems in image processing and VLSI. J. ACM (JACM) **32**(1), 130–136 (1985)
2. Gonzalez, T.F.: Covering a set of points in multidimensional space. Inf. Process. Lett. **40**(4), 181–188 (1991)
3. Charikar, M., Chekuri, C., Feder, T., Motwani, R.: Incremental clustering and dynamic information retrieval. SIAM J. Comput. **33**(6), 1417–1440 (2004)
4. Brönnimann, H., Goodrich, M.T.: Almost optimal set covers in finite VC-dimension. Discrete Comput. Geometry **14**(4), 463–479 (1995). https://doi.org/10.1007/BF02570718
5. Franceschetti, M., Cook, M., Bruck, J.: A geometric theorem for approximate disk covering algorithms (2001)
6. Liu, P., Lu, D.: A fast 25/6-approximation for the minimum unit disk cover problem. arXiv preprint arXiv:1406.3838 (2014)
7. Biniaz, A., Liu, P., Maheshwari, A., Smid, M.: Approximation algorithms for the unit disk cover problem in 2D and 3D. Comput. Geom. **60**, 8–18 (2017)
8. Dumitrescu, A., Ghosh, A., Toth, C.D.: Online unit covering in Euclidean space. Theoret. Comput. Sci. **809**, 218–230 (2020)
9. Panov, S.: Finding the minimum number of disks of fixed radius needed to cover a set of points in the plane by MaxiMinMax approach. In: 2022 International Conference Automatics and Informatics (ICAI). IEEE (2022)
10. Friederich R., Ghosh A., Graham M., Hicks B., Shevchenko R.: Experiments with unit disk cover algorithms for covering massive pointsets. Computat. Geometry **109**, 101925 (2023). https://doi.org/10.1016/j.comgeo.2022.101925
11. Fowler, R.J., Paterson, M.S., Tanimoto, S.L.: Optimal packing and covering in the plane are NP-complete. Inf. Process. Lett. **12**(3), 133–137 (1981)
12. Chong, E.K.P., Zak, S.H.: An introduction to optimization, 3rd ed., Wiley (2011)
13. Martins, J.R.R.A., Ning, A.: Engineering design optimization. Cambridge University Press (2021)
14. Griva, I., Nash, S.G., Sofer, A.: Linear and nonlinear optimization. Society For Industrial Mathematics, 2nd ed. (2009)
15. Kirkpatrick, S., Gelatt, C.D., Jr., Vecchi, M.P.: Optimization by Simulated Annealing. Science **220**(4598), 671–680 (1983). https://doi.org/10.1126/science.220.4598.671

16. Bertsimas, D., Tsitsiklis, J.: Simulated annealing. Statistical Sci. **8**(1), 10–15 (1993). https://www.jstor.org/stable/2246034
17. Filipov, S.M., Panov, S.M., Tomova, F.N., Kuzmanova, V.D.: Maximal covering of a point set by a system of disks via simulated annealing. In: 2021 International Conference Automatics and Informatics (ICAI), pp. 261–269. IEEE (2021). https://doi.org/10.1109/ICAI52893.2021.9639536
18. Frenkel, D., Smit, B.: Understanding molecular simulation: from algorithms to applications (2002). https://doi.org/10.1016/B978-0-12-267351-1.X5000-7

Mathematical Modelling of Nonlinear Heat Conduction with Relaxing Boundary Conditions

Stefan M. Filipov[1]($\boxtimes$)(iD), István Faragó[2](iD), and Ana Avdzhieva[3](iD)

[1] Department of Computer Science, Faculty of Chemical and System Engineering, University of Chemical Technology and Metallurgy, Sofia, Bulgaria
sfilipov@uctm.edu
[2] Department of Differential Equations, Institute of Mathematics, Budapest University of Technology and Economics, & MTA-ELTE Research Group, Budapest, Hungary
[3] Faculty of Mathematics and Informatics, "St. Kl. Ohridski" University of Sofia, Sofia, Bulgaria
aavdzhieva@fmi.uni-sofia.bg

Abstract. This paper considers one dimensional unsteady heat condition in a media with temperature dependent thermal conductivity. When the thermal conductivity depends on the temperature, the corresponding heat equation is nonlinear. At one or both boundaries, a relaxing boundary condition is applied. It is a time dependent condition that approaches continuously, as time increases, a certain time independent condition. Such behavior at the boundaries arises naturally in some physical systems. As an example, we propose a simple model system that can give rise to either Dirichlet or convective relaxing boundary condition. Due to the dependence of the thermal conductivity on the temperature, the convective condition is nonlinear. For the solution of the problem, we propose a new numerical approach that first discretizes the heat equation in time, whereby a sequence of two-point boundary value problems (TPBVPs) is obtained. We use implicit time discretization, which provides for unconditional stability of the method. If the initial condition is given, we can solve consecutively the TPBVPs and get approximations of the temperature at the different time levels. For the solution of the TPBVPs, we apply the finite difference method. The arising nonlinear systems are solved by Newton's method. A number of example problems are solved demonstrating the efficiency of the proposed approach.

Keywords: PDE · ODE · Nonlinear heat equation · Relaxing boundary condition · Finite difference method

S. M. Filipov and A. Avdzhieva were supported by the Sofia University Research Fund under Contract No. 80-10-156/23.05.2022. I. Faragó was supported by the Hungarian Scientific Research Fund OTKA SNN125119 and also OTKA K137699.

I. Georgiev et al. (Eds.): NMA 2022, LNCS 13858, pp. 146–158, 2023.
https://doi.org/10.1007/978-3-031-32412-3_13

1   Introduction

This work considers unsteady one dimensional heat conduction in a solid body with temperature dependent thermal conductivity. The body occupies a finite space region. The exchange of energy within the body and at the boundaries is assumed to be due to heat transfer only. It is assumed that no energy is exchanged through radiation and that there are no internal heat sources. Since we consider the one dimensional case, the temperature u within the body is a function of one space variable only. This variable is denoted by x. The temperature at any position x may depend on the time t, hence $u = u(x, t)$. Heat transfer processes for which the temperature changes with time are called unsteady. If we are given some initial distribution of the temperature within the body, i.e. a function $u(x, 0)$, and certain appropriate boundary conditions, we should be able, in principle, to predict how the temperature distribution evolves with time, i.e. to find the function $u(x, t)$.

From the energy balance in a differentially small element of the body, we can derive a partial differential equation (PDE) for the temperature $u(x, t)$. This equation is called a heat equation and must hold at any point x within the considered space interval and any time $t > 0$. Solving this equation, subject to the given initial and boundary conditions, would yield the sought function $u(x, t)$ When the thermal conductivity of the media is constant, this equation is a linear parabolic equation. This PDE has been studies extensively over the years and many analytical and numerical methods for its solution have been developed [1]–[6]. However, when the thermal conductivity of the media depends on the temperature, which is the considered case, the heat equation is nonlinear. This nonlinear equation is quite significant in science and engineering but is much more difficult to solve. Because of that various methods for particular cases and approximate techniques have been developed [7]–[14]. An important numerical method for solving the equation is the method of lines (MOL). The method was originally introduced in 1965 by Liskovetz [15]–[17] for partial differential equations of elliptic, parabolic, and hyperbolic type. The method first discretizes the equation in space, whereby, adding the boundary conditions, a system of first order ordinary differential equations (ODEs) is obtained. The system, together with the initial condition, constitutes an initial value problem (Cauchy problem). It can be solved using various numerical approaches. However, when explicit time-discretization schemes are applied, the method is only conditionally sable.

To overcome this and other drawbacks of the MOL, the authors have recently proposed a method [18] that first discretizes the heat equation in time using implicit discretization scheme. The discretized PDE is a sequence of second order ODEs which, together with the boundary conditions, forms a sequence of two-point boundary value problems (TPBVPs). Using the initial condition, we can solve the first TPBVP from the sequence and get an approximation of the temperature $u(x, t)$ at the first time level $t = t_1$. Then, using the obtained solution, we can solve the second TPBVP, and so on, getting approximations of the temperature at each time level $t = t_1, t_2, \ldots$ For the solution of the TPBVPs, we employ the finite difference method (FDM) with Newton iteration. An important feature of the proposed method is that it is unconditionally stable. The technique was applied successfully for solving nonlinear heat and mass transfer problems [18]–[19]. Originally, the method was developed for Dirichlet boundary conditions. These boundary conditions are linear and time-independent. However, many important physical situations require the application of time-dependent or even nonlinear time-dependent boundary conditions. This paper extends the proposed method to incorporate such boundary conditions. To motivate their use, we propose a simple physical model that naturally gives rise to the so called relaxing boundary conditions. They have been introduced first for certain diffusion processes [20] but, as shown in this work, can be applied successfully in certain heat transfer problems. Roughly speaking, a relaxing boundary condition is a time-dependent condition that approaches continuously, as time increases, a certain time-independent condition. The relaxing boundary condition can be a Dirichlet type, which is linear, but also a Neumann or a Robin type, which, given the nonlinear nature of the considered process, lead to nonlinear time-dependent conditions at the boundary.

2 Physical System and Mathematical Model

This section presents a continuous mathematical model of a simple physical system that gives rise to one dimensional nonlinear heat transfer with relaxing boundary condition. A silicon rod is situated along the x-axis between $x = a$ and $x = b$ (Fig. 1). The temperature in the silicon is denoted by $u(x,t)$. Because of the symmetry of the problem, the temperature in the rod is a function of one spatial variable only. We have chosen the substance silicon because its thermal conductivity depends on the temperature. The silicon rod is laterally insulated so that no energy, in any form, can be transferred through the lateral surface. At the point $x = a$ the silicon rod is in thermal contact with a tank filled with liquid. Heat can flow freely through the contact surface in both directions. Apart from the thermal contact with the silicon, the tank is well insulated. The temperature in the tank is denoted by $T(t)$. It is homogenous inside the tank. A possible exception could be a thin layer close to the silicon contact surface, where the temperature may not be equal to $T(t)$. There are two pipes connected

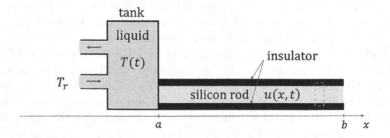

Fig. 1. Physical system

to the tank. Through one of them liquid at temperature T_r is being pumped into the tank at a constant volume flow rate. Through the other pipe liquid at temperature $T(t)$ is going out of the tank at the same volume flow rate. We assume that the liquid in the tank is well stirred throughout the process so that the incoming liquid is mixed with the liquid inside the tank fast enough. We also assume that the density and heat capacity do not depend on the temperature. How to find the function $T(t)$ and how to use it to set the boundary condition at $x = a$ will be shown in the next section. In this section we will derive the heat equation for the silicon rod and set the other necessary conditions.

To derive the heat equation we consider a differentially small volume element in the silicon rod, e.g. the element shown with dashed line in Fig. 1. Energy can enter or leave the volume element only via heat transfer through the two vertical surfaces. Let the left surface be at position x and the right at position $x+\Delta x$. Let $\kappa = \kappa(u)$ be the thermal conductivity of the silicon. The energy flux through the left surface is $-\kappa(u)\frac{\partial u}{\partial x}|_x$. It tells us how much energy per unit time per unit area enters the volume element. Note that, by definition, a positive value of the flux means that the energy flows in the positive direction of the axis, i.e. from left to right in Fig. 1. The sign minus in the flux expression ensures that the energy moves from regions with high temperature to regions with low temperature. The energy flux through the right surface is $-\kappa(u)\frac{\partial u}{\partial x}|_{x+\Delta x}$. It tells us how much energy per unit time per unit area leaves the volume element. Let A be the area of the cross section of the silicon rod. Then, the energy that enters the element per unit time minus the energy that leaves the element per unit time divided by the volume of the element is

$$\frac{A\left(-\kappa(u)\frac{\partial u}{\partial x}|_x\right) - A\left(-\kappa(u)\frac{\partial u}{\partial x}|_{x+\Delta x}\right)}{A\Delta x}. \tag{1}$$

Since there are no internal heat sources within the silicon rod, we can equate the limit of (1) as $\Delta x \to 0$ to the rate of change of the energy density (energy per unit volume) at point x:

$$\rho c_p \frac{\partial u}{\partial t} = -\frac{\partial}{\partial x}\left(-\kappa(u)\frac{\partial u}{\partial x}\right). \tag{2}$$

In this equation ρ is the density of the silicon, and c_p is its heat capacity at constant pressure. In our model they are assumed to be constant. The left hand side of the equation represents the rate of change of the energy density. The quantity in the brackets on the right hand side is the energy flux, while the right hand side itself can be viewed as the negative divergence of the energy flux. Equation (2) is the heat equation for the silicon rod. It is a PDE for the unknown function $u(x,t)$ and must hold for any $x \in (a,b)$ and any $t > 0$. If κ were a constant, the equation would be a linear parabolic PDE. However, since $\kappa = \kappa(u)$, the equation is a nonlinear PDE.

To define the particular process that is taking place, however, in addition to (2), we need to provide the initial condition and the boundary conditions at $x = a$ and $x = b$. Let the temperature distribution in the silicon rod at time $t = 0$ be $u_0(x)$. Then, the initial condition is

$$u(x, 0) = u_0(x), x \in [a, b]. \tag{3}$$

At the boundary $x = b$, we consider a simple Dirichlet boundary condition, namely

$$u(b, t) = \beta, t > 0. \tag{4}$$

Other boundary conditions can also be applied.

In the next section, we discuss possible approaches to define the boundary condition at $x = a$ and show that the considered physical system leads naturally to the so called relaxing boundary condition.

3   Relaxing Boundary Condition

Let the temperature in the tank at some initial time $t = 0$ be T_0. Obviously, if the temperature T_r of the incoming liquid is different from T_0, the temperature in the tank $T(t)$ will be changing. Now we proceed to find the temperature in the tank $T(t)$. Let Q be the volume flow rate of the incoming liquid (bottom pipe in Fig. 1). The rate at which energy is entering the tank with the incoming liquid is $\rho_l c_{p,l} T_r Q$, where ρ_l is the density of the liquid and $c_{p,l}$ is its heat capacity at constant pressure. The volume flow rate of the outgoing liquid (top pipe in Fig. 1) is also Q. Therefore, the rate at which energy is leaving the tank with the outgoing liquid is $\rho_l c_{p,l} T(t) Q$. Note that the mechanism of energy transfer through the pipes is convective, i.e. liquid with some energy density moves in and replaces (pushes out) liquid with different energy density. As a result the total energy contained in the tank is changing. Since in our model we assume that the tank is well insulated and no heat transfer between the tank and its surroundings occur, conservation of energy tells us that

$$\rho_l c_{p,l} \frac{dT(t)}{dt} V = \rho_l c_{p,l} T_r Q - \rho_l c_{p,l} T(t) Q, \tag{5}$$

where V is the volume of the tank. The left hand side of the equation is the rate of change of the tank energy. Note that we have neglected the heat transfer through the contact surface separating the liquid from the silicon rod. This is justifiable as long as the contact surface area A is small enough. In fact, since (1) is independent of A, as long as A is not zero, we can choose it to be as small as necessary without causing a change in the heat equation (2). Simplifying (5), we get

$$\frac{dT(t)}{dt} = -\frac{Q}{V}(T(t) - T_r). \tag{6}$$

Equation (6) is a first order linear ordinary differential equation for the unknown function $T(t)$. Solving the equation and taking into account the initial condition $T(0) = T_0$, we get

$$T(t) = T_r + (T_0 - T_r)e^{-\frac{Q}{V}t}. \tag{7}$$

Hence, the temperature in the tank $T(t)$ is an exponentially relaxing function. As t approaches infinity, the temperature approaches the finite value T_r. For $T_r > T_0$, the function $T(t)$ is monotonically increasing and looks like the one shown with blue in Fig. 2.

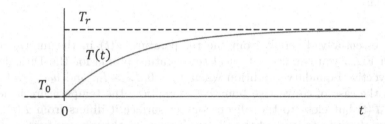

Fig. 2. The temperature at $x = a$ as a function of time for the Dirichlet relaxing boundary condition (solid blue line) and the traditional Dirichlet boundary condition (dashed line).

Now we are ready to set the boundary condition at $x = a$. The first approach is to equate the temperature of the silicon at $x = a$ to the temperature in the tank

$$u(a,t) = T(t), t > 0. \tag{8}$$

This boundary condition is called Dirichlet relaxing boundary condition [20]. The main feature of this condition is that the temperature value at the boundary increases (or decreases) gradually with time and approaches some finite value. In the traditional Dririchlet boundary condition, the temperature at the boundary changes abruptly at $t = 0$ and stays constant for all $t > 0$, i.e. it is a Heaviside function of time (the dashed line in Fig. 2).

The second approach to set the boundary condition at $x = a$ is to apply the so called convective boundary condition (= convection boundary condition) wherein we equate the energy flux at $x = a$ expressed through the thermal conductivity and the temperature gradient in the silicon to the same energy flux expressed through the transport properties and the state of the liquid system:

$$- \kappa(u(a,t))\frac{\partial u(x,t)}{\partial x}\Big|_{x=a} = c(T(t) - u(a,t)), t > 0, \tag{9}$$

where c is the mean convection heat transfer coefficient. Condition (9) holds, to a good approximation, for solid-liquid contact surfaces where the heat transport mechanism is mainly due to convection in the liquid system. For this type of boundary condition, unlike in (8), the temperature at the surface $u(a,t)$ is

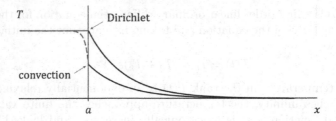

Fig. 3. The temperature profile in the tank ($x < a$) and along the silicon rod ($x \geq a$) at time t for (i) Dirichlet boundary condition and (ii) convective boundary condition (convection).

deemed essentially different from the temperature $T(t)$ in the interior of the tank. In Fig. 3 you can see a typical temperature profile for the Dirichlet and the convective boundary condition when $T_0 = 0, T_r > T_0$, and $u_0(x) = 0$. Note that, in the case of convective boundary condition, the temperature inside the tank is $T(t)$ but close to the silicon contact surface it differs from $T(t)$. Since the thermal conductivity κ of the silicon depends on the temperature, condition (9) is nonlinear. The numerical method proposed for solving the heat equation (2) was originally developed for linear boundary conditions. In this paper, we show how the method can be changed to incorporate the nonlinear condition (9). Condition (9), besides being nonlinear, is time dependent. As time increases, the function $T(t)$ approaches the constant value T_r, hence the condition approaches a certain time independent steady state condition. Thus, the unsteady state is transient and will approach a steady state. Any time dependent boundary condition that approaches continuously some time independent condition can be considered a relaxing boundary condition. Hence, condition (9) will be called *convective relaxing* boundary condition.

4   Numerical Method

In this section we briefly introduce the numerical method proposed in [18] and then we show how it can altered in order to incorporate the nonlinear condition (9). Discretizing Eq. (2) on the time mesh $t_n = n\tau, n = 1, 2, \ldots$, we get

$$\rho c_p \frac{u_n - u_{n-1}}{\tau} = \partial_u \kappa(u_n) \left(\frac{du_n}{dx} \right)^2 + \kappa(u_n) \frac{d^2 u_n}{dx^2}, \tag{10}$$

where $u_n = u_n(x)$ approximates the unknown function $u(x, t)$ at time $t = t_n$. Equation (10) is a second order ODE for the unknown function $u_n(x)$. In a more compact form (10) can be written as

$$\frac{d^2 u_n}{dx^2} = f(u_n, v_n; u_{n-1}), \tag{11}$$

where $f(u_n, v_n; u_{n-1}) = \phi_n / \kappa(u_n), \phi_n = \rho c_p (u_n - u_{n-1}) / \tau - \partial_u \kappa(u_n) v_n^2, v_n = du_n / dx$. Equation (11), together with the given boundary conditions, is a

TPBVP [21]. If $u_{n-1}(x)$ is given, we can solve the problem and obtain $u_n(x)$. Therefore, starting from the initial condition $u_0(x)$, we can solve consecutively the TPBVP for $n = 1, 2, \ldots$ and get $u_1(x), u_2(x), \ldots$ To solve the problem, the finite difference method (FDM) is used [21]–[23]. We introduce a uniform mesh on $x \in [a, b] : x_i = a + (i - 1)h, i = 1, 2, \ldots, N, h = (b - a)/(N - 1)$. The ODE (11) is discretized, using the central difference approximation

$$\frac{u_{n,i+1} - 2u_{n,i} + u_{n,i-1}}{h^2} = f(u_{n,i}, v_{n.i}; u_{n-1,i}), \ i = 2, 3, \ldots, N - 1, \qquad (12)$$

where $u_{n,i}$ is an approximation of $u_n(x_i)$ and $v_{n,i} = (u_{n,i+1} - u_{n,i-1})/(2h)$. Equations (12), together with the two equations for the boundary conditions, form a nonlinear system of N equations for the N unknowns $u_{n,i}, i = 1, 2, \ldots, N$. To solve the system, we apply the Newton method. To implement the method, we need the partial derivatives of the function f with respect to u_n and v_n. Let $f_n = f(u_n, v_n; u_{n-1})$, then:

$$q_n = \frac{\partial f_n}{\partial u_n} = \frac{1}{\kappa(u_n)}\left(-f_n \partial_u \kappa(u_n) + \frac{\partial \phi_n}{\partial u_n}\right), p_n = \frac{\partial f_n}{\partial v_n} = \frac{1}{\kappa(u_n)}\frac{\partial \phi_n}{\partial v_n}, \qquad (13)$$

where $\partial \phi_n/\partial u_n = \rho c_p/\tau - \partial_{uu}^2 \kappa(u_n)v_n^2, \partial \phi_n/\partial v_n = -2\partial_u \kappa(u_n)v_n$. Now we can implement the Newton method. First, we introduce the vector

$$\mathbf{G}_n = [G_{n,1}, G_{n,2}, \ldots G_{n,N}]^T$$

where

$$G_{n,i} = u_{n,i+1} - 2u_{n,i} + u_{n,i-1} - h^2 f_{n,i}, i = 2, 3, \ldots, N - 1. \qquad (14)$$

The components $G_{n,1}$ and $G_{n,N}$ come from the boundary condition. They are given later on in the section. The nonlinear system can be written as $\mathbf{G}_n(\mathbf{u}_n) - 0$, where $\mathbf{u}_n = [u_{n,1}, u_{n,2}, \ldots, u_{n,N}]^T$. Using $\mathbf{u}_{n-1}$ as initial guess $\mathbf{u}_n^{(0)}$, we apply Newton iteration for $k = 0, 1, \ldots$

$$\mathbf{u}_n^{(k+1)} = \mathbf{u}_n^{(k)} - (\mathbf{L}_n^{(k)})^{(-1)}\mathbf{G}_n(\mathbf{u}_n^{(k)}), \mathbf{L}_n^{(k)} = \frac{\partial \mathbf{G}_n}{\partial \mathbf{u}_n}(\mathbf{u}_n^{(k)}). \qquad (15)$$

The nonzero elements of the Jacobian $\mathbf{L}_n^{(k)}$ for rows $i = 2, 3, \ldots, N - 1$ are

$$L_{n(i,i-1)}^{(k)} = 1 + \frac{1}{2}hp_{n,i}^{(k)}, L_{n(i,i)}^{(k)} = -2 - h^2 q_{n,i}^{(k)}, L_{n(i,i+1)}^{(k)} = 1 - \frac{1}{2}hp_{n,i}^{(k)}. \qquad (16)$$

The elements of the first row are determined from the first boundary condition, and the elements of row N from the second boundary condition.

First, we show how to apply the Dirichlet relaxing boundary condition (8). The boundary condition at time level n is $u(a, t_n) - T(t_n) = 0$. Replacing $u(a, t_n)$ by $u_n(a)$ and then $u_n(a)$ by $u_{n,1}$ as required by the FDM, we get $u_{n,1} - T(t_n) = 0$. Let

$$G_{n,1} = u_{n,1} - T(t_n). \qquad (17)$$

Then, the only nonzero element of the first row of the Jacobian is

$$L_{n(1,1)}^{(k)} = \frac{\partial G_{n,1}}{\partial u_{n,1}}\Big|_{\mathbf{u}_n = \mathbf{u}_n^{(k)}} = 1. \tag{18}$$

Analogously, the convective relaxing boundary condition (9) yields

$$G_{n,1} = \kappa(u_{n,1})(u_{n,2} - u_{n,1}) + hcT(t_n) - hcu_{n,1}, \tag{19}$$

where we have used the forward finite difference to approximate $\partial u/\partial x$. The two nonzero elements of the first row of the Jacobian are

$$L_{n(1,1)}^{(k)} = \frac{\partial G_{n,1}}{\partial u_{n,1}}\Big|_{\mathbf{u}_n = \mathbf{u}_n^{(k)}} = \partial_u \kappa(u_{n,1}^{(k)})(u_{n,2}^{(k)} - u_{n,1}^{(k)}) - \kappa(u_{n,1}^{(k)}) - hc, \tag{20}$$

$$L_{n(1,2)}^{(k)} = \frac{\partial G_{n,1}}{\partial u_{n,2}}\Big|_{\mathbf{u}_n = \mathbf{u}_n^{(k)}} = \kappa(u_{n,1}^{(k)}). \tag{21}$$

For the silicon rod, where $\kappa = \kappa_0 e^{\chi u}$ [24], formula (19) becomes

$$G_{n,1} = \kappa_0 e^{\chi u_{n,1}}(u_{n,2} - u_{n,1}) + hcT(t_n) - hcu_{n,1}, \tag{22}$$

and formulas (20)–(21) become

$$L_{n(1,1)}^{(k)} = \frac{\partial G_{n,1}}{\partial u_{n,1}}\Big|_{\mathbf{u}_n = \mathbf{u}_n^{(k)}} = \chi\kappa_0 e^{\chi u_{n,1}^{(k)}}(u_{n,2}^{(k)} - u_{n,1}^{(k)}) - \kappa_0 e^{\chi u_{n,1}^{(k)}} - hc, \tag{23}$$

$$L_{n(1,2)}^{(k)} = \frac{\partial G_{n,1}}{\partial u_{n,2}}\Big|_{\mathbf{u}_n = \mathbf{u}_n^{(k)}} = \kappa_0 e^{\chi u_{n,1}^{(k)}}. \tag{24}$$

Finally, we apply the right boundary condition (4). At time level n the condition is $u(b,t_n) - \beta = 0$. Replacing $u(b,t_n)$ by $u_n(b)$ and then $u_n(b)$ by $u_{n,N}$, we get $u_{n,N} - \beta = 0$. Let

$$G_{n,N} = u_{n,N} - \beta. \tag{25}$$

Then, the only nonzero element of the last row of the Jacobian is

$$L_{n(N,N)}^{(k)} = \frac{\partial G_{n,N}}{\partial u_{n,N}}\Big|_{\mathbf{u}_n = \mathbf{u}_n^{(k)}} = 1. \tag{26}$$

Of course, at $x = b$, instead of the simple condition (4), we can apply Dirichlet relaxing boundary condition or convective relaxing boundary condition. The approach is analogous to the one described for $x = a$. In the next section we show examples with Dirichlet relaxing condition at the left or right boundary, and also at both boundaries.

5   Computer Experiments

In this section the proposed numerical approach is applied for the solution of several nonlinear heat transfer problems with relaxing boundary condition. Example 1 compares the Dirichlet with the convective relaxing boundary condition. In Example 2 and Example 3, Dirichlet relaxing boundary conditions are applied at one or both boundaries. In all examples thermal conductivity $\kappa = 0.1e^{0.5u}$ is considered. This dependence is similar to the one exhibited by silicon but the numerical values are different [24]. The density and the heat capacity are chosen to be $\rho = 1$ and $c_p = 1$. The solid rod is placed between $x = 1$ and $x = 3$. The heat equation is

$$\frac{\partial u}{\partial t} = \frac{\partial}{\partial x}\left(0.1e^{0.5u}\frac{\partial u}{\partial x}\right), x \in [1,3]. \tag{27}$$

The initial condition is $u(x,0) = 1, x \in [1,3]$.

Example 1. We choose the following parameters for the liquid containing tank system: $T_r = 2, T_0 = 1, Q = 1, V = 1$, hence $T(t) = 2 - e^{-t}$. At the right boundary the temperature is kept fixed at 1, i.e. $u(3,t) = 1$. At the left boundary, we first apply the Dririchlet relaxing boundary condition (8):

$$u(1,t) = 2 - e^{-t}, t > 0. \tag{28}$$

Then, we apply the convective relaxing boundary condition (9) with $c = 0.1$:

$$-e^{0.5u(1,t)}\frac{\partial u(x,t)}{\partial x}\Big|_{x=1} = 2 - e^{-t} - u(1,t), t > 0. \tag{29}$$

Results are shown in Fig. 4.

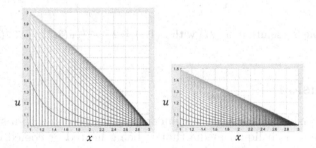

Fig. 4. Example 1 - solution of (27) for the Dirichlet relaxing BC (28) (left) and the convective relaxing BC (29) (right).

Example 2. Equation (27) is solved with the same initial condition and a Dirichlet relaxing boundary condition at the left, right, and both boundaries. First we apply $T(t) = 2 - e^{-t}$ (Fig. 5 - top line) and then $T(t) = e^{-t}$ (Fig. 5 - bottom line).

Example 3. Equation (27) is solved with the same initial condition and Dirichlet relaxing boundary conditions $u(1,t) = 2 - e^{-t}, u(3,t) = e^{-t}$. Then, the problem is solved with $u(1,t) = e^{-t}, u(3,t) = 2 - e^{-t}$. Results are shown in Fig. 6.

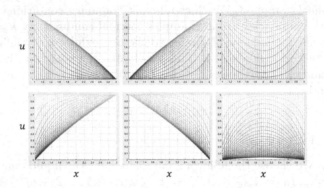

Fig. 5. Example 2 - solution of (27) with Dirichlet relaxing BC at the left, right, and both boundaries for $T(t) = 2 - e^{-t}$ (top) and $T(t) = e^{-t}$ (bottom).

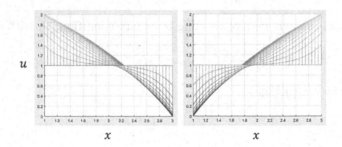

Fig. 6. Example 3 - solution of (27) with $u(1,t) = 2 - e^{-t}, u(3,t) = e^{-t}$ (left) and vice versa (right).

6   Conclusion

This paper considered unsteady nonlinear heat transfer in a thin silicon rod in thermal contact with liquid media that is being heated or cooled convectively. The mathematical model describing the system consists of nonlinear heat equation with initial condition and a Dirichlet relaxing or convective relaxing boundary condition. The convective condition is nonlinear. To solve the problem a new numerical approach was proposed that first discretizes the heat equation in time. The discretization scheme is implicit, which results in unconditional stability of the method. The resulting TPBVPs were solved by FDM with Newton method. Computer experiments were performed demonstrating the suitability of the approach for solving the nonlinear heat equation with variety of time-dependent and nonlinear conditions at the boundaries.

References

1. Carslaw, H.S., Jaeger, J.C.: Conduction of heat in solids. Oxford University Press, 2nd edn. (1986)
2. Friedman, A.: Partial differential equations of parabolic type. Prentice-Hall (1964)
3. Cannon, J.R.: The one-dimensional heat equation. Addison-Wesley (1984)
4. Evans, L.C.: Partial differential equations. Graduate Studies in Mathematics, vol. 19, 2nd edn. Providence, R.I.: American Mathematical Society (2010)
5. Powers, D.L.: Boundary value problems and partial differential. Equations, 6th edn. Boston: Academic Press (2010)
6. Smith, G.D.: Numerical solution of partial differential equations: finite difference methods. Oxford Applied Mathematics and Computing Science Series, 3rd edn. (1986)
7. Polyanin, A.D., Zhurov, A.I., Vyazmin, A.V.: Exact solutions of nonlinear heat- and mass-transfer equations. Theoret. Found. Chem. Eng. **34**(5), 403–415 (2000)
8. Sadighi, A., Ganji, D.D.: Exact solutions of nonlinear diffusion equations by variational iteration method. Comput. Math. Appl. **54**, 1112–1121 (2007)
9. Hristov, J.: An approximate analytical (integral-balance) solution to a nonlinear heat diffusion equation. Therm. Sci. **19**(2), 723–733 (2015)
10. Hristov, J.: Integral solutions to transient nonlinear heat (mass) diffusion with a power-law diffusivity: a semi-infinite medium with fixed boundary conditions. Heat Mass Transf. **52**(3), 635–655 (2016)
11. Fabre, A., Hristov, J.: On the integral-balance approach to the transient heat conduction with linearly temperature-dependent thermal diffusivity. Heat Mass Transf. **53**(1), 177–204 (2017)
12. Fabre, A., Hristov, J., Bennacer, R.: Transient heat conduction in materials with linear power-law temperature-dependent thermal conductivity: integral-balance approach. Fluid Dyn. Mater. Process. **12**(1), 69–85 (2016)
13. Grarslan, G.: Numerical modelling of linear and nonlinear diffusion equations by compact finite difference method. Appl. Math. Comput. **216**(8), 2472–2478 (2010)
14. Grarslan, G., Sari, G.: Numerical solutions of linear and nonlinear diffusion equations by a differential quadrature method (DQM). Commun. Numer. Meth, En (2009)
15. Liskovets, O.A.: The method of lines. J. Diff. Eqs. **1**, 1308 (1965)
16. Zafarullah, A.: Application of the method of lines to parabolic partial differential equations with error estimates. J. ACM **17**(2), 294–302 (1970)
17. Marucho, M.D., Campo, A.: Suitability of the Method Of Lines for rendering analytic/numeric solutions of the unsteady heat conduction equation in a large plane wall with asymmetric convective boundary conditions. Int. J. Heat Mass Transf. **99**, 201–208 (2016)
18. Filipov, S.M., Faragó, I.: Implicit Euler time discretization and FDM with Newton method in nonlinear heat transfer modeling. Math. Model. **2**(3), 94–98 (2018)
19. Faragó, I., Filipov, S.M., Avdzhieva, A., Sebestyén, G.S.: A numerical approach to solving unsteady one-dimensional nonlinear diffusion equations. In book: A Closer Look at the Diffusion Equation, pp. 1–26. Nova Science Publishers (2020)
20. Hristov, J.: On a non-linear diffusion model of wood impregnation: analysis, approximate solutions and experiments with relaxing boundary conditions. In: Singh, J., Anastassiou, G.A., Baleanu, D., Cattani, C., Kumar, D. (eds.) Advances in Mathematical Modelling, Applied Analysis and Computation. LNNS, vol. 415. Springer, Singapore (2023). https://doi.org/10.1007/978-981-19-0179-9_2

21. Ascher, U.M., Mattjeij, R.M.M., Russel, R.D.: Numerical solution of boundary value problems for ordinary differential equations. Classics in Applied Mathematics 13. SIAM (1995)
22. Filipov, S.M., Gospodinov, I.D., Faragó, I.: Replacing the finite difference methods for nonlinear two-point boundary value problems by successive application of the linear shooting method. J. Comput. Appl. Math. **358**, 46–60 (2019)
23. Faragó, I., Filipov, S.M.: the linearization methods as a basis to derive the relaxation and the shooting methods. In: A Closer Look at Boundary Value Problems, Nova Science Publishers, pp. 183–210 (2020)
24. Lienemann, J., Yousefi, A., Korvink, J.G.: Nonlinear heat transfer modeling. In: Benner, P., Sorensen, D.C., Mehrmann, V. (eds.), Dimension Reduction of Large-Scale Systems. Lecture Notes in Computational Science and Engineering, vol. 45, pp. 327–331. Springer, Heidelberg (2005). https://doi.org/10.1007/3-540-27909-1_13

Random Sequences in Vehicle Routing Problem

Mehmet Emin Gülşen[✉] and Oğuz Yayla[iD]

Institute of Applied Mathematics, Middle East Technical University,
Çankaya, 06800 Ankara, Türkiye
mehmet.gulsen@metu.edu.tr

Abstract. In this paper, we study the Capacitated Vehicle Routing
Problem (CVRP) and implemented a simulation-based algorithm with
different random number generators. The Binary-CWS-MCS algorithm
has been integrated with six different random number generators and
their variations. The random number generators used in this study gath-
ered with respect to two perspectives, the first is to compare the mostly
known and used RNGs in simulation-based studies which are Linear Congruential Generator (LCG) and its shift variant, Multiple Recursive Generator (MRG) and its shift variant and the second perspective is based
on the improvements in the random number generator algorithms which
are Mersenne Twister Pseudo Random Generator (MT) and Permuted
Congruential Generator (PCG). The results of experiments showed that
the PCG and MT pseudo random generators can generate better results
than the other random number generators.

Keywords: Monte-Carlo simulation · Random number generators ·
Capacitated vehicle routing problem

1 Introduction

Development of supply chain technologies and the increase in demand for goods
made distribution centers must arrange optimal vehicle routes to ensure the
possible lowest transportation and logistic costs. In this regard, planning efficient
routes for the distribution of goods to customers can generate important savings
for companies. Consequently, VRP is can be accepted as an important task in
many private and public concerns.

VRP has become one of the most studied problems in the field of combinatorial optimization. The first example of the problem was introduced by Dantzig
[6] in 1959 and is applied to the design of optimal routes, which seeks to serve a
number of customers with a fleet of vehicles. The objective of this problem is to
determine the optimal route to serve multiple clients, using a group of vehicles
to minimize the overall transportation cost.

VRP and its variants are classified as NP-hard problems. Because of this
reason, for finding an optimal solution, many factors are has to be considered

The original version of this chapter was revised: the last name of Mehmet Emin Gülşen
was misspelled. The correction to this chapter is available at
https://doi.org/10.1007/978-3-031-32412-3_32

I. Georgiev et al. (Eds.): NMA 2022, LNCS 13858, pp. 159–170, 2023.
https://doi.org/10.1007/978-3-031-32412-3_14

with many possibilities of permutation and combinations. VRP becomes more complex as constraints and the number of customers increase. There are multiple variations of VRP with different type of constraints such as Capacitated VRP (CVRP), Multi-depot VRP (MDVRP), Periodic VRP (PVRP), Stochastic VRP (SVRP), and VRP with Time Windows (VRPTW), among others [5].

In order to test the performance of pseudo random generators on Monte-Carlo simulation for solving CVRP instances, the Binary-CWS-MCS algorithm from the study [13] adopted. Linear Congruential Generator (LCG), Multiple Recursive Generator (MRG), Inversive Congruential Generator (ICG), Permuted Congruential Generator (PCG) and Mersenne-Twister (MT) pseudo random generators are employed for a comparison of their cost performances. The results show that the LCG and MRG pseudo random number generators are better performing than the others. In addition, variations of MRG can affect the cost performance of the solutions in the use case of Monte Carlo simulations. This is a comprehensive extension of the methods given in [7], in which only limited number of RNGs and VRP instances were tested and analysed.

The outline of the paper is as follows. In the next section, the methods including random number generators used in our implementation of pseudo random generators on Monte Carlo simulation for solving CVRP instances will be presented. Then, we will give our implementation results in Sect. 3, where we also discuss the consequences of our implementation. We conclude our paper in Sect. 4.

2   Methodology

In this section, the Monte-Carlo method and the Monte Carlo Tree Search will be expressed firstly with the employed method Binary-CWS-MCS [13] and the random number generators used in this study will be given.

2.1   Monte-Carlo Method

Monte-Carlo method first introduced by John von Neumann and Stanislaw Ulam increases the accuracy of decision making. Monte Carlo method has a wide range of applications in the industry including computational physics, computer graphics to artificial intelligence for games [2]. More details about Monte Carlo simulations and the theory can be seen in the study [3].

Monte Carlo Tree Search (MCTS). Monte Carlo Tree Search (MCTS) is one of the earliest methods in the field of artificial intelligence. MCTS is mostly used in NP-Hard problems since using deterministic models is not applicable. The summary of MCTS steps can be seen in Fig. 1.

The steps of MCTS can be examined in four different steps Selection, Expansion, Simulation and Back-propagation. The selection process starts from a root node and spans the tree based on an evaluation function. The selection of nodes iterates through the tree by getting the value of the evaluation function and continues with the maximum value generating option. The expansion step adds child nodes to the candidate solution and the process continues with the simulation step that performs strategies until a feasible solution is obtained under the given constraints of the problem.

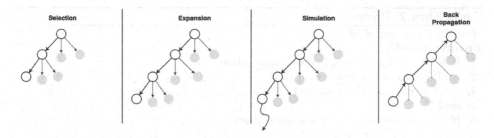

Fig. 1. Steps of Monte-Carlo Tree Search

Algorithm 1. Binary-CWS Algorithm

Require: Savings List (S), probability threshold value (p)
Ensure: routeList
1: routeList = []
2: Initialize tempList as [1,2,...,length of S]
3: **while** tempList is not empty **do**
4:     rand = Generate random number
5:     rand = rand (mod 100)
6:     **for do** i in tempList
7:         **if** rand $\geq p$*100 **then**
8:             Process (S[i],routelist)
9:             Remove i from templist
10:        **end if**
11:    **end for**
12: **end while**
13: **return** routeList

2.2 Monte-Carlo Simulation Applied on CWS

In order to compare the random number generators on CVRP instances, the binary version of the Clarke and Wright's Savings (CWS) algorithm with Monte-Carlo simulation approach called Binary-CWS-MCS is given in [13]. In order to compare the random number generators on CVRP instances, the binary version of CWS algorithm with Monte Carlo simulation approach was implemented which is given in [13]. In the CWS algorithm, the nodes and distances between them represented on the form of a matrix called as distance matrix, c_{ij}. A new matrix called the savings matrix is derived from the distance matrix with the following equation, $s_{ij} = c_{0i} + c_{0j} - c_{ij}$. The c_{ij} represents the distance cost of travelling from node i to j and s_{ij} represents the savings obtained from passing through nodes without returning to the depot. The binary version of the CWS algorithm arises from the probability threshold p of the process step from the CWS, see Algorithm 1. The process step of the CWS algorithm is detailed in Algorithm 2. As given in Algorithm 1, the required inputs are savings list S, which is built such that within S the edges are ordered by descending saving value, and probability p. The S is the representation of the savings matrix in

---

Algorithm 2. Process

Require: i, j, routeList
Ensure: routeList
 1: **if** both i and j not assigned to a route **then**
 2:     Initialize a new route with (i, j)
 3:     Add new route to routeList
 4: **end if**
 5: **if** i or j exists at the end of a route **then**
 6:     Link (i, j) in that route
 7: **end if**
 8: **if** both i and j exists at the end of route **then**
 9:     Merge two routes into one route
10:     Remove the old routes from routeList
11: **end if**
12: **return** routeList

---

the form of (i, j, s_{ij}) and the p value represents the probability threshold. The Binary-CWS-MCS algorithm iterates through the savings matrix and for each iteration it computes the average of bulk operation for two different directions to manage the selection of the next customer on the route. The simulation continues until the demand of each node is satisfied.

Algorithm 3. Binary-CWS-MCS Algorithm

Require: Distance Matrix C, number of nodes n
Ensure: routeList
 1: Generate Savings Matrix(C, n)
 2: listOrdered = order list descending with respect to savings s_{ij}
 3: routeList = []
 4: **while** listOrdered is not empty **do**
 5:     t_1: average of 1000 calls of score Binary-CWS(s,list-ordered)
 6:     t_2: average of 1000 calls of score Binary-CWS($s + 1$,list-ordered)
 7:     **if** $t_1 \geq t_2$ **then**
 8:         **Process(list-ordered[s], route list)**
 9:         Remove list-ordered[s] from list-ordered
10:     **end if**
11: **end while**
12: **return** routeList

The critical decision for the Binary-CWS method which is presented in Algorithm 1, is to set the correct range for the p value. Because if the range for the p value is set to close 0, the cost value of the Binary-CWS will converges to the cost value of CWS algorithm and reduces the improvement for the simulation, and if the range for the p value set to larger values both the time consumption of the algorithm will increase and the cost value of the routes will be resulted on unsatisfactory performance compared to CWS cost/distance value.

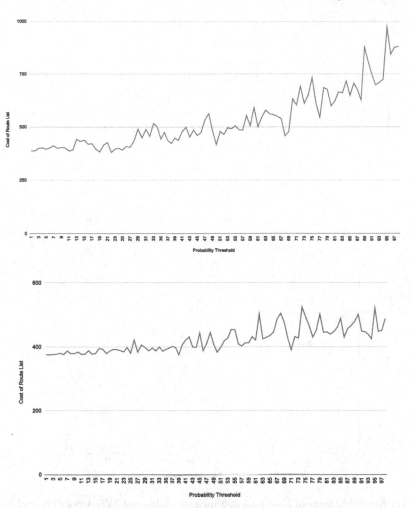

Fig. 2. Probability tests of E-n22-k4 instance with Binary-CWS Algorithm and Binary-CWS-MCS Algorithm, respectively

Two different CVRP instances E-n22-k4 and E-n51-k5 are chosen to test the p-value in order to distinguish a better range. The E-n22-k4 (1 depot, 21 clients, 4 vehicles, vehicle capacity = 100) and E-n51-k5 (1 depot, 50 clients, 5 vehicles, vehicle capacity = 160) instances have been tested with both Binary-CWS and Binary-CWS-MCS algorithms. As can be seen in Figs. 2 and 3, the cost/distance value is mostly biased to the p value. Also, it can be seen from the figures, when p-value increases, the cost of the routes created increases. Thus, the p-value has been chosen such that $0.05 \leq p \leq 0.26$.

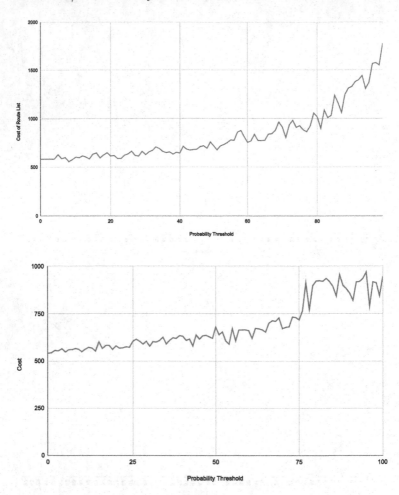

Fig. 3. Probability tests of E-n51-k5 instance with Binary-CWS Algorithm and Binary-CWS-MCS Algorithm, respectively

2.3   Random Number Generators

Random numbers are utilized in many different applications, including modelling, encryption, sampling, numerical analysis and simulation-based algorithms. There are different types of random number generators for different purposes of usages. Five different random number generators with different parameters have been employed in this study to test the quality of random number generators in the Monte Carlo simulation.

Linear Congruential Generator (LCG). Firstly introduced by W.E.Thomson in 1958 [14]. LCG is the most widespread and easiest to comprehend and implement pseudo random number generator. It is defined by recursion as given in Eq. 1.

$$X_{n+1} = (a.X_n + c) \mod m \tag{1}$$

the modulus $m, 0 < m$,

the multiplier $a, 0 < a < m$,

the increment $c, 0 \le c < m$,

seed of the generator $X_0, 0 \le X_0 < m$, $\tag{2}$

Although LCGs can generate pseudo random numbers that pass formal criteria for randomness, the output quality is highly dependent on the values of the parameters m and a. For instance, $a = 1$ and $c = 1$ generate a basic modulo-m counter with a lengthy period but it is not a random sequence. The detailed analysis and the tables of known good parameters can be found in [9,12]. The period and parameters used for LCG employed in this study are given in Table 1, which are taken from [9].

Table 1. LCG Parameters [9]

a	b	$Period$
18145460002477866997	1	2^{64}

LCG with Shift Transformation (LCGS). Applying shift transformation is a way to improve the randomness of sequences. If we assume that the output of a given LCG is r and the output of transformation operation is q, the operation of LCG in combination with shift transformation can be expressed as the following operations. The $x >> n$ operation denotes for n bit-shift of x to the right and $x << n$ operations denote for n bit-shift x to the left.

$$t_0 = r$$
$$t_1 = t_0 \oplus (t_0 >> 17)$$
$$t_2 = t_1 \oplus (t_0 << 31)$$
$$t_3 = t_2 \oplus (t_0 >> 8)$$
$$q = t_3$$

The LCG parameters used in this study for LCG with shift transformation are the same as LCG and can be seen in Table 1.

Multiple Recursive Generator (MRG). A common type of random number generator for simulation based methods, are based on a linear recurrence of the form given in (3). The prime modulus m was either chosen as Mersenne-Prime

Table 2. MRG Parameters with Mersenne-Prime Modulus [8]

	a_1	a_2	a_3	a_4	a_5	$Period$	m
$mrg2$	1498809829	1160990996				2^{62}	$2^{31} - 1$
$mrg3$	2021422057	1826992351	1977753457			2^{93}	$2^{31} - 1$
$mrg4$	2001982722	1412284257	1155380217	1668339922		2^{124}	$2^{31} - 1$
$mrg5$	107374182	0	0	0	104480	2^{155}	$2^{31} - 1$

Table 3. MRG Parameters with Sophie-Germain Prime Modulus [8]

	a_1	a_2	a_3	a_4	a_5	$Period$	m
$mrg3s$	2025213985	1112953677	2038969601			2^{93}	$2^{31} - 21069$
$mrg5s$	1053223373	1530818118	1612122482	133497989	573245311	2^{155}	$2^{31} - 22641$

or Sophie-Germain Prime. The parameters used for generating random numbers with MRG with chosen Mersenne-Prime modulus can be seen in Table 2 and the chosen Sophie-Germain Prime modulus can be seen in Table 3.

$$X_n = (a_1.X_{n-1} + a_2.X_{n-2} + ... + a_k.X_{n-k}) \mod m \qquad (3)$$

There are several studies to find good parameters for both MRGs and LCGs, the parameters used in this study were obtained from the research that L'Ecuyer et al. made [8].

Mersenne Twister Pseudo Random Number Generator (MT).
Mersenne twister algorithm was invented in 1997 by Makoto Matsumoto and Takuji Nishimura [10]. It has proven that the period of $2^{19937} - 1$ can be established with good parameters and is proven to be equidistributed in (up to) 623 dimensions (for 32-bit values), and runs faster than other statistically reasonable generators. The Mersenne twister algorithm mostly used and chosen for simulations and models that requires random number generators. The Mersenne Twister algorithm is used as the default random number generator for many software and operating systems. The most widely used version is based on the Mersenne prime $2^{19937} - 1$ with the parameters given in Table 4.

Permuted Congruential Generator (PCG). The PCG family of random number generators were developed by O'Neill in 2014 [11]. PCG variants have a common approach for generating random numbers, an RNG (LCG or MCG) is employed for internal state generation and the permutation function is applied to the output of the internal state generator with truncation of the value. The variants of PCG are named with the permutation function applied to the internal generator. The members of PCG family algorithms are expressed as PCG-XSH-RR (XorShift, Random Rotation), PCG-XSH-RR (XorShift, Random Rotation),

Table 4. Mersenne-Twister Parameters [10]

w	32
n	624
m	397
r	31
a	$(9908B0DF)_{16}$
u	11
d	$(FFFFFFFF)_{16}$
s	7
b	$(9D2C5680)_{16}$
t	15
c	$(EFC6000)_{16}$
l	18

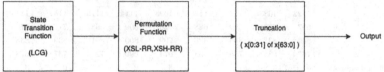

Fig. 4. Permuted Congruential Generator

PCG-XSH-RS (XorShift, Random Shift) and PCG-XSL-RR (XorShift Low bits, Random Rotation). The random number generation process of the PCG algorithm can be generalized as given in Fig. 4.

The LCG function is used as the internal state generator of PCG-XSL-RR and the permutation function of the PCG is given in the Eq. 4. The >>> operation denotes the right rotation and the $\oplus$ denotes the XOR operation.

$$S_{i+1} = aS_i + c \mod 2^{128} \tag{4}$$

$$X_i = \left(S_i[0:64] \oplus S_i[64:128] >>> S_i[122:128] \right) \tag{5}$$

3   Results and Experiments

The Binary-CWS method has been integrated into the MCTS method as the selection process, where the simulation is accomplished with the bulk 1000-trials given in Algorithm 3. This method is called as Binary-CWS-MCS. Then, the comparison was made based on the cost performance of the Binary-CWS-MCS under different instances and random number generators. LCG, LCGS, MRG, MRGS, MT and PCG pseudo random number generators have been employed. The PRNGs have been implemented from two different well-known PRNG libraries, and more information about the PRNGs can be obtained from [1]. The

Table 5. Comparison of Pseudo Random Number Generators

Instance	LCG	LCGS	MRG2	MRG3	MRG4	MRG5	MRG3S	MRG5S	MT	PCG64
E-n22-k4	388	376	392	381	388	379	376	379	393	379
E-n33-k4	844	853	864	847	865	877	882	921	863	854
E-n31-k7	509	471	505	535	491	552	535	538	540	469
E-n13-k4	257	257	257	257	257	257	248	257	257	275
E-n30-k3	514	548	511	550	510	533	531	525	505	508
E-n23-k3	595	618	636	593	591	607	594	610	587	583
E-n51-k5	563	578	539	550	560	558	577	567	585	567
E-n76-k7	715	743	763	754	756	750	754	750	745	736
E-n76-k8	803	809	803	782	802	774	799	826	796	847
B-n39-k5	562	567	566	578	580	574	582	577	571	583
B-n35-k5	989	976	988	974	965	992	974	989	992	976
B-n31-k5	682	684	682	681	685	685	694	688	691	681
B-n43-k6	762	770	784	781	777	777	771	768	796	776
B-n45-k5	782	764	769	775	769	756	790	796	799	792
A-n63-k10	1382	1400	1370	1343	1366	1350	1397	1394	1343	1391
A-n60-k9	1394	1374	1383	1398	1377	1417	1424	1415	1390	1394
A-n55-k9	1122	1083	1118	1129	1098	1114	1116	1152	1148	1134
A-n33-k5	697	702	699	688	726	699	698	716	676	707
A-n32-k5	842	832	808	836	799	800	861	859	835	850
A-n37-k5	702	688	735	698	674	699	724	737	678	725
Number of best solutions	4	3	1	2	3	2	2	0	3	3
Number of worst solutions	0	1	2	1	1	2	3	3	5	3
Avg. loss against best	-2.36%	-2.76%	-3.07%	-3.07%	-2.44%	-3.45%	-3.82%	-4.54%	-3.69%	-2.97%
Avg. loss against median	0.72%	0.31%	0.02%	0.07%	0.64%	-0.30%	-0.66%	-1.37%	-0.53%	0.11%
Avg. loss against avg.	0.82%	0.42%	0.13%	0.16%	0.74%	-0.20%	-0.57%	-1.27%	-0.44%	0.21%

consideration for choosing he PRNGs is to both test the mostly used PRNGs on Monte-Carlo Simulation and adapt new methods into the field. Since MCTS on larger problem instances takes long time, and comparing the PRNGs also requires to run the method multiple times, we choose A, B and E instance sets from [4]. The comparison is made by using five different cases: the number of best solutions, the number of worst solutions, the loss against the median cost value (Avg. loss against median), the loss against the average cost value (Avg. loss against average) and the average loss against to best solution (Avg. loss against best). As seen in Table 5, the best performing random number generator is the MRG4 followed by LCG by concerning the number of best solutions. If the number of worst solutions considered as the performance indicator, it can be seen that the minimum value is 0 obtained with usage of LCG and followed with value of 1 which appears on three different PRNG:, MRG2, MRG3 and MRG4. If we examine the average loss against to best solution parameter, minimum value appears in MRG4 and LCG tests, also it is important to note that MRG4 performs better compared to LCG considering the number of best solutions. MRG4 also performs better than LCG considering the average loss against median solution and the average loss against the average solution. The table also shows that shift transformation variants of LCG and MRG are not as good as classical LCG and MRG. It can also can be examined that using different parameters for generating pseudo random numbers using MRG can produce better results if we compare with MRG2, MRG3, MRG4 and MRG5. It is also important to note that while both MRG3S and MRG5S use a shift transformation, MRG3S generates better solutions based on the cost function.

4   Conclusion

In conclusion, the experiments are conducted based on the pseudo random number generators (PRNGs) and their variants on three different CVRP instance sets and on twenty different instances. The random number generators used in this study are chosen with respect to two perspective. The first one is to compare the mostly known and used PRNGs on simulation based studies. For this purpose, we have implemented the library known as Tina's PRNG library [1]. We have also implemented MT and PCG for their performance comparison. Our results have shown that the choice of PRNG affects the performance of the solutions. As a result of experiments, the best performing PRNG for the given set of instances is the LCG. If we compare the number of best solutions for LCG and LCG with shift transformation, LCG concluded on better results than the LCG with shift transformation; in addition, the average loss values of LCG are better. On the other hand, MRG generator has produced better results than the MRG with shift transformation variants. PCG and MT were as good as other PRNGs on the concern of both number of best solutions value and average loss parameters.

The research in this manner can be continued by both adding a parallel computing mechanism to the simulation process for improving the time consumption. Also, other benchmark instances with larger problem size can be tested with different random number generators.

References

1. Bauke, H., Mertens, S.: Random numbers for large-scale distributed monte Carlo simulations. Phys. Rev. E: Stat., Nonlin, Soft Matter Phys. **75**(6), 066701 (2007)
2. Bird, G.: Monte-Carlo simulation in an engineering context. Prog. Astronaut. Aeronaut. **74**, 239–255 (1981)
3. Chan, W.K.V.: Theory and applications of monte Carlo simulations. BoD-Books on Demand (2013)
4. Charles, P.: Capacitated vehicle routing problem library. http://vrp.galgos.inf.puc-rio.br/index.php/en/updates (2014)
5. Cordeau, J.F., Laporte, G., Savelsbergh, M.W., Vigo, D.: Vehicle routing. Handbooks Oper. Res. Management Sci. **14**, 367–428 (2007)
6. Dantzig, G.: Linear programming and extensions. Princeton University Press (2016)
7. Demirci, I.E., Özdemir, Ş.E., Yayla, O.: Comparison of randomized solutions for constrained vehicle routing problem. In: 2020 International Conference on Electrical, Communication, and Computer Engineering (ICECCE), pp. 1–6. IEEE (2020)

8. L'ecuyer, P.: Tables of linear congruential generators of different sizes and good lattice structure. Math. Comput. **68**(225), 249–260 (1999)
9. L'Ecuyer, P., Blouin, F., Couture, R.: A search for good multiple recursive random number generators. ACM Trans. Model. Comput. Simul. (TOMACS) **3**(2), 87–98 (1993)
10. Matsumoto, M., Nishimura, T.: Mersenne twister: a 623-dimensionally equidistributed uniform pseudo-random number generator. ACM Trans. Model. Comput. Simul. (TOMACS) **8**(1), 3–30 (1998)
11. O'Neill, M.E.: PCG: a family of simple fast space-efficient statistically good algorithms for random number generation. Tech. Rep. HMC-CS-2014-0905, Harvey Mudd College, Claremont, CA (2014)
12. Steele, G.L., Vigna, S.: Computationally easy, spectrally good multipliers for congruential pseudorandom number generators. Softw. Pract. Exper. **52**(2), 443–458 (2022)
13. Takes, F., Kosters, W.A.: Applying monte Carlo techniques to the capacitated vehicle routing problem. In: Proceedings of 22th Benelux Conference on Artificial Intelligence (BNAIC 2010) (2010)
14. Thomson, W.: A modified congruence method of generating pseudo-random numbers. Comput. J. **1**(2), 83 (1958)

A Database of High Precision Trivial Choreographies for the Planar Three-Body Problem

I. Hristov[1]([✉])[iD], R. Hristova[1,2][iD], I. Puzynin[3][iD], T. Puzynina[3][iD],
Z. Sharipov[3][iD], and Z. Tukhliev[3][iD]

[1] Faculty of Mathematics and Informatics, Sofia University "St. Kliment Ohridski",
Sofia, Bulgaria
`ivanh@fmi.uni-sofia.bg`
[2] Institute of Information and Communication Technologies, Bulgarian Academy
of Sciences, Sofia, Bulgaria
[3] Meshcheryakov Laboratory of Information Technologies, Joint Institute
for Nuclear Research, Dubna, Russia

Abstract. Trivial choreographies are special periodic solutions of the planar three-body problem. In this work we use a modified Newton's method based on the continuous analog of Newton's method and a high precision arithmetic for a specialized numerical search for new trivial choreographies. As a result of the search we computed a high precision database of 462 such orbits, including 397 new ones. The initial conditions and the periods of all found solutions are given with 180 correct decimal digits. 108 of the choreographies are linearly stable, including 99 new ones. The linear stability is tested by a high precision computing of the eigenvalues of the monodromy matrices.

Keywords: Three-body problem · Trivial choreographies · Modified Newton's method · High precision arithmetic

1 Introduction

A choreography is a periodic orbit in which the three bodies move along one and the same trajectory with a time delay of $T/3$, where T is the period of the solution. A choreography is called trivial if it is a satellite (a topological power) of the famous figure-eight choreography [1,2]. Trivial choreographies are of special interest because many of them are expected to be stable like the figure-eight orbit. About 20 new trivial choreographies with zero angular momentum and bodies with equal masses are found in [3–5], including one new linearly stable choreography, which was the first found linearly stable choreography after the famous figure-eight orbit. Many three-body choreographies (345) are also found in [6], but they are with nonzero angular momentum and an undetermined topological type.

In our recent work [7] we made a purposeful numerical search for figure-eight satellites (not necessarily choreographies). A purposeful search means, that we

I. Georgiev et al. (Eds.): NMA 2022, LNCS 13858, pp. 171–180, 2023.
https://doi.org/10.1007/978-3-031-32412-3_15

do not consider the entire domain of possible initial velocities, but a small part of it, where there is a concentration of figure-eight satellites and we use a finer search grid. For numerical search we used a modification of Newton's method with a larger domain of convergence. The three-body problem is well known with the sensitive dependence on the initial conditions. To overcome the obstacle of dealing with this sensitivity and to follow the trajectories correctly for a long time, we used as an ODE solver the high order multiple precision Taylor series method [8–10]. As a result we found over 700 new satellites with periods up to 300 time units, including 45 new choreographies. 7 of the newly found choreographies are shown to be linearly stable, bringing the number of the known linearly stable choreographies up to 9.

This work can be regarded as a continuation of our recent work [7]. Now we make a specialized numerical search for new trivial choreographies by using the permuted return proximity condition proposed in [4]. We consider the same searching domain and the same searching grid step as in [7]. Considering pretty long periods (up to 900 time units), which are much longer than those in the previous research, allows us to compute a high precision database of 462 trivial choreographies, including 397 new ones. 99 of the newly found choreographies are linearly stable, so the number of the known linearly stable choreographies now rises to 108.

2   Differential Equations Describing the Bodies Motion

The bodies are with equal masses and they are treated as point masses. A planar motion of the three bodies is considered. The normalized differential equations describing the motion of the bodies are:

$$\ddot{r}_i = \sum_{j=1, j \neq i}^{3} \frac{(r_j - r_i)}{\|r_i - r_j\|^3}, \quad i = 1, 2, 3. \tag{1}$$

The vectors r_i, $\dot{r}_i$ have two components: $r_i = (x_i, y_i)$, $\dot{r}_i = (\dot{x}_i, \dot{y}_i)$. The system (1) can be written as a first order one this way:

$$\dot{x}_i = vx_i, \quad \dot{y}_i = vy_i, \quad \dot{vx}_i = \sum_{j=1, j \neq i}^{3} \frac{(x_j - x_i)}{\|r_i - r_j\|^3}, \quad \dot{vy}_i = \sum_{j=1, j \neq i}^{3} \frac{(y_j - y_i)}{\|r_i - r_j\|^3}, \quad i = 1, 2, 3. \tag{2}$$

We solve numerically the problem in this first order form. Hence we have a vector of 12 unknown functions $X(t) = (x_1, y_1, x_2, y_2, x_3, y_3, vx_1, vy_1, vx_2, vy_2, vx_3, vy_3)^\top$. Let us mention that this first order system actually coincides with the Hamiltonian formulation of the problem.

We search for periodic planar collisionless orbits as in [3,4]: with zero angular momentum and symmetric initial configuration with parallel velocities:

$$(x_1(0), y_1(0)) = (-1, 0), \quad (x_2(0), y_2(0)) = (1, 0), \quad (x_3(0), y_3(0)) = (0, 0),$$
$$(vx_1(0), vy_1(0)) = (vx_2(0), vy_2(0)) = (v_x, v_y), \tag{3}$$
$$(vx_3(0), vy_3(0)) = -2(vx_1(0), vy_1(0)) = (-2v_x, -2v_y).$$

The velocities $v_x \in [0,1], v_y \in [0,1]$ are parameters. We denote the periods of the orbits with T. So, our goal is to find triplets (v_x, v_y, T) for which the periodicity condition $X(T) = X(0)$ is fulfilled.

3   Numerical Searching Procedure

The numerical searching procedure consists of three stages. During the first stage we search for candidates for correction with the modified Newton's method, i.e. we compute initial approximations of the triplets (v_x, v_y, T). In what follows we will use the same notation for v_x, v_y, T and their approximations. We introduce a square 2D searching grid with stepsize $1/4096$ for the parameters v_x, v_y in the same searching domain as in [7] (the domain will be shown later in Sect. 5). We simulate the system (2) at each grid point (v_x, v_y) up to a prefixed time $T_0 = 300$. For an ODE solver we use the high order multiple precision Taylor series method [8–10] with a variable stepsize strategy from [11]. Because we concentrate only on choreographies (that is why we call this search specialized), we take as in [4] the candidates to be the triplets (v_x, v_y, T) for which the cyclic permutation return proximity function R_{cp} (the minimum):

$$R_{cp}(v_x, v_y, T_0) = \min_{1 < t \le T_0} \|\hat{P} X(t) - X(0)\|_2$$

is obtained at $t = T/3$ and is less than 0.1. We also set the constraint that R_{cp} has a local minimum on the grid for v_x, v_y. Here $\hat{P}$ is a cyclic permutation of the bodies' indices. Using cyclic permutation return proximity instead of the standard return proximity, reduces three times the needed integration time at the first stage (now we obtain candidates with periods T up to 900).

During the second stage we apply the modified Newton's method, which has a larger domain of convergence than the classic Newton's method. Convergence during this stage means that a periodic orbit is found. The following linear algebraic system with a 12×3 matrix for the corrections $\Delta v_x, \Delta v_y, \Delta T$ has to be solved at each iteration step [12].

$$
\begin{pmatrix}
\frac{\partial x_1}{\partial v_x}(T) & \frac{\partial x_1}{\partial v_y}(T) & \dot{x}_1(T) \\
\frac{\partial y_1}{\partial v_x}(T) & \frac{\partial y_1}{\partial v_y}(T) & \dot{y}_1(T) \\
\frac{\partial x_2}{\partial v_x}(T) & \frac{\partial x_2}{\partial v_y}(T) & \dot{x}_2(T) \\
\frac{\partial y_2}{\partial v_x}(T) & \frac{\partial y_2}{\partial v_y}(T) & \dot{y}_2(T) \\
\frac{\partial x_3}{\partial v_x}(T) & \frac{\partial x_3}{\partial v_y}(T) & \dot{x}_3(T) \\
\frac{\partial y_3}{\partial v_x}(T) & \frac{\partial y_3}{\partial v_y}(T) & \dot{y}_3(T) \\
\frac{\partial vx_1}{\partial v_x}(T) - 1 & \frac{\partial vx_1}{\partial v_y}(T) & \dot{vx}_1(T) \\
\frac{\partial vy_1}{\partial v_x}(T) & \frac{\partial vy_1}{\partial v_y}(T) - 1 & \dot{vy}_1(T) \\
\frac{\partial vx_2}{\partial v_x}(T) - 1 & \frac{\partial vx_2}{\partial v_y}(T) & \dot{vx}_2(T) \\
\frac{\partial vy_2}{\partial v_x}(T) & \frac{\partial vy_2}{\partial v_y}(T) - 1 & \dot{vy}_2(T) \\
\frac{\partial vx_3}{\partial v_x}(T) + 2 & \frac{\partial vx_3}{\partial v_y}(T) & \dot{vx}_3(T) \\
\frac{\partial vy_3}{\partial v_x}(T) & \frac{\partial vy_3}{\partial v_y}(T) + 2 & \dot{vy}_3(T)
\end{pmatrix}
\begin{pmatrix}
\Delta v_x \\
\Delta v_y \\
\Delta T
\end{pmatrix}
=
\begin{pmatrix}
x_1(0) - x_1(T) \\
y_1(0) - y_1(T) \\
x_2(0) - x_2(T) \\
y_2(0) - y_2(T) \\
x_3(0) - x_3(T) \\
y_3(0) - y_3(T) \\
vx_1(0) - vx_1(T) \\
vy_1(0) - vy_1(T) \\
vx_2(0) - vx_2(T) \\
vy_2(0) - vy_2(T) \\
vx_3(0) - vx_3(T) \\
vy_3(0) - vy_3(T)
\end{pmatrix}
\tag{4}
$$

For the classic Newton's method we correct to obtain the next approximation this way:

$$v_x := v_x + \Delta v_x, \quad v_y := v_y + \Delta v_y, \quad T := T + \Delta T.$$

For the modified Newton's method based on the continuous analog of Newton's method [13] we introduce a parameter $\tau_k : 0 < \tau_k \leq 1$, where k is the number of the iteration. Now we correct this way:

$$v_x := v_x + \tau_k \Delta v_x, \quad v_y := v_y + \tau_k \Delta v_y, \quad T := T + \tau_k \Delta T.$$

Let R_k be the residual $\|X(T) - X(0)\|_2$ at the k-th iteration. With a given τ_0, the next τ_k is computed by the following adaptive algorithm [13]:

$$\tau_k = \begin{cases} \min(1, \ \tau_{k-1} R_{k-1}/R_k), & R_k \leq R_{k-1}, \\[2mm] \max(\tau_0, \ \tau_{k-1} R_{k-1}/R_k), & R_k > R_{k-1}. \end{cases} \tag{5}$$

The value $\tau_0 = 0.2$ is chosen in this work. We iterate until the value R_k at some iteration becomes less than some tolerance or the number of the iterations becomes greater than some number *maxiter* to detect divergence. The modified Newton's method has a larger domain of convergence than the classic Newton's method, allowing us to find more choreographies for a given search grid. To compute the matrix elements in (4), a system of 36 ODEs (the original 12 differential equations plus the 24 differential equations for the partial derivatives with respect to the parameters v_x and v_y) has to be solved. The high order multiple precision Taylor series method is used again for solving this 36 ODEs system. The linear algebraic system (4) is solved in linear least square sense using QR decomposition based on Householder reflections [14].

During the third stage we apply the classic Newton's method with a higher precision in order to specify the solutions with more correct digits (180 correct digits in this work). This stage can be regarded as some verification of the found periodic solutions since we compute the initial conditions and the periods with many correct digits and we check the theoretical quadratic convergence of the Newton's method.

During the first and the second stage the high order multiple precision Taylor series method with order 154 and 134 decimal digits of precision is used. During the third stage we make two computations. The first computation is with 242-nd order Taylor series method and 212 digits of precision and the second computation is for verification - with 286-th order method and 250 digits of precision. The most technical part in using the Taylor series method is the computations of the derivatives for the Taylor's formula, which are done by applying the rules of automatic differentiation [15]. We gave all the details for the Taylor series method, particularly we gave all needed formulas, based on the rules of the automatic differentiation in our work [16].

4   Linear Stability Investigation

The linear stability of a given periodic orbit $X(t), 0 \le t \le T$, is determined by the eigenvalues λ of the 12×12 monodromy matrix $M(X;T)$ [17]:

$$M_{ij}[X;T] = \frac{\partial X_i(T)}{\partial X_j(0)}, \quad M(0) = I.$$

The elements M_{ij} of the monodromy matrix M are computed in the same way as the partial derivatives in the system (4) - with the multiple precision Taylor series method using the rules of automatic differentiation (see for details [16]). The eigenvalues of M come in pairs or quadruplets: $(\lambda, \lambda^{-1}, \lambda^*, \lambda^{*-1})$. They are of four types:

1) Elliptically stable - $\lambda = \exp(\pm 2\pi i \nu)$, where $\nu > 0$ (real) is the stability angle. In this case the eigenvalues are on the unit circle. Angle ν describes the stable revolution of adjacent trajectories around a periodic orbit.
2) Marginally stable - $\lambda = \pm 1$.
3) Hyperbolic - $\lambda = \pm \exp(\pm \mu)$, where $\mu > 0$ (real) is the Lyapunov exponent.
4) Loxodromic - $\lambda = \exp(\pm \mu \pm i\nu)$, μ, ν (real).

Eight of the eigenvalues of M are equal to 1 [17]. The other four determine the linear stability. Here we are interested in elliptically stable orbits, i.e. the four eigenvalues to be $\lambda_j = \exp(\pm 2\pi i \nu_j), \nu_j > 0, j = 1, 2$. For computing the eigenvalues we use a Multiprecision Computing Toolbox [18] for MATLAB [19]. First the elements of M are obtained with 130 correct digits and then two computations with 80 and 130 digits of precision are made with the toolbox. The four eigenvalues under consideration are verified by a check for matching the first 30 digits of them and the corresponding condition numbers obtained by the two computations (with 80 and 130 digits of precision).

5   Numerical Results

To classify the periodic orbits into topological families we use a topological method from [20]. Each family corresponds to a different conjugacy class of the free group on two letters (a, b). Satellites of figure-eight correspond to free group elements $(abAB)^k$ for some natural power k. For choreographies the power k can not be divisible by 3 (see [3]). We use "the free group word reading algorithm" from [21] to obtain the free group elements. Together with the triplet (v_x, v_y, T), we compute the scale-invariant period T^*. T^* is defined as $T^* = T|E|^{\frac{3}{2}}$, where E is the energy of our initial configuration: $E = -2.5 + 3(v_x{}^2 + v_y{}^2)$. Equal T^* for two different initial conditions means two different representations of the same solution (the same choreography). Some of the choreographies are presented by two different initial conditions.

We found 462 trivial choreographies in total (including Moore's figure-eight orbit and old choreographies, and counting different initial conditions as different solutions). 397 of the solutions are new (not included in [3–5,7]). For each found solution we computed the power k and the four numbers (v_x, v_y, T, T^*) with 180 correct digits. This data can be seen in [22] together with the plots of the trajectories in the real $x - y$ plane.

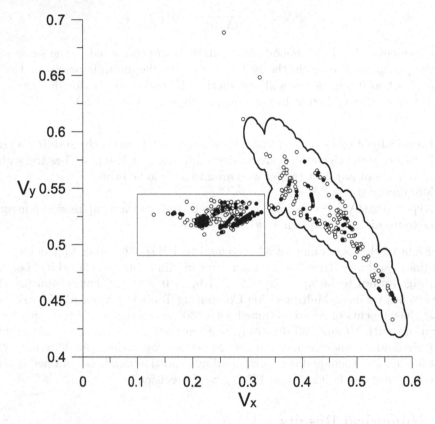

Fig. 1. Initial velocities for all found solutions, black points are the linearly stable solutions, white points - unstable (more precisely, not confirmed to be linearly stable).

As a result of computing the eigenvalues of the monodromy matrices we obtain that 108 of the choreographies are linearly stable (99 new ones). These 108 choreographies correspond to 150 initial conditions (some of the choreographies are presented by two different initial conditions and have the same T^*). All stability angles $\nu_{1,2}$ with 30 correct digits for the linearly stable solutions are given in a table in [22]. The distribution of the initial condition points can be seen in Fig. 1 (the black points are the linearly stable solutions, the white points – the unstable ones (more precisely, not confirmed to be linearly stable)). The searching domain is the same as in [7] and consists of the rectangle $[0.1, 0.33] \times [0.49, 0.545]$ plus the domain with the curved boundary. The three points out of the searching domain are previously found solutions.

Table 1. Data with 17 correct digits for a pair of linearly stable and hyperbolic-elliptic solutions.

N	v_x	v_y	T	T^*	k
119	0.41817368353651279e0	0.54057212735770067e0	0.52133539095545824e3	0.600424230253006803e3	65
120	0.26562094559259036e0	0.5209803403964781e0	0.33548942966568876e3	0.600424230253006829e3	65

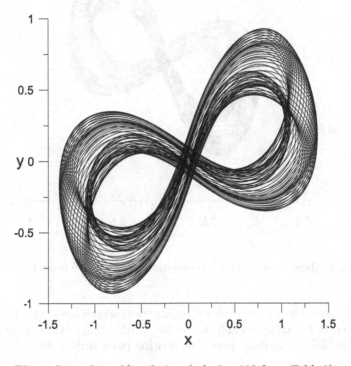

Fig. 2. Linearly stable solution (solution 119 from Table 1).

Analyzing the linear stability data, we observe 13 pairs of solutions with the same power k and very close T^*, where one is linearly stable and the other is of hyperbolic-elliptic type. The pairs have the following property: "The stability angle of the elliptic eigenvalues of the hyperbolic-elliptic type solution is very close to the larger stability angle of the linearly stable solution, the smaller stability angle of the linearly stable solution is close to zero and the larger hyperbolic eigenvalue of the hyperbolic-elliptic type solution is greater than one but very close to one". For example, there exists a pair of solutions with $\nu_1 = 0.2550119442211337538756666925693$, $\nu_2 = 2.19223274459622941216216635818e-05$ and $\nu = 0.255011941995861150357102898351$, $\lambda = 1.0001377515325471896758$ 5223182. The initial conditions and the periods for this pair can be seen in Table 1. The solutions are those with numbers 119 and 120 in [22]. The trajectories of the three bodies in the real $x - y$ plane can be seen in Fig. 2 and Fig. 3. A table with the linear stability data for all the 13 pairs can be seen in [22].

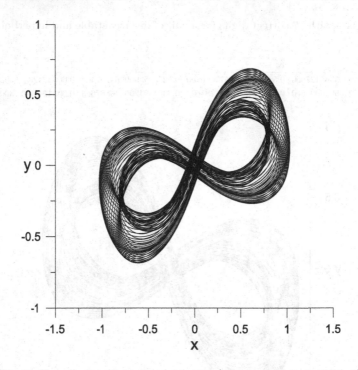

Fig. 3. Hyperbolic-elliptic type solution (solution 120 from Table 1).

The extensive computations for the numerical search are performed in "Nestum" cluster, Sofia, Bulgaria [23], where the GMP library (GNU multiple precision library) [24] for multiple precision floating point arithmetic is installed.

6   Conclusions

A modified Newton's method with high precision is successfully used for a specialized numerical search of new trivial choreographies for the planar three-body problem. Considering pretty long periods (up to 900) allows us to compute a high precision database of 462 solutions (397 new ones). 99 of the newly found solutions are linearly stable, bringing the number of the known linearly stable trivial choreographies from 9 to 108.

Acknowledgements. We greatly thank for the opportunity to use the computational resources of the "Nestum" cluster, Sofia, Bulgaria. We would also like to thank Veljko Dmitrashinovich from Institute of Physics, Belgrade University, Serbia for a valuable e-mail discussion and advice, and his encouragement to continue our numerical search for new periodic orbits.

References

1. Moore, Christopher: Braids in classical gravity. Phys. Rev. Lett **70**(24), 3675–3679 (1993)
2. Chenciner, A., Montgomery, R.: A remarkable periodic solution of the three-body problem in the case of equal masses. Ann. Math. **152**, 881–901 (2000)
3. Shuvakov, Milovan: Numerical search for periodic solutions in the vicinity of the figure-eight orbit: slaloming around singularities on the shape sphere. Celest. Mech. Dyn. Astron. **119**(3), 369–377 (2014)
4. Shuvakov, Milovan, Shibayama, Mitsuru: Three topologically nontrivial choreographic motions of three bodies. Celest. Mech. Dyn. Astron. **124**(2), 155–162 (2016)
5. Dmitrashinovich, V., Hudomal, A., Shibayama, M., Sugita, A.: Linear stability of periodic three-body orbits with zero angular momentum and topological dependence of Kepler's third law: a numerical test. J. Phys. Mathe. Theor. **51**(31), 315101 (2018)
6. Simo, C.: Dynamical properties of the figure eight solution of the threebody problem, Celestial Mechanics, dedicated to Donald, S., for his 60th birthday, Chenciner, A., Cushman, R., Robinson, C., and Xia, Z.J., eds., Contemporary Mathematics **292**, 209–228 (2002)
7. Hristov, I., Hristova, R., Puzynin, I., et al.: Hundreds of new satellites of figure-eight orbit computed with high precision. (2022) arXiv preprint arXiv:2203.02793
8. Roberto, B., et al.: Breaking the limits: the Taylor series method. Appl. Mathe. Comput., 217.20, 7940–7954 (2011)
9. Li, Xiaoming, Liao, Shijun: Clean numerical simulation: a new strategy to obtain reliable solutions of chaotic dynamic systems. Appl. Mathe. Mech. **39**(11), 1529–1546 (2018)
10. Li, XiaoMing, Liao, ShiJun: More than six hundred new families of Newtonian periodic planar Collisionless three-body orbits. Sci. China Phys. Mech. Astron. **60**(12), 1–7 (2017)
11. Jorba, Angel, Zou, Maorong: A software package for the numerical integration of ODEs by means of high-order Taylor methods. Exp. Mathe. **14**(1), 99–117 (2005)
12. Abad, Alberto, Barrio, Roberto, Dena, Angeles: Computing periodic orbits with arbitrary precision. Phys. Rev. E **84**(1), 016701 (2011)
13. Puzynin, I.V., et al.: The generalized continuous analog of Newton's method for the numerical study of some nonlinear quantum-field models. Phys. Part. Nucl. **30**(1), 87–110 (1999)
14. James, W.: Applied numerical linear algebra. Society for Industrial and Applied Mathematics, (1997)
15. Barrio, Roberto: Sensitivity analysis of ODEs/DAEs using the Taylor series method. SIAM J. Sci. Comput. **27**(6), 1929–1947 (2006)
16. Hristov, I., Hristova, R., Puzynin, I., et al.: Newton's method for computing periodic orbits of the planar three-body problem. (2021) arXiv preprint arXiv:2111.10839
17. Roberts, Gareth E.: Linear stability analysis of the figure-eight orbit in the three-body problem. Ergodic Theor. Dyn. Syst. **27**(6), 1947–1963 (2007)
18. Advanpix, L.L.C.: Multiprecision Computing Toolbox for MATLAB. http://www.advanpix.com/. Version 4.9.0 Build 14753, 2022-08-09
19. MATLAB version 9.12.0.1884302 (R2022a), The Mathworks Inc, Natick, Massachusetts, 2021

20. Montgomery, Richard: The N-body problem, the braid group, and action-minimizing periodic solutions. Nonlinearity **11**(2), 363 (1998)
21. Shuvakov, Milovan, Dmitrashinovich, Veljko: A guide to hunting periodic three-body orbits. Am. J. Phys. **82**(6), 609–619 (2014)
22. http://db2.fmi.uni-sofia.bg/3bodychor462/
23. http://hpc-lab.sofiatech.bg/
24. https://gmplib.org/

Evaluation of the Effective Material Properties of Media with Voids Based on Numerical Homogenization and Microstructure Data

Roumen Iankov[1], Ivan Georgiev[2], Elena Kolosova[3], Mikhail Chebakov[3], Miriana Raykovska[2], Gergana Chalakova[1], and Maria Datcheva[1,2]($\boxtimes$)

[1] Institute of Mechanics, Bulgarian Academy of Sciences, 1113 Sofia, Bulgaria
datcheva@imbm.bas.bg, iankovr@abv.bg
[2] Institute of Information and Communication Technologies, Bulgarian Academy of Sciences, 1113 Sofia, Bulgaria
ivan.georgiev@parallel.bas.bg
[3] Institute of Mathematics, Mechanics and Computer Sciences, Southern Federal University, 344090 Rostov-on-Don, Russia
{kolosova,mchebakov}@sfedu.ru

Abstract. This work is devoted to a 3D hybrid numerical-experimental homogenization strategy for determination of elastic characteristics of materials with closed voids. The performed homogenization procedure employs micro-computed tomography (micro–CT) and instrumented indentation testing data (IIT). Based on the micro–CT data a 3D geometrical model of a cubic representative elementary volume (RVE) is created assuming periodic microstructure of the material with closed voids. Creating the RVE respects the following principle of equivalence: the porosity assigned to the RVE is the same as the porosity calculated based on the micro CT images. Next, this geometrical model is used to generate the respective finite element model where, for simplicity, the voids are considered to have a spherical form. The numerical homogenization technique includes proper periodic boundary conditions with unit force applied in normal and shear directions. The employed constitutive model for the solid phase is the linear elastic model whose parameters are determined based on IIT data. It is performed a validation and verification study using simplified geometries for the RVE and under different assumptions for modelling the voids.

Keywords: media with voids · numerical homogenization · micro–CT data analysis

1 Introduction and Motivation

Porous materials occupy an important place in modern technologies. They are used both as a construction material with specific properties and also as a material that exhibits certain properties under certain impacts and conditions (functional graded material). Generally, these materials are considered to be composed

I. Georgiev et al. (Eds.): NMA 2022, LNCS 13858, pp. 181–187, 2023.
https://doi.org/10.1007/978-3-031-32412-3_16

of a solid skeleton and empty or filled with fluids pores. In the preparation of these materials, different technologies are used, in which the skeleton of the porous material is obtained from a certain starting bulk material. This is related to the fact that usually the material characteristics of the skeleton differ from those of the starting material. Of practical interest for industrial applications of porous materials is their characterization and determination of their mechanical characteristics, which allows their wider application in various fields. The aim of the research in the present work is to develop an effective methodology for the characterization of porous materials based on contemporary experimental approaches and the application of modern numerical methods.

2   Methodology

The proposed in this work methodology for determining the effective mechanical properties of a medium with voids consists of combining experimental data and numerical simulations as it is schematically depicted in Fig. 1.

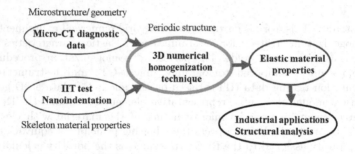

Fig. 1. Flow chart of the proposed methodology.

2.1   Microstructure Diagnostic: Micro–CT Testing

The capabilities of modern computer tomography and particularly of the micro–CT allow to investigate the 3D–structure of porous materials [1]. The micro–CT imaging and 3D–structure reconstruction are particularly useful for porous materials with closed pores where the use of other experimental techniques for characterizing the pore system is not possible. For the purposes of the proposed here homogenization methodology, the micro–CT diagnosis is applied to gain the following information about the microstructure of the material:

* pore volume fraction;
* minimum pore size;
* maximum pore size.

Figure 2 shows results of micro–CT diagnosis of a closed-pore material, namely, sintered ceramic foam made of metallurgical slag and industrial sand, [2]. It also depicts the region of interest (ROI) selected for quantification of the pore system at microlevel.

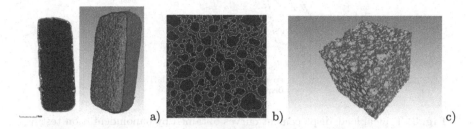

a)   b)   c)

Fig. 2. micro–CT images of the selected porous material: a) the whole specimen covered with x-ray contrast coating; b) 2D aspect of smaller ROI from the centre of the specimen, c) reconstructed 3D RVE from the same ROI.

2.2   Instrumented Indentation Testing

For the purposes of the mathematical modeling and the homogenization procedure, it is necessary to have the material characteristics of the porous material and in particular of the solid skeleton. To this end, we use the instrumented indentation technique (IIT) to determine the elastic characteristics of the considered here ceramic foam skeleton. The instrumented indentation (IIT), also known as nanoindentation is a relatively new but well established and very intensively developing and promising experimental technique used for characterization of materials in a local area [3, 4]. Moreover, nanoindentation has become a commonly approved tool for the measurement of mechanical properties at small scales, but it may have even greater importance as a technique for experimental study of fundamental material behavior of composites, functionally graded materials and foams. Figure 3 depicts typical set of load–displacements curves after nanoindentation testing at selected 5 locations onto the surface of the ceramic foam. The prescribed maximum applied load was 10 mN and the average value of 35 GPa for the elastic modulus (indentation modulus, E_{sk}) was determined using the descending part of the load–displacement curves and applying the Oliver&Pharr method [3] assuming the Poisson's ratio to be $\nu_{sk} = 0.33$.

3   Numerical Homogenization Technique

In this study a numerical homogenization is used to determine the elastic characteristics of the porous medium. This technique requires solving 6 boundary value problems in a region representing the structure of the non-homogeneous (RVE). The basic assumption behind the applied here homogenization technique is that the porous medium has a periodic structure whose unit element is the selected

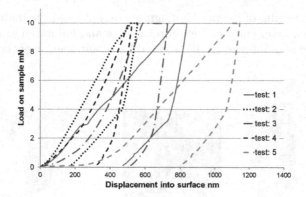

Fig. 3. Typical load–displacement curves obtained by nanoindentation testing.

RVE. The micro–CT data is used for building the geometrical model of the RVE (Fig. 4) with additional simplification assuming the pores can be considered to be idle spheres. Furthermore, periodic boundary conditions are imposed on the RVE consisting in three boundary value problems with applied normal to the paired parallel RVE faces unit forces and three boundary value problems with applied unit shear forces on the RVE faces. This sequence of RVE boundary value problems (BVPs) is given schematically in Fig. 5 in general case, and the geometry of the RVE inclusions depicted in Fig. 5 is illustrative and not tied to the specific RVE geometry considered here.

RVE geometry and FE discretization
Spherical voids, randomly distributed
Void volume fraction
Void size variation interval (min,max)
Material model:
Two phases: air and solid skeleton
Material: linear elastic
Skeleton's elastic properties: from IIT data
Boundary conditions:
Periodic boundary conditions

Fig. 4. RVE–geometry, finite element model and material model assumptions.

Fig. 5. BVPs for 3D homogenization: 6 loading cases – 3 tensile tests and 3 shear tests.

3.1 Effective Elastic Material Properties

As a result of the numerical homogenization, the elastic tensor of the relationship between stresses and strains describing the behavior of the porous material is obtained. We assumed for generality that the foam is not isotropic and suggested the material behavior may be explained employing the orthotropic elastic model given with Eq. (1). There are total of 6 independent elastic parameters that have been identified using the numerical homogenization procedure.

$$
\begin{bmatrix} \epsilon_x \\ \epsilon_y \\ \epsilon_z \\ \gamma_{xy} \\ \gamma_{xz} \\ \gamma_{yz} \end{bmatrix} = \begin{bmatrix} \frac{1}{E_x} & -\frac{\nu_{yx}}{E_y} & -\frac{\nu_{zx}}{E_z} & 0 & 0 & 0 \\ -\frac{\nu_{xy}}{E_x} & \frac{1}{E_y} & -\frac{\nu_{zy}}{E_z} & 0 & 0 & 0 \\ -\frac{\nu_{xz}}{E_x} & -\frac{\nu_{yz}}{E_y} & \frac{1}{E_z} & 0 & 0 & 0 \\ 0 & 0 & 0 & \frac{1}{G_{xy}} & 0 & 0 \\ 0 & 0 & 0 & 0 & \frac{1}{G_{xz}} & 0 \\ 0 & 0 & 0 & 0 & 0 & \frac{1}{G_{yz}} \end{bmatrix} \begin{bmatrix} \sigma_x \\ \sigma_y \\ \sigma_z \\ \tau_{xy} \\ \tau_{xz} \\ \tau_{yz} \end{bmatrix} \tag{1}
$$

It has been pointed out that considering solely the pore volume fraction may lead to underestimation of the effective material properties [5,6]. Accounting the uncertainty in the micro-CT image analysis as well as to have data to compare the explored here numerical homogenization with the analytical models for deriving the effective material properties using solely the volume fraction of voids, we did a parametric study to identify the influence of the void size distribution used in the homogenization procedure on the obtained effective material characteristics. The pore volume fraction was estimated based on the average pore volume obtained from the micro–CT image analysis. Two types of void size distribution are considered with fixed pore volume fraction of 0.8, namely:

(a) *log*-normal distribution with mean void diameter of 350 um and a standard deviation of 150 um;
(b) constant void size distribution with a void diameter of 280 um.

Figure 6 presents the obtained effective elastic material parameters as follows: in blue color for case (a) and in red color for case (b). The resulting effective material properties can be directly used for structural analysis of elements or constructions involving foam materials or materials with empty voids showing inherent difference in the void size characteristics, e.g. in the apparent void diameter.

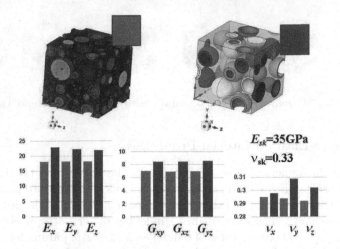

Fig. 6. Parametric study results for two types of pore size distribution: *log*–normal distribution (blue colour); constant distribution (red colour). (Color figure online)

4   Conclusions

The main contribution in this study is the developed and verified against real data methodology for determination of the effective properties of foam materials (materials with empty voids). It is shown that the consideration of the void size distribution, in our study this was the size distribution of the void diameter, can give difference in the derived effective elastic material parameters as compared to the effective parameters values obtained using solely the average pore diameter size and porosity. This fact has to be taken into account also in case if using effective material properties obtained by employing solely the volume fraction of the voids.

Acknowledgements. The authors are grateful for the financial support of the Bulgarian National Science Fund, grants KP-06-H27/6 from 08.12.2018 (I.G.), KP-06-Russia-1 from 27.09.2019 (R.I., G.C.) and by the Ministry of Science and Higher Education of the Russian Federation (State task in the field of scientific activity, scientific project No. FENW-2023-0012) (M.C.,E.K.). The financial support by the Science and Education for Smart Growth Operational Program (2014-2020) and the ESIF through grant BG05M2OP001-1.001-0003 is also acknowledged (M.D.).

References

1. Iassonov, P., Gebrenegus, T., Tuller, M.: Segmentation of X-ray computed tomography images of porous materials: A crucial step for characterization and quantitative analysis of pore structures. Water Resources Res. **45**(9), W09415 (2009). https://doi.org/10.1029/2009WR008087

2. Jordanov, N., Georgiev, I., Karamanov, A.: Sintered glass-ceramic self-glazed materials and foams from metallurgical waste slag. Materials **14**(9), 2263 (2021). https://doi.org/10.3390/ma14092263
3. Oliver, W., Pharr, G.: An improved technique for determining hardness and elastic modulus using load and displacement sensing indentation experiments. J. Mater. Res. **7**(6), 1564–1583 (1992). https://doi.org/10.1557/JMR.1992.1564
4. Iankov, R., Cherneva, S., Stoychev, D.: Investigation of material properties of thin copper films through finite element modelling of microindentation test. Appl. Surf. Sci. **254**(17), 5460–5469 (2008). https://doi.org/10.1016/j.apsusc.2008.02.101
5. Parashkevova, L.: Characterization of porous media by a size sensitive homogenization approach. J. Theor. Appl. Mech. **50**, 338–353 (2020). https://doi.org/10.7546/JTAM.50.20.04.04
6. Parashkevova, L., Drenchev, L.: Effects of porosity and pore size distribution on the compressive behaviour of Pd based glass foams. J. Theor. Appl. Mech. **51**(3), 335–351 (2021)

Variable Neighborhood Search Approach to Community Detection Problem

Djordje Jovanović[1]($\boxtimes$), Tatjana Davidović[1], Dragan Urošević[1],
Tatjana Jakšić Krüger[1], and Dušan Ramljak[2]

[1] Mathematical Institute, Serbian Academy of Sciences and Arts, Belgrade, Serbia
giorgaki.jovanovic@gmail.com, {tanjad,draganu,tatjana}@mi.sanu.bg.ac.rs
[2] School of Professional Graduate Studies at Great Valley,
The Pennsylvania State University, Malvern, PA, USA
dusan@psu.edu

Abstract. Community detection on graphs can help people gain insight into the network's structural organization, and grasp the relationships between network nodes for various types of networks, such as transportation networks, biological networks, electric power networks, social networks, blockchain, etc. The community in the network refers to the subset of nodes that have greater similarity, i.e. have relatively close internal connections. They should also have obvious differences with members from different communities, i.e. relatively sparse external connections. Solving the community detection problem is one of long standing and challenging optimization tasks usually treated by *metaheuristic methods*. Thus, we address it by basic variable neighborhood search (BVNS) approach using *modularity* as the score for measuring quality of solutions. The conducted experimental evaluation on well-known benchmark examples revealed the best combination of BVNS parameters. Preliminary results of applying BVNS with thus obtained parameters are competitive in comparison to the state-of-the-art methods from the literature.

Keywords: Optimization on graphs · Social networks · Metaheuristics · Modularity maximization

1 Introduction

Many complex systems can be described by networks, such as social, biological, citation, scientific collaboration, blockhain transaction networks. As with all such systems, to better understand their structure, it is common to discover the underlying communities. Community detection refers to the procedure of

This work has been funded by the Serbian Ministry of Education, Science and Technological Development, Agreement No. 451-03-9/2021-14/200029 and by the Science Fund of Republic of Serbia, under the project "Advanced Artificial Intelligence Techniques for Analysis and Design of System Components Based on Trustworthy BlockChain Technology (AI4TrustBC)".

I. Georgiev et al. (Eds.): NMA 2022, LNCS 13858, pp. 188–199, 2023.
https://doi.org/10.1007/978-3-031-32412-3_17

identifying groups of interacting nodes in a network depending upon their structural properties. More precisely nodes in a community have more interactions among members of the same community than with the remainder of the network. Number of communities may or may not be known *a priori*.

Community detection algorithms inspired by different disciplines have been developed as there is a wide variety of complex networks generated from different processes. Those algorithms might not perform well on all types of networks [15], but metaheuristics like (VNS) thrive under these circumstances. In this paper we are focused on finding a set of parameters that would allow the Basic VNS (BVNS) perform competitively in comparison to the state-of-the-art methods on well-known benchmark examples from the literature.

In the remainder of this paper, we present a problem formulation in Sect. 2 and review of relevant literature in Sect. 3. Section 4 contains the description of the methodology. Discussion of experimental setup and results are presented in Sect. 5. Finally, Sect. 6 contains a summary and conclusions of the performed work, as well as suggestions for future research.

2   Problem Formulation

The concept of community detection has emerged in network science as a method for finding groups within complex systems represented on a graph. It helps us to reveal the hidden relations among the nodes in the network. In contrast to more traditional decomposition problems in which a strict block diagonal or block triangular structure has to be found, community detection problems refers to the determination of sub-networks with statistically significantly more links between nodes in the same group than between nodes in different groups [10,16]. Community detection is know as hard optimization problem [24] that is usually addressed by various (meta)heuristic methods [2]. To evaluate the performance of the applied optimization method, various measures have been proposed in the literature: Conductance, Internal Density, Normalized Cut, etc. [17]. However, the most commonly used are Modularity Q that evaluates the quality, with respect to the connectivity, of communities formed by an algorithm and Normalized Mutual Information (NMI) that measures the community structure similarity between two networks [24]. The modularity value is most often [3,8,21] defined by Eq. (1).

$$Q = \frac{1}{2m} \sum_{ij} \left(A_{ij} - \frac{d_i d_j}{2m} \right) \delta(g_i, g_j) \qquad (1)$$

Here, A represents the adjacency matrix (A_{ij} takes value 1 if there exists an edge between nodes i and j and 0 otherwise), d_i denotes the degree of node i, m is the total number of edges in graph. The Kronecker delta function $\delta(g_i, g_j)$ evaluates to one if nodes i and j belong to the same community, and zero otherwise. The array g is used to store the community indices of each node. More precisely, g_i is equal to the index of community to which node i belongs.

Modularity Q represents the criterion for deciding how to partition a network. The edge cases, networks that form a single community and those that place every node in its own community will, both have Q equal to zero. The goal of community detection, then, is to find communities that maximize the value of Q. Figure 1 illustrates the change in Q value when the edges are re-distributed for the same number of communities.

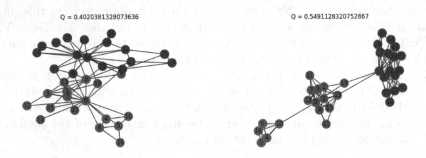

Q = 0.4020381328073636   Q = 0.5491128320752867

Fig. 1. Change in modularity values with respect to graph edges distribution.

Given that U and V are two partitions of a network, NMI is defined in the Eq. 2 [4]:

$$NMI(U,V) = \frac{-2 \sum_{i=1}^{|U|} \sum_{j=1}^{|V|} C_{ij} \log \frac{nC_{ij}}{C_{i.}C_{.j}}}{\sum_{i=1}^{|U|} C_{i.} \log(C_{i.}/n) + \sum_{j=1}^{|V|} C_{.j} \log(C_{.j}/n)} \tag{2}$$

where n is the number of nodes of the network, C is a confusion matrix. C_{ij} counts the nodes shared between community i in partition U and community j in partition V. $|U|$ and $|V|$ denote the number of communities in partition U and V, respectively. $C_{i.}$ and $C_{.j}$ denote the sum of elements of C in row i and column j, respectively. The reliability of NMI as a similarity measure has been proved in [7].

The most common usage of these metrics is to find communities by maximization of Q and then to apply NMI to compare the identified communities with ground truth or results obtained by different methods. Although maximization of modularity Q is a NP-hard optimization problem, many efficient algorithms exist to find sub-optimal solutions, including spectral clustering [21], fast unfolding [3], and VNS metaheuristic [1,8]. We also adopted the modularity maximization as the objective and apply BVNS metaheuristic to the community detection on undirected graphs. After the review of relevant literature, we describe the implementation of the proposed BVNS.

3   Related Work

In this section we review the state-of-the-art algorithms for community detection and application of VNS metaheuristics for this problem. A detailed review, including the classification of existing algorithms, challenges, open issues and future trends related to community detection could be found in [15]. An overview of quality measures and their experimental comparison on some graph examples is provided in [17].

Community detection problem formulated using a mathematical programming model can be found in [24] along with a branch and bound algorithm that is developed for solving it. The first heuristic algorithm for detecting community structure based on the idea of modularity is proposed in [20]. The experiments have shown that it runs to completion in reasonable times for networks of up to a million vertices.

The most common approach to community detection problem in the literature is the application of metaheuristic methods [2,4,5,9,15]. Evolutionary algorithms for community detection are surveyed in [5] and divided into two categories: the single- and multi-objective optimisation based algorithms. In [4] Particle Swarm Optimization (PSO) for the community detection algorithm is proposed. The authors propose discrete PSO where position and velocity are redefined to discrete (instead usual continuous) values. In addition, updating rules for position and velocity of each particle have been reformulated in such a way as to explore the topology of the network. To compare the proposed approach with other methods from the literature, NMI quality measure is used.

In [14] the Ant Colony Optimization (ACO) method is proposed for community detection based on modularity optimization approach. This method does not need to have a priori knowledge about how many communities are there in the network. The authors employ the following instances: Zachary's karate club, the bottlenose dolphins social network, the American College Football, and the Krebs' books on US politics. For the measure of performance the authors utilize modularity and NMI. In [9] the adaptive ant colony clustering i.e. a hybrid between ACO and honeybee hive optimization algorithm is proposed. They compare 14 different algorithms on three real world networks, Karate club, Bottlenose dolphin and American college football. The measures of performance is NMI, Fraction of Vertices Classified Correctly (FVCC) and number of clusters.

Several papers propose VNS-based approaches to community detection problem. In [1] a Variable Neighborhood Decomposition Search (VNDS) is developed for the community detection problem. The decomposition is used due to the size of certain graph instances. In the perturbation phase various operations involving the whole set of communities are performed, while the LPAm+ algorithm is used for local improvement of perturbed solutions. In [8], an additional improvement step is added that decides whether or not to search the whole solution space in local search phase.

In [22] the Community Detection Problem is considered as multi–objective optimization problem using VNS framework to minimize two objective: the Negative Ratio Association (NRA) and the Ratio Cut (RC). The initial set of solutions for the VNS is generated by using the Greedy Randomized Adaptive Search Procedure (GRASP). In [23] GRASP for maximizing the modularity is proposed. The method consists of two phases: the constructive procedure and the local optimization. The constructive procedure begins with $n = |V|$ communities each of which contains only one vertex and at each iteration selects at random from the restricted candidate list community C_1 and the best candidate community C_2 for merging. Restricted candidate list contains communities having modularity larger than previously determined threshold. If modularity of clustering after merging communities C_1 and C_2 is increased then these two communities are merges otherwise community C_1 is removes from candidate list. Local optimization consists of moving selected vertex from current community into another such that total modularity increases. But the local search method in each step selects the vertex v with the smallest ratio r between number of edges in the same community and the total number of edges incident to v.

4 VNS Implementation

In this section, a brief description of the VNS algorithm is given first. Afterwards, the details of the implementation of BVNS algorithm for the problem of community detection is presented. Finally, the process of parameter tuning of the proposed BVNS algorithm is described.

4.1 Brief Description of VNS

VNS is a trajectory based metaheuristic method proposed in [19]. It explores some problem-specific local search procedure. VNS uses one or more neighborhood structures to efficiently search the solution space of a considered optimization problem. If necessary, it explores also a systematic change of neighborhoods in a local search phase. In addition, VNS contains the diversification (perturbation) phase to ensure escaping from local optima traps. In this phase, it may use both neighborhood change and increasing of distances within a single neighborhood. Basic VNS consists of three main steps: Shaking, Local Search, and Neighborhood Change (see Algorithm 1). The role of Shaking step is to ensure the diversification of the search. It performs a random perturbation of the current best solution in the given neighborhood and provides a starting solution to the next step (the local search procedure). Local Search step takes the shaken solution as an input and tries to improve it by visiting its neighbors in one or more neighborhoods. After the Local Search, VNS performs Neighborhood Change step in which it examines the quality of the obtained local optimum. If the newly found local optimum is better than the current best solution, the search is concentrated around it (the global best solution and the neighborhood index are updated properly). Otherwise, only the neighborhood index for further Shaking is changed. The three main steps are repeated until a pre-specified stopping criterion is satisfied [12,13].

4.2   Implementation Details

In this subsection, the solution implementation, as well as all of the basic steps of the VNS algorithm will be presented.

The VNS algorithm is comprised of three basic steps, as described in the previous subsection: Shaking, Local Search and Neighborhood change, until a stopping criterion is met. In this implementation, the stopping criterion is 100 iterations of Shaking-Local Search operations while the neighborhood changes from k_{mim} to k_{max}, at a step of k_{step}. Algorithm 1 gives an overview of the basic structure of the BVNS algorithm in the case of optimization problems that require maximization of the objective function value.

Algorithm 1. Pseudo-code for BVNS method

---

procedure BVNS(Problem input data, k_{min}, k_{step}, k_{max} STOP)

    $x \leftarrow$ INITSOLUTION

    $x_{best} \leftarrow x$

    repeat

        $k \leftarrow k_{min}$

        repeat

            $x' \leftarrow$ RANDOMSOLUTION($x_{best}, \mathcal{N}_k$)   ▷ Shaking

            $x'' \leftarrow$ LS(()x')   ▷ Local Search

            if $(f(x'') > f(x_{best}))$ **then**   ▷ Neighborhood Change

                $x_{best} \leftarrow x''$

                $k \leftarrow k_{min}$

            else

                $k \leftarrow k + k_{step}$

            end if

            $STOP \leftarrow$ STOPPINGCRITERION

        until $(k > k_{max} \vee STOP)$

    until (STOP)

    return $(x_{best}, f(x_{best}))$

end procedure

A solution of the community detection problem is represented by an array of n integers containing, for each node i, the index of the corresponding community. More precisely, $solution[i] = c_i$, $c_i \in \{1, 2, \ldots, c\}$, with c being the current number of communities.

The initial solution is constructed by randomly assigning the community index to each node. The input data for this construction are number of nodes n and the initial number of communities c. The i-th element of the $solution$ array is determined as follows $solution[i] = rand(1, c)$. The resulting $solution$ array is used to calculate the modularity Q for the initial solution.

Our BVNS explores a single neighborhood structure that consists of moving nodes from the current communities to some others. This neighborhood is explored randomly at distance k, $k = 1, 2, \ldots, k_{max}$ in the shaking phase. In addition, the neighborhood is explored systematically at distance 1 in the local search phase. The term distance here, counts the number of nodes involved in the transformation. Once the node (or a subset of k nodes) change communities, the quality of the obtained solution has to be determined by calculating the value of modularity Q.

In the local search procedure, an attempt to put each of the nodes in all the other communities, save for its own, is made, calculating the improvement of modularity at each such attempt. If the modularity is better than the current best modularity, the index of the node and the community which designates such movement. After checking all nodes in this manner, the memorized best move is made, and the process starts again from the first node. The local search is terminated when, after checking all of the nodes, there exist no more moves which increase the modularity. Algorithm 2 shows all the mentioned details of local search.

Algorithm 2. Pseudo-code for Local Search

---

procedure LS(*solution*)
    visited ← 0   ▷ Counter for nodes
    nodeId ← 0
    improved ← 0
    temp1 ← *solution*   ▷ *temp1* stores solution modified by local search
    bestQ ← $Q(solution)$
    bestmoveNode ← −1
    bestmoveCommunity ← −1
    while *visited* < *n* **do**
        visited ← *visited* + 1
        communityId ← 0
        while (*communityId* < *c*) ∧ (¬*improved*) **do**   ▷ For each community
            temp2 ← *temp1*
            temp2 ← MOVENODE(*temp2*, *visited*, *communityId*)
            if $Q(temp2) > Q(solution)$ **then**   ▷ If better, note the solution
                bestQ ← $Q(temp2)$
                bestmoveNode ← *visited*
                bestmoveCommunity ← *communityId*
                improved ← 1
            end if
            communityId ← *communityId* + 1
        end while
        if (*visited* = *n*) ∧ (*improved*) **then**   ▷ Make a move if solution is improved
            temp1 ← MOVENODE(*temp1*, *bestmoveNode*, *bestmoveCommunity*)
            improved ← 0   ▷ Loop variable reset
            visited ← 0
        end if
    end while
    solution ← *temp1*
    return (*solution*, *bestQ*)
end procedure

4.3   Parameter Tuning

Parameter tuning focuses on tuning the three parameters of the BVNS algorithm k_{min}, k_{step}, and k_{max} by two different methods "manually", and by using the iRace library [18].

The "manual" method consists of an exhaustive search through parameter space. Each of these parameters is given a set of discrete values to be tested on a set of selected problem instances with fixed number of communities. Each parameter combination was tested five times, using different random seeds. The results of these runs were integrated in the following manner: if the runs gave the same solution, the maximum time since the last improvement was selected as the measure of goodness for the parameter combination. Otherwise, the maximum time of the whole run was selected. After parameter combination aggregation, a set of p parameter combinations with minimal time was selected for each combination of a given graph instance and fixed number of communities. Parameter combination with the highest occurrence was selected as the best.

iRace is a statistical library which uses statistical tests in order to perform hyperparameter tuning for combinatorial optimization algorithms including metaheuristics [18]. It offers iterated racing procedures, which have been successful in automatic configurations of various state-of-the-art algorithms. A stopping criterion can either be a number of experiments to be performed, or a given budget of time. iRace could be defined as a supervised machine learning algorithm for learning good parameter values.

5   Experimental Evaluation

In this section a detailed outline of the conducted experiments is given, along with the tested graph set, tested values for the parameters, and the results of the algorithm for the best found configuration of parameter values for both manual and iRace tuning.

5.1   Experimental Setup

As stated in the previous section, the primarily goal of these experiments is to determine the best possible values for k_{min}, k_{step}, and k_{max} parameters for a given set of graph examples. The significance of such an experiment is in the improvement the diversification step (Shaking). This step is realized by random perturbation of the current best solution and should enable BVNS to escape the local optima traps. This is especially important for large graphs, containing tens of thousands or even more nodes.

All of the experiments are performed in a Linux Mint 20.04 virtual machine, running on two cores of Intel Core i5 1035G1 processor and 8GB of RAM, while the algorithm is programmed using the Python programming language. The graph libraries NetworkX [11] for the graph representation, and the igraph [6] library for implementation of modularity score are used. In the case of the iRace package, an R script was written to load the scenario and the set of parameters, and the Python implementation is used as the runner program.

The experiments were performed for a well-known set of 8 benchmark graphs, namely the karate, chesapeakcese, dolphins, lesmis, polbooks, jazz, adjnoun, and football instan. More information on these instances can be found in [1]. Usually, the number of communities is not specified in advance, instead, it has to be determined in such a way to correspond to the maximal value of Q. The referent number of communities are provided in [1]. For all of the given graphs, except for the adjnoun and football instances, number of communities is less than seven, and we calculate the value of modularity for two to seven communities. For the adjnoun and football instances, we let the number of communities to range from six to twelve. An instance of the considered community detection problem is defined by the combination of graph and the number of communities. Therefore, having 6 graphs combined with 6 values for the number of communities and 2 graphs with 7 possible values for the number of communities, we and up with 50 problem instances to be evaluated.

Stopping criterion for BVNS is defined as 100 iterations, where iteration is represented by a single combination of Shake and Local Search steps. For manual tuning, we set $p = 10$, and then identified 3 combinations that perform well for most of the instances. In the parameter tuning process, for each of the parameters, minimum and maximum values are specified. As we are addressing combinatorial optimization problem and parameters can take only discrete values, for each parameter we assign also a constant step for changing the current value. During the preliminary experimental analysis, we noticed that for small values of BVNS parameters algorithm gets stuck in the local optima. Therefore, we foreseen large perturbations of the current best solution in each Shaking step, which correspond to the parameter values summarized in Table 1.

Table 1. Minimal, maximal and step values for each of the parameters.

Parameter Name	Minimum	Step	Maximum
k_{min}	0.05	0.02	0.15
k_{step}	0.05	0.02	0.15
k_{max}	0.20	0.05	0.7

In the case of iRace experiments, default settings are used and 1000 iterations are given as budget. The experiments are run with the training set in such a way that both the training and test set contain the same instances.

5.2   Parameter Settings

The results of the manual tuning are shown in the first three columns of Table 2. As stated in the Parameter tuning section, a list of three parameter combinations with the highest occurrence is presented. The best three combinations according to iRace are given in the columns 4-6 of the same table. iRace actually returned a single combination as a results of ceiling operation.

Table 2. Results of the manual (columns 1-3) and iRace (columns 4-6) tuning.

k_{min}	k_{step}	k_{max}	k_{min}	k_{step}	k_{max}
0.09	**0.09**	**0.65**	0.07	0.15	0.7
0.09	0.09	0.55	0.07	0.15	0.7
0.11	0.09	0.65	0.07	0.15	0.7

As all the reported combination generate more or less the same BVNS results, we decided to trust manual parameter tuning and selected its best combination for the comparison with the state-of-the-art methods.

5.3   Comparison Results

In this subsection, we provide BVNS results for benchmark graphs with different numbers of communities. In Table 3 the results for graphs with 2-7 communities are presented, and Table 4 contains the results for graphs with 6-12 communities. However, in the relevant literature, values for Q are provided only for a single number of communities, referred to as the *optimal* number of communities. Usually, it is ground truth stated by the instance creator or the number for which maximum Q value is obtained. The obtained values of Q corresponding to the optimal number of communities are presented in bold in Tables 3 and 4.

Table 3. BVNS results for graphs with smaller number of communities.

Graph	n	m	Number of communities					
			2	3	4	5	6	7
karate	34	78	0.371794	0.402038	**0.419789**	0.419789	0.419789	0.419789
chesapeake	39	170	0.233217	**0.265795**	0.265795	0.265795	0.265795	0.265795
dolphins	62	159	0.402733	0.494185	0.526798	**0.528519**	0.528519	0.528519
lesmis	77	254	0.382788	0.497372	0.542834	0.556195	**0.560008***	0.560008
polbooks	105	441	0.456874	0.522074	0.526938	**0.527236**	0.527236	0.527236
jazz	198	2742	0.320609	0.444469	**0.445143**	0.445143	0.445143	0.445143

* The maximum found in literature for this graph is 0.566

As can be seen from Tables 3 and 4, in all the cases, except for the lesmis and adjnoun graph, the obtained Q value coincides with the corresponding value provided in [1] and declared as optimum. The execution times of our BVNS are negligible for graphs with less than 100 nodes. For the remaining graphs, the required CPU time is 2-4 min.

Table 4. BVNS results for graphs with larger number of communities.

Graph	n	m	Number of communities						
			6	7	8	9	10	11	12
adjnoun	112	425	0.309735	**0.311283**[*]	0.309544	0.309522	0.313367	0.309057	0.306148
football	115	615	0.592785	0.594547	0.603161	0.604407	**0.604407**	0.604569	0.604569

[*] The maximum found in literature for this graph is 0.313367

6 Conclusion

Understanding complex networks through community detection has been an important topic in many research areas in recent years. In this paper we focused on parameter tuning that would allow the Basic Variable Neighbourhood Search (BVNS) to efficiently address community detection problem for a set of benchmark networks from literature. We have described the BVNS algorithm and a procedure for finding the best set of parameters. Finally, after obtaining parameters for BVNS the conducted experimental evaluation shows that the proposed BVNS is competitive in comparison with the state-of-the-art algorithms from the literature.

The future work includes extensions of the algorithm to address large scale networks. As the proposed algorithm is only suited for undirected networks, another focus of future research could be the adjustment of VNS to the community discovery of attribute networks and multilayer networks.

References

1. Aloise, D., Caporossi, G., Hansen, P., Liberti, L., Perron, S., Ruiz, M.: Modularity maximization in networks by variable neighborhood search. Graph Partition. Graph Clust. **588**, 113 (2012)
2. Bara'a, A.A., et al.: A review of heuristics and metaheuristics for community detection in complex networks: current usage, emerging development and future directions. Swarm Evol. Comput. **63**, 100885 (2021)
3. Blondel, V.D., Guillaume, J.L., Lambiotte, R., Lefebvre, E.: Fast unfolding of communities in large networks. J. Stat. Mech: Theory Exp. **2008**(10), P10008 (2008)
4. Cai, Q., Gong, M., Shen, B., Ma, L., Jiao, L.: Discrete particle swarm optimization for identifying community structures in signed social networks. Neural Netw. **58**, 4–13 (2014)
5. Cai, Q., Ma, L., Gong, M., Tian, D.: A survey on network community detection based on evolutionary computation. Int. J. Bio-Inspired Comput. **8**(2), 84–98 (2016)
6. Csardi, G., Nepusz, T.: The Igraph software package for complex network research. InterJ. **Complex Systems**, 1695 (2006). https://igraph.org
7. Danon, L., Diaz-Guilera, A., Duch, J., Arenas, A.: Comparing community structure identification. J. Stat. Mech: Theory Exp. **2005**(09), P09008 (2005)

8. Džamić, D., Aloise, D., Mladenović, N.: Ascent-descent variable neighborhood decomposition search for community detection by modularity maximization. Ann. Oper. Res. **272**(1), 273–287 (2019)
9. Ghasabeh, A., Abadeh, M.S.: Community detection in social networks using a hybrid swarm intelligence approach. Int. J. Knowl. -Based Intell. Eng. Syst. **19**(4), 255–267 (2015)
10. Girvan, M., Newman, M.E.: Community structure in social and biological networks. Proc. Natl. Acad. Sci. **99**(12), 7821–7826 (2002)
11. Hagberg, A.A., Schult, D.A., Swart, P.J.: Exploring network structure, dynamics, and function using networkx. In: Varoquaux, G., Vaught, T., Millman, J. (eds.) Proceedings of the 7th Python in Science Conference, pp. 11–15. Pasadena, CA USA (2008)
12. Hansen, P., Mladenović, N.: Variable neighborhood search. In: Martí, R., Pardalos, P.M., Resende, M.G.C. (eds.) Handbook of Heuristics, pp. 759–787. Springer, Cham (2018). https://doi.org/10.1007/978-3-319-07124-4_19
13. Hansen, P., Mladenović, N., Todosijević, R., Hanafi, S.: Variable neighborhood search: basics and variants. EURO J. Comput. Optimiz. **5**(3), 423–454 (2017)
14. Honghao, C., Zuren, F., Zhigang, R.: Community detection using ant colony optimization. In: 2013 IEEE Congress on Evolutionary Computation, pp. 3072–3078. IEEE (2013)
15. Javed, M.A., Younis, M.S., Latif, S., Qadir, J., Baig, A.: Community detection in networks: a multidisciplinary review. J. Netw. Comput. Appl. **108**, 87–111 (2018)
16. Leicht, E.A., Newman, M.E.J.: Community structure in directed networks. Phys. Rev. Lett. **100**(11), 118703:1–4 (2008)
17. Leskovec, J., Lang, K.J., Mahoney, M.: Empirical comparison of algorithms for network community detection. In: Proceedings of the 19th International Conference on World Wide Web, pp. 631–640 (2010)
18. López-Ibáñez, M., Dubois-Lacoste, J., Cáceres, L.P., Birattari, M., Stützle, T.: The irace package: Iterated racing for automatic algorithm configuration. Oper. Res. Persp. **3**, 43–58 (2016)
19. Mladenović, N., Hansen, P.: Variable neighborhood search. Comput. Oper. Res. **24**(11), 1097–1100 (1997)
20. Newman, M.E.: Fast algorithm for detecting community structure in networks. Phys. Rev. E **69**(6), 066133 (2004)
21. Newman, M.E.: Finding community structure in networks using the eigenvectors of matrices. Phys. Rev. E **74**(3), 036104 (2006)
22. Pérez-Peló, S., Sánchez-Oro, J., Gonzalez-Pardo, A., Duarte, A.: On the analysis of the influence of the evaluation metric in community detection over social networks. Appl. Soft Comput. **112**, 107838 (2021)
23. Pérez-Peló, S., Sanchez-Oro, J., Martin-Santamaria, R., Duarte, A.: On the analysis of the influence of the evaluation metric in community detection over social networks. Electronics **8**(1), 23 (2018)
24. Srinivas, S., Rajendran, C.: Community detection and influential node identification in complex networks using mathematical programming. Expert Syst. Appl. **135**, 296–312 (2019)

Simulation Investigations of Pyramidal Cells Layer Effect on Conscious Visual Perception

Petia Koprinkova-Hristova$^{(\boxtimes)}$ (ORCID) and Simona Nedelcheva

Institute of Information and Communication Technologies, Bulgarian Academy of Sciences, Acad. G. Bonchev str. bl.25A, 1113 Sofia, Bulgaria
{petia.koprinkova,simona.nedelcheva}@iict.bas.bg
http://iict.bas.bg

Abstract. The aim of presented research is to investigate one of the latest theories of consciousness and its neural correlates in the brain - the dendrite theory. It states that the conscious perception depends on the connectivity between the thalamus and sensory cortex mediated via neural pyramidal cells. In present research we upgraded our spike timing neural network model of visual information perception with an elaborated thalamus structure (first and second order thalamus) as well as more realistic structure of the visual cortex including pyramidal neurons. Conducted simulation experiment demonstrated the important role of the second order thalamus. It was shown also that when the L5 pyramidal cells layer is omitted the propagation of the perceived visual information is destructed.

Keywords: Consciousness · Visual Perception · Pyramidal Cells

1 Introduction

The debates about the link between sensation, perception, attention, and consciousness date since the earliest days of psychophysiology. The scientists were interested in answering the question what happens to a sensory signal in the brain and how to distinguish its conscious and outside of awareness processing. Research in this direction give the origin of the concept called "Neural correlates of consciousness" denoting a set of neuronal events and mechanisms generating a specific conscious perception. Thus in [3,4] consciousness is viewed as a

This work was supported by the Bulgarian Science Fund, grant No KP-06-COST/9 from 2021 "Supercomputer simulation investigation of a neural model of brain structures involved in conscious visual perception", co-funding scheme of COST action CA18106 "Neural architectures of consciousness". S. Nedelcheva was partially supported by the Bulgarian Ministry of Education and Science under the National Research Programme "Young scientists and postdoctoral students -2" approved by DCM 206 / 07.04.2022.

I. Georgiev et al. (Eds.): NMA 2022, LNCS 13858, pp. 200–208, 2023.
https://doi.org/10.1007/978-3-031-32412-3_18

state-dependent property of some complex, adaptive, and highly interconnected biological structures in the brain. In [23] a theoretical description that relates brain phenomena such as fast irregular electrical activity and widespread brain activation to expressions of consciousness was called a model of consciousness.

Recent studies on neural correlates of consciousness continue the research initiated at the end of the 19th century [15]. Because the visual system was already intensively investigated [2–4,10] in the 1990s an empirical approach focusing on visual awareness was developed. Since then, consciousness research becomes more diverse but its link to visual perception continued [20,21]. Irrespective of the intense interest and research efforts in studying consciousness, a scientific consensus which brain regions are essential for conscious experience and what are the minimal neural mechanisms sufficient for conscious perception is still not reached [3].

Recently, a dominant trend is to view consciousness as emerging from the global activity patterns of corticocortical and thalamocortical loops [5–7,16,25–27]. In [18] a hypothesis supported by numerous works was proposed that the thalamus is the primary candidate for the location of consciousness since it has been referred to as the gateway of nearly all sensory inputs to the corresponding cortical areas. Recent findings [8] support the important role of the lateral geniculate nucleus in the emergence of consciousness and provide a more complex view of its connections to the other parts of the thalamus and the visual cortex. In [1] it was suggested that the hallmark of conscious processing is the flexible integration of bottom-up and top-down thalamocortical streams and a novel neurobiological theory of consciousness called "Dendritic Information Theory", was proposed. It is based on the findings that the cortex also has a hierarchical structure in which feed forward and feedback pathways have a layer-specific pattern. Recent advances in measuring subcellular activity in axons and dendrites as well as high-resolution fMRI allowed for carrying out studies on how cognitive events arise from the laminar distribution of cortex inputs [14]. It was hypothesised that a specific type of neural cells in the cortex - neocortical pyramidal cells presented mainly in fifth layer of the cortex - selectively amplify sensory signals and their contribution to the consciousness have to be distinguished from that of thalamocortical system [17,22]. A view of thalamocortical processing is proposed in [24] where two types of thalamic relays are defined: first-order relays receiving subcortical driver input, e.g. retinal input to the lateral geniculate nucleus, and higher-order relays, receiving driver input from layer 5 of the cortex, that participate in corticothalamo-cortical circuits.

We have already developed a hierarchical spike timing model of visual information perception and decision making including detailed structure of thalamic relay and laretal geniculate nucleus [11]. In our previous work [12] we investigated the influence of thalamo-cortical connectivity on the conscious perception of visual stimuli by changing the bottom-up and top-down connections between thalamic relay (TRN) and lateral geniculate nucleus (LGN) and primary visual cortex (V1). The model was implemented on NEST Simulator [9] on the supercomputer Avitohol.

In present research we investigate the role of pyramidal cells in 5th layer of the cortex [17] as well as of the first and high order thalamus [24] on conscious visual perception by further upgrade of our model using literature information.

The structure of the rest of the paper is as follows: next section presents briefly our model structure and its parameters under investigation; simulation results are presented in Sect. 3 followed by discussion; the paper finishes with concluding remarks pointing out directions of our further work.

2   Model Structure

For the aims of planed here simulation investigations we used only a part of our model [11], presented on the left side in the Fig. 1. Each rectangle represents a group of neurons, called further layer, positioned on a regular two-dimensional grid that corresponds to a brain structures involved in visual information perception as follows: retinal ganglion cells (RGC) in eyes that transform the light to electrical signal fed into the brain via the optic nerve; Lateral geniculate nucleus (LGN), Thalamic reticulate nucleus (TRN) and Interneurons (IN) are parts of the thalamus - the relay structure forwarding the information to the visual cortex; Primary visual cortex (V1) which detects orientation of visual stimulus and whose neurons are organized in orientation columns as shown on the right side of the Fig. 1. The details of connectivity, described in our previous works [11,19], are based on literature information [8,13]. Briefly, each neuron has its own receptive field - area of neurons from a given layer that it is connected to - depending on the function of the layer it belongs to as well as on its position inside its own layer.

RGC were modelled by spatio-temporal filters while the rest of the neural cells were modeled by leaky integrate and fire model (for the details see [11,19]).

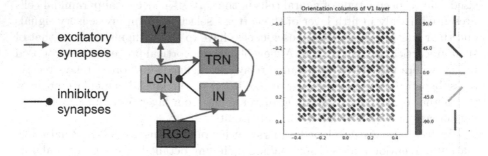

Fig. 1. Thalamo-cortical part of our model.

Based on literature information we extended the model including first (FO) and high (HO) order thalamus as well as four layers of the visual cortex as shown in Fig. 2.

Each layer within V1 has the same orientation structure as in Fig. 1. According to the literature the sensory input from the retina influences mainly the FO thalamus while HO is much less influenced. However, only HO is connected to the fifth layer (L5) of the cortex thus influencing the conscious perception as the literature supposes. The sensory signal is fed into the layers 4 and 6 of V1 via both FO and HO thalamus. The connections between V1 layers are based on literature information [17] as is shown in Fig. 1.

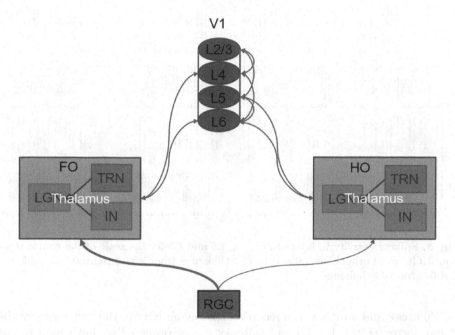

Fig. 2. Thalamo-cortical connectivity.

3   Simulation Results

In order to investigate the role of both thalamus parts (FO and HO) as well as of pyramidal cells layer L5 in conscious visual perception, the following simulation experiments were carried out:

- Strength of the connections from the RGC to the FO and HO thalamus.
- Presence/absence of the pyramidal cells layer L5.

The model input is a moving dots stimulus presenting 50 dots expanding from an imaginary center on the visual field as it was described in our previous work [11].

3.1   Connectivity from RGC to FO and HO Thalamus

Since the theories suppose much stronger connectivity from RGC to FO than to HO, we investigated two scaling values of the connections to FO (1.0 and 2.0) and three smaller values for connections to HO (0.005, 0.05 and 0.5). Thus the sensory signal portion send to both thalamus parts are as follows: 200:1, 20:1, 2:1, 400:1, 40:1 and 4:1 respectively.

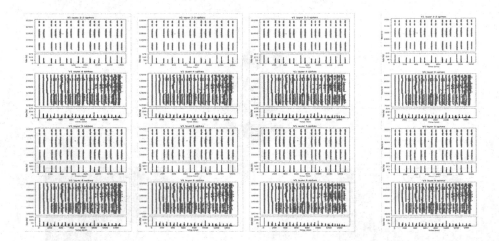

Fig. 3. Spiking activity in layers L2/3, L4, L5 and L5 for the scale of the connections from RGC to FO thalamus 1.0 and to HO thalamus from left to right: 0.005, 0.05, 0.5 and 0.0 (no HO thalamus).

Figures 3 and 4 present the recorded spiking activity in the four layers of the visual cortex (L2/3, L4, L5 and L6). As it was observed on both figures, the spiking activity in all V1 layers for the smallest scaling values of the connections from RGN to HO thalamus (0.005 and 0.5) is the same as in the case without HO thalamus (zero scaling value). Only in the case of the highest scaling value (0.5) for the connections to HO thalamus led to visible changes the spiking activity in the visual cortex and especially in the layer L2/3 which sends processed visual information to other brain areas.

The presence of strong enough connections from sensory input to the HO thalamus (ratio of FO:HO connections 2:1 and 4:1 respectively) influences the spiking frequencies in L2/3 and L5 layers as follows:

– In the first case (ratio 2:1) spiking frequency in L5 slightly increased while that in L2/3 slightly decreased.
– In the second case (ratio 4:1) in both L2/3 and L5 layers spiking frequency was slightly decreased.

These results confirm the theory that since the HO thalamus is responsible also for the propagation of the sensory signal to the deeper brain structures via L5 layer, its presence should not be neglected.

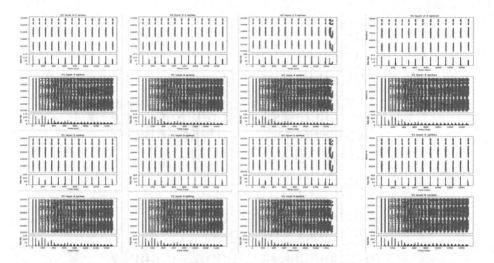

Fig. 4. Spiking activity in layers L2/3, L4, L5 and L5 for the scale of the connections from RGC to FO thalamus 2.0 and to HO thalamus from left to right: 0.005, 0.05, 0.5 and 0.0 (no HO thalamus).

3.2   Presence/Absence of the Pyramidal Cells Layer L5

Based on the previous simulation experiment we choose to investigate further the combinations of 1.0/2.0 and 0.5 scaling factors for the connections to FO and HO thalamus respectively (ratios 2:1 and 4:1 of the connections strength).

Simulations with present/missing L5 layer of the model using chosen connectivity ratios are shown in Figs. 5 and 6 respectively.

This experiment proofs the significant role of the pyramidal cells layer in propagation of the processed visual information. Although the spiking activity in the last (L2/3 layer) of the V1 did not disappeared, its frequency is significantly lower in the case of missing pyramidal cells layer (L5) and connections ratio 2:1. Although in the second case (ratio 4:1) the mean firing rate in L2/3 is a little bit higher in case of missing L5 layer, the spiking activity in a part of L2/3 neurons disappeared especially at the end of stimulation.

The spiking patterns in all V1 layers were changed in both considered cases. This effect was most clearly observable in the L2/3 layer. This result is inline with the theory that anesthesia drugs block the pyramidal cells dendrite projections on the brain surface thus preventing the propagation of sensory signal which a mechanism of consciousness loss.

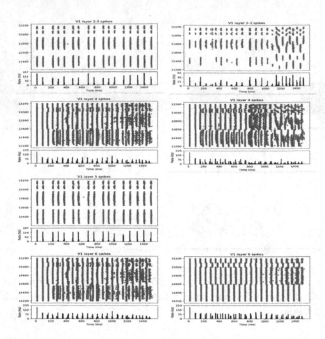

Fig. 5. Spiking activity in V1 for RGC-FO thalamus connections scale 1.0, RGC-HO thalamus connection scale 0.5 and present (left)/missing (right) pyramidal cells layer.

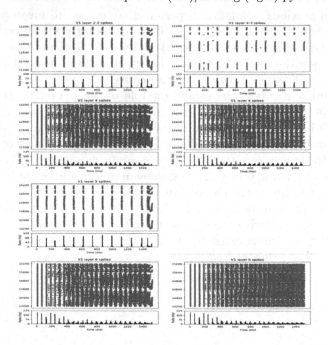

Fig. 6. Spiking activity in V1 for RGC-FO thalamus connections scale 2.0, RGC-HO thalamus connection scale 0.5 and present (left)/missing (right) pyramidal cells layer.

4   Conclusions

The simulation results reported above are inline with the consciousness theories as follows:

- The role of HO is important although it receives neglectfully small sensory information in comparison with the FO thalamus. Its presence enhances the sensory signal propagation to deeper brain areas
- Missing L5 layer destroys perceptual ability of the brain thus influencing conscious sensory information processing.

Further investigations will include more detailed dendrite three structure of the L5 pyramidal cells since the theory that anesthesia blocks the dendrite branches thus suppressing the consciousness.

References

1. Aru, J., Suzuki, M., Larkum, M.E.: Cellular mechanisms of conscious processing trends in cognitive sciences, vol. 24, no. 10 (2020). https://doi.org/10.1016/j.tics.2020.07.006
2. Block, N.: How many concepts of consciousness? Behav. Brain Sci. **18**(2), 272–287 (1995). https://doi.org/10.1017/S0140525X00038486
3. Crick, F., Koch, C.: Towards a neurobiological theory of consciousness. Semin. Neurosci. **2**, 263–275 (1990)
4. Crick, F., Koch, C.: The problem of consciousness. Sci. Am. **267**, 152–159 (1992)
5. Dehaene, S., et al.: A neuronal network model linking subjective reports and objective physiological data during conscious perception. PNAS **100**(14), 8520–8525 (2003). https://doi.org/10.1073/pnas.1332574100
6. Dehaene, S., et al.: Conscious, preconscious, and subliminal processing: a testable taxonomy. Trends Cogn. Sci. **10**, 204–211 (2006)
7. Dehaene, S., Changeux, J.-P.: Experimental and theoretical approaches to conscious processing. Neuron **70**, 200–227 (2011)
8. Ghodratia, M., et al.: Recent advances in understanding its role: towards building a more complex view of the lateral geniculate nucleus. Prog. Neurobiol. **156**, 214–255 (2017)
9. Jordan, J., et al.: NEST 2.18.0, Zenodo (2019). https://doi.org/10.5281/zenodo.2605422
10. Koch, C.: The quest for consciousness: a neurobiological approach. Roberts Publishers (2004)
11. Koprinkova-Hristova, P., Bocheva, N.: Brain-inspired spike timing model of dynamic visual information perception and decision making with STDP and reinforcement learning. In: Nicosia, G., et al. (eds.) LOD 2020. LNCS, vol. 12566, pp. 421–435. Springer, Cham (2020). https://doi.org/10.1007/978-3-030-64580-9_35
12. Koprinkova-Hristova, P., Nedelcheva, S.: Spike timing neural network model of conscious visual perception. Biomath **11**, 2202258 (2022). https://doi.org/10.55630/j.biomath.2022.02.258
13. Kremkow, J., et al.: Push-pull receptive field organization and synaptic depression: mechanisms for reliably encoding naturalistic stimuli in V1. Front. Neural Circuits (2016). https://doi.org/10.3389/fncir.2016.00037

208   P. Koprinkova-Hristova and S. Nedelcheva

14. Larkum, M.E., Petro, L.S., Sachdev, R.N.S., Muckli, L.: A perspective on cortical layering and layer-spanning neuronal elements. Front. Neuroanat. **12**, 56 (2018). https://doi.org/10.3389/fnana.2018.00056
15. LeDoux, J.E., Michel, M., Lau, H.: A little history goes a long way toward understanding why we study consciousness the way we do today. PNAS **117**(13), 6976–6984 (2020)
16. Mashour, G.A., et al.: Conscious processing and the global neuronal workspace hypothesis. Neuron **105**, 776–798 (2020)
17. Mease, R.A., Gonzalez, A.J.: Corticothalamic pathways from layer 5: emerging roles in computation and pathology. Front. Neural Circuits **15**, 730211 (2021). https://doi.org/10.3389/fncir.2021.730211
18. Min, B.K.: A thalamic reticular networking model of consciousness. Theor. Biol. Med. Model. **7**(1), 10 (2010). https://doi.org/10.1186/1742-4682-7-10
19. Nedelcheva, S., Koprinkova-Hristova, P.: Orientation selectivity tuning of a spike timing neural network model of the first layer of the human visual cortex. In: Georgiev, K., Todorov, M., Georgiev, I. (eds.) BGSIAM 2017. SCI, vol. 793, pp. 291–303. Springer, Cham (2019). https://doi.org/10.1007/978-3-319-97277-0_24
20. Palva, J.M., Monto, S., Kulashekhar, S., Palva, S.: Neuronal synchrony reveals working memory networks and predicts individual memory capacity. Proc. Natl. Acad. Sci. USA **107**, 7580–7585 (2010)
21. Panagiotaropoulos, T.I., Kapoor, V., Logothetis, N.K.: Subjective visual perception: from local processing to emergent phenomena of brain activity. Philosop. Trans. Royal Soc. B: Biol. Sci. **369**, 1641 (2014). https://doi.org/10.1098/rstb.2013.0534
22. Phillips, W.A., Bachmann, T., Storm, J.F.: Apical function in neocortical pyramidal cells: a common pathway by which general anesthetics can affect mental state. Front. Neural Circuits **12**, 50 (2018). https://doi.org/10.3389/fncir.2018.00050
23. Seth, A.K.: Explanatory correlates of consciousness: theoretical and computational challenges. Cogn. Comput. **1**(1), 50–63 (2009)
24. Sherman, S.M.: Thalamus plays a central role in ongoing cortical functioning. Nat. Neurosci. **19**(4), 533–541 (2016)
25. Tononi, G., Edelman, G.M.: Consciousness and complexity. Science **282**, 1846–1851 (1998)
26. Tononi, G.: Consciousness as integrated information: a provisional manifesto. Biol. Bull. **215**, 216–242 (2008)
27. Tononi, G., et al.: Integrated information theory: from consciousness to its physical substrate. Nat. Rev. Neurosci. **17**, 450–461 (2016)

On Decision Making Under Uncertainty

Mikhail I. Krastanov[1,2] and Boyan K. Stefanov[1,2(✉)]

[1] Department of Mathematics and Informatics, University of Sofia,
5 James Bourchier Blvd, Sofia 1164, Bulgaria
bojanks@fmi.uni-sofia.bg
[2] Institute of Mathematics and Informatics, Bulgarian Academy of Sciences,
Acad. Georgi Bonchev Str., Bl. 8, Sofia 1113, Bulgaria

Abstract. Decision-making under uncertainty has recently received an increasing amount of interest in the context of dynamical games. Here we study a discrete dynamical game on an infinite-time horizon. This game is used for modelling optimal control problems in the presence of unknown disturbances. The main result provides a necessary condition of Pontryagin's maximum principle type. The presented example illustrates the possible practical applications.

Keywords: optimal control problems · disturbances · discrete-time games · Pontryagin maximum principle

1 Introduction

Uncertainty plays a significant role in a wide range of fields such as physics, biology, political science, communications, environment, and econometrics, to name a few, where it interacts in a non-cooperative manner and has a direct impact on the management of dynamic systems modelling the processes in these areas. Very often, decision makers work with models that are only approximations of the studied reality, generating the data. Sometimes it is not feasible to assign probabilities to the various scenarios, which renders the expected utility approach inapplicable. Faced with the problem of misspecification, the decision maker seeks rules ensuring the robustness of the proposed solution. A possible interpretation is that of a game against a hypothetical malevolent agent (nature), which chooses the disturbances such as to maximize the loss that the policy maker is trying to minimize. So, we accept the dynamic game's approach to the problem of robustness and consider a two-person non-cooperative discrete game on an infinite-time horizon. The aim is to create an analytical framework

This work has been partially supported by the Sofia University "St. Kliment Ohridski" Fund "Research & Development" under contract 80-10-180/27.05.2022, by the Bulgarian Ministry of Science and Higher Education National Fund for Science Research under contract KP-06-H22/4/ 04.12.2018, by the Center of Excellence in Informatics and ICT, Grant No. BG05M2OP001-1.001-0003 (financed by the Science and Education for Smart Growth Operational Program (2014-2020) and co-financed by the European Union through the European structural and investment funds).

© The Author(s), under exclusive license to Springer Nature Switzerland AG 2023
I. Georgiev et al. (Eds.): NMA 2022, LNCS 13858, pp. 209–220, 2023.
https://doi.org/10.1007/978-3-031-32412-3_19

using game theory to represent the decision maker's non-cooperative behavior in response to the intervention of these disturbances.

This paper is concerned with finding optimality conditions of Pontryagin's maximum principle type for discrete control systems in the presence of disturbances (uncertainty). Discrete-time systems are often more appropriate when the nature of the studied problem is discrete. Optimality is understood in the sense of a Nash equilibrium. So, the optimal control minimizes the "worst" value of the criterion.

The considered optimal control problem belongs to the class of optimization problems of the min-max type. The main result is a necessary optimality condition of Pontryagin's maximum principle type for a problem formulated in the context of a determined discrete zero-sum game. The proof of this result is based on a relatively recent result (Theorem 2.2) obtained in [1]. The main assumption (Assumption A) in this cited paper guarantees that the definition of the adjoint sequence $\psi := \{\psi_k\}_{k=1}^\infty$ produces a finite vector. Of course, this assumption rules out abnormal optimal processes for which the maximum principle cannot have a normal form. On the other hand, it is not restrictive for a large class of problems, including rather challenging ones, where the known optimality conditions do not hold or are not informative. We have to point out that the adjoint sequence ψ is explicitly defined. And this is a basic fact we use in the proof of our main result.

The paper is structured as follows: The next section contains some assumptions, definitions, and preliminaries needed for our exposition. The main result, a necessary optimality condition, and its proof are presented in Sect. 3. Section 4 presents an example illustrating the obtained results, which is a version of the linear quadratic controller often used for modelling and solving problems in various fields.

2   Statement of the Problem and Preliminaries

Denote by $\mathbb{R}^n$ the n-dimensional Euclidean space and by $\mathbb{N}$ – the set of all nonnegative integers. Let us fix a vector x_0 in an open subset $G \subset \mathbb{R}^n$. We consider a two-person discrete-time dynamic game on an infinite time horizon. The dynamics of the game is described by the following discrete-time control system:

$$x_{k+1} = f_k(x_k, u_k, v_k), \ u_k \in U_k, \ v_k \in V_k, \ k \in \mathbb{N}. \tag{1}$$

Here, x_0 is the initial state, and x_k denotes the state of the system at the moment of time k. Also, U_k and V_k, $k \in \mathbb{N}$, are non-empty, closed, and convex subsets of $\mathbb{R}^{m_u}$ and $\mathbb{R}^{m_v}$, respectively. We assume that each of the functions $f_k : G \times \tilde{U}_k \times \tilde{V}_k \to \mathbb{R}^n$, $k \in \mathbb{N}$, is continuously differentiable, where $\tilde{U}_k$ and $\tilde{V}_k$ are open sets satisfying the relations $\tilde{U}_k \supset U_k$ and $\tilde{V}_k \supset V_k$ for each $k \in \mathbb{N}$.

Both players influence the system through their choice of functions $u : \mathbb{N} \to \mathbb{R}^{m_u}$ for the first player and $v : \mathbb{N} \to \mathbb{R}^{m_v}$ for the second player.

We call a pair of control sequences $(\mathbf{u}, \mathbf{v})$, where $\mathbf{u} := \{u_k\}_{k=0}^{\infty}$, $\mathbf{v} := \{v_k\}_{k=0}^{\infty}$ *admissible* if for every $k \in \mathbb{N}$, the inclusions $u_k \in U_k$ and $v_k \in V_k$ hold true. By $\mathcal{U}$ and $\mathcal{V}$, we denote the sets of all admissible strategies of the first and second players, respectively.

For a given admissible control pair $(\mathbf{u}, \mathbf{v})$, the equality (1) generates the trajectory $x_0, x_1, \ldots$. Note that this trajectory may be extended either to infinity or to the minimal number k, for which the following relation holds true: $f_k(x_k, u_k, v_k) \notin G$ (if such a k exists). In the former case, we call the triple $(\mathbf{x}, \mathbf{u}, \mathbf{v})$, where $\mathbf{x} := \{x_k\}_{k=0}^{\infty}$ an *admissible process*.

Given an admissible process $(\mathbf{x}, \mathbf{u}, \mathbf{v})$, the state trajectory $\{x_k\}_{k=0}^{\infty}$ can be represented from (1) as

$$x_{k+1} := f_k^{(u_k, v_k)} \circ f_{k-1}^{(u_{k-1}, v_{k-1})} \circ \cdots \circ f_0^{(u_0, v_0)}(x_0), \quad k \in \mathbb{N}. \tag{2}$$

Here, $\circ$ denotes the composition of the corresponding maps and $f_k^{(u,v)}(x) := f_k(x, u, v)$.

Given the dynamics (1), we consider the following infinite-time horizon discrete dynamic game:

$$\min_{\mathbf{u} \in \mathcal{U}} J(x_0, \mathbf{u}, \mathbf{v}), \quad \max_{\mathbf{v} \in \mathcal{V}} J(x_0, \mathbf{u}, \mathbf{v}), \tag{3}$$

whose criterion is defined as:

$$J(x_0, \mathbf{u}, \mathbf{v}) = \sum_{k=0}^{\infty} g_k(x_k, u_k, v_k). \tag{4}$$

It is evident from (3) that the first player strives to "minimize" this criterion, while the aim of the second player is to "maximize" it. Here, the functions $g_k : G \times \tilde{U}_k \times \tilde{V}_k \to \mathbb{R}$, $k \in \mathbb{N}$, are assumed to be continuously differentiable.

For this antagonistic zero-sum game, the problem of looking for a *Nash equilibrium* is equivalent to the problem of finding a saddle point $(\bar{\mathbf{u}}, \bar{\mathbf{v}})$ of the criterion, i.e., an admissible control pair $(\bar{\mathbf{u}}, \bar{\mathbf{v}}) \in \mathcal{U} \times \mathcal{V}$ satisfying the inequalities

$$J(x_0, \bar{\mathbf{u}}, \mathbf{v}) \leq J(x_0, \bar{\mathbf{u}}, \bar{\mathbf{v}}) \leq J(x_0, \mathbf{u}, \bar{\mathbf{v}}) \text{ for every } (\mathbf{u}, \mathbf{v}) \in \mathcal{U} \times \mathcal{V}.$$

The triple $(\bar{\mathbf{x}}, \bar{\mathbf{u}}, \bar{\mathbf{v}})$, formed by a Nash equilibrium $(\bar{\mathbf{u}}, \bar{\mathbf{v}})$ and the corresponding state trajectory $\bar{\mathbf{x}}$, we call an *optimal process*.

Let $(\bar{\mathbf{x}}, \bar{\mathbf{u}}, \bar{\mathbf{v}})$ be an optimal process. For every $k \in \mathbb{N}$ and for each vector ξ, we denote by $x^{k,\xi} = (x_k, x_{k+1}, \ldots)$ the trajectory induced by (1) with an initial state of $x_k = \xi$ at time k, i.e.,

$$x_{s+1}^{k,\xi} := f_s^{(\bar{u}_s, \bar{v}_s)} \circ f_{s-1}^{(\bar{u}_{s-1}, \bar{v}_{s-1})} \circ \cdots \circ f_k^{(\bar{u}_k, \bar{v}_k)}(\xi), \quad s = k, k+1, \ldots .$$

Clearly, $x^{k,\xi}$ may happen to be an infinite sequence or may terminate at the minimal $s > k$ such that $f_s(x_s^{k,\xi}, \bar{u}_s, \bar{v}_s) \notin G$.

As in [1], we make the following assumption:

Assumption 1. *For every* $k = 0, 1, \ldots,$ *there exist* $\alpha_k > 0$ *and a sequence* $\{\beta_s^k\}_{s=k}^\infty$ *with* $\sum_{s=k}^\infty \beta_s^k < \infty$, *such that* $\mathbf{B}(\bar{x}_k; \alpha_k) \subset G$ *for every* $\xi \in \mathbf{B}(\bar{x}_k; \alpha_k)$, *the sequence* $x^{k,\xi}$ *is infinite and*

$$\sup_{\xi \in \mathbf{B}(\bar{x}_k; \alpha_k)} \left\| \frac{\partial}{\partial \xi} g_s(x_s^{k,\xi}, \bar{u}_s, \bar{v}_s) \right\| \le \beta_s^k,$$

where $\mathbf{B}(\bar{x}_k; \alpha_k)$ *is a closed ball in* $\mathbb{R}^n$ *centered at* $\bar{x}_k$ *with a radius* α_k.

Assumption 1 implies that the series

$$\sum_{s=k}^\infty \frac{\partial}{\partial \xi} g_s(x_s^{k,\xi}, \bar{u}_s, \bar{v}_s), \quad k = 1, 2, \ldots$$

is absolutely convergent, uniformly with respect to $\xi \in \mathbf{B}(\bar{x}_k; \alpha_k)$.

By the identity

$$g_s(x_s^{k,\xi}, \bar{u}_s, \bar{v}_s) = g_s(f_{s-1}^{(\bar{u}_{s-1}, \bar{v}_{s-1})} \circ f_{s-2}^{(\bar{u}_{s-2}, \bar{v}_{s-2})} \circ \cdots \circ f_k^{(\bar{u}_k, \bar{v}_k)}(\xi), \bar{u}_s, \bar{v}_s)$$

and the chain rule we have that

$$\frac{\partial}{\partial \xi} g_s(x_s^{k,\xi}, \bar{u}_s, \bar{v}_s) = \frac{\partial}{\partial x} g_s(x_s^{k,\xi}, \bar{u}_s, \bar{v}_s) \prod_{i=s-1}^k \frac{\partial}{\partial x} f_i(x_i^{k,\xi}, \bar{u}_i, \bar{v}_i), \tag{5}$$

where the following notation is used:

$$\prod_{i=s-1}^k A_i := \begin{cases} A_{s-1} A_{s-2} \ldots A_k & \text{if } s > k, \\ E & \text{if } s \le k. \end{cases}$$

Here, $A_i := \frac{\partial}{\partial x} f_i(x_i^{k,\xi}, \bar{u}_i, \bar{v}_i)$ and E denotes the identity matrix of dimension $n \times n$. Also, here we use the symbol $\prod_{i=s}^k$ instead of the usual symbol $\prod_{i=s}^k$ to indicate that the "increment" of the running index i is -1 (since $s > k$).

Following [1], we define the *adjoint sequence* $\psi := \{\psi_k\}_{k=1}^\infty$ as:

$$\psi_k = \sum_{s=k}^\infty \frac{\partial}{\partial \xi} g_s(x_s^{k,\xi}, \bar{u}_s, \bar{v}_s)_{|\xi=\bar{x}_k}, \quad k = 1, 2, \ldots. \tag{6}$$

Remark 1. Assumption 1 actually implies that $\|\psi_k\| < \infty$, $k = 1, 2, \ldots.$ Furthermore, we obtain from (5) that

$$\psi_k = \sum_{s=k}^\infty \frac{\partial}{\partial \xi} g_s(x_s^{k,\xi}, \bar{u}_s, \bar{v}_s)_{|\xi=\bar{x}_k}$$

(taking into account that $x_s^{k,\bar{x}_k} = \bar{x}_s$) $\tag{7}$

$$= \sum_{s=k}^\infty \frac{\partial}{\partial x} g_s(\bar{x}_s, \bar{u}_s, \bar{v}_s) \prod_{t=s-1}^k \frac{\partial}{\partial x} f_t(\bar{x}_t, \bar{u}_t, \bar{v}_t).$$

Due to the second equality in (7), it turns out that the so-defined adjoint sequence solves the *adjoint equation*

$$\psi_k = \psi_{k+1}\frac{\partial}{\partial x}f_k(\bar{x}_k,\bar{u}_k,\bar{v}_k) + \frac{\partial}{\partial x}g_k(\bar{x}_k,\bar{u}_k,\bar{v}_k), \quad k = 1,2,\dots. \tag{8}$$

In order to formulate a corollary of the main result (Theorem 2.2) in [1], we denote by $T_S(\bar{y})$ the *Bouligand tangent cone* to the closed subset $S \subset \mathbb{R}^m$ at the point $\bar{y} \in S$. We remind that $T_S(\bar{y})$ consists of all $w \in \mathbb{R}^m$ such that there exist a sequence of positive real numbers $\{t_\mu\}_{\mu=1}^\infty$ convergent to 0 and a sequence $\{w_\mu\}_{\mu=1}^\infty \in \mathbb{R}^m$ convergent to w such that $\bar{y} + t_\mu w_\mu \in S$ for each $\mu = 1,2,\dots$ (see [2], Chap. 4.1). Also, for each $k \in \mathbb{N}$, we define the matrix Z_k as

$$Z_k :=$$

$$= \frac{\partial}{\partial x}f_{k-1}(\bar{x}_{k-1},\bar{u}_{k-1},\bar{v}_{k-1})\frac{\partial}{\partial x}f_{k-2}(\bar{x}_{k-2},\bar{u}_{k-2},\bar{v}_{k-2})\dots\frac{\partial}{\partial x}f_0(\bar{x}_0,\bar{u}_0,\bar{v}_0). \tag{9}$$

Corollary 1. *Let the triple* $(\bar{\mathbf{x}},\bar{\mathbf{u}},\bar{\mathbf{v}})$ *be an optimal process. Let the adjoint sequence* $\psi = \{\psi_k\}_{k=1}^\infty$ *be defined by (6) and Assumption 1 be fulfilled. Then, for every* $k \in \mathbb{N}$, *the following local maximum condition holds true:*

$$\left(\frac{\partial}{\partial v}g_k(\bar{x}_k,\bar{u}_k,\bar{v}_k) + \psi_{k+1}\frac{\partial}{\partial v}f_k(\bar{x}_k,\bar{u}_k,\bar{v}_k)\right)w \leq 0 \quad \textit{for every } w \in T_{V_k}(\bar{v}_k) \tag{10}$$

as well as the transversality condition

$$\lim_{k\to+\infty}\psi_k Z_k = 0. \tag{11}$$

Remark 2. The function $\mathcal{H}_k : G \times \mathbb{R}^{m_u} \times \mathbb{R}^{m_v} \times \mathbb{R}^n \to \mathbb{R}$, defined as

$$\mathcal{H}_k(x,u,v,\psi) := g_k(x,u,v) + \psi f_k(x,u,v),$$

is called a *Hamiltonian.* Using it, the relations (1), (8), and (10) can be written as follows:

$$\bar{x}_{k+1} = \frac{\partial}{\partial \psi_{k+1}}\mathcal{H}_k(\bar{x}_k,\bar{u}_k,\bar{v}_k,\psi_{k+1}), \tag{12}$$

$$\psi_k = \frac{\partial}{\partial x_k}\mathcal{H}_k(\bar{x}_k,\bar{u}_k,\bar{v}_k,\psi_{k+1}), \tag{13}$$

and

$$\frac{\partial}{\partial v}\mathcal{H}_k(\bar{x}_k,\bar{u}_k,\bar{v}_k,\psi_{k+1})w \leq 0 \quad \text{for every } w \in T_{V_k}(\bar{v}_k). \tag{14}$$

3   A Necessary Optimality Condition for the Existence of a Saddle Point

Let the triple $(\bar{\mathbf{x}}, \bar{\mathbf{u}}, \bar{\mathbf{v}})$ be an optimal process and let the adjoint sequence $\psi = \{\psi_k\}_{k=1}^{\infty}$ be defined by (7). For our main result, we need the following assumption:

Assumption 2. *The following conditions hold true:*

(i) *The function $g_k(\bar{x}_k, \cdot, \bar{v}_k) : U_k \to \mathbb{R}$ is convex. When $\psi_{k+1}^j > 0$ (the j-component of ψ_{k+1}), the j-component $f_k^j(\bar{x}_k, \cdot, \bar{v}_k) : U_k \to \mathbb{R}$ of the vector function $f_k(\bar{x}_k, \cdot, \bar{v}_k) : U_k \to \mathbb{R}^n$ is convex and when $\psi_{k+1}^j \leq 0$, it is concave.*

(ii) *The function $g_k(\bar{x}_k, \bar{u}_k, \cdot) : V_k \to \mathbb{R}$ is concave; When $\psi_{k+1}^j \geq 0$ (the j-component of ψ_{k+1}), the j-component $f_k^j(\bar{x}_k, \bar{u}_k, \cdot) : V_k \to \mathbb{R}$ of the vector function $f_k(\bar{x}_k, \bar{u}_k, \cdot) : V_k \to \mathbb{R}^n$ is concave and when $\psi_{k+1}^j < 0$, it is convex.*

Remark 3. Assumption 2 implies that for every $k \in \mathbb{N}$, the Hamiltonian function $\mathcal{H}_k(\bar{x}_k, \cdot, \bar{v}_k, \psi_{k+1}) : U_k \to \mathbb{R}$ is convex, while the function $\mathcal{H}_k(\bar{x}_k, \bar{u}_k, \cdot, \psi_{k+1}) : V_k \to \mathbb{R}$ is concave.

Theorem 1. *Let the triple $(\bar{\mathbf{x}}, \bar{\mathbf{u}}, \bar{\mathbf{v}})$ be an optimal process and let the adjoint sequence $\psi = \{\psi_k\}_{k=1}^{\infty}$ be defined by (7). Let the Assumptions 1 and 2 hold true. Then, the adjoint sequence $\psi = \{\psi_k\}_{k=1}^{\infty}$ solves the adjoint system (13), and for every $k \in \mathbb{N}$, the following conditions are satisfied:*

(i) $\min\limits_{u \in U_k} \mathcal{H}_k(\bar{x}_k, u, \bar{v}_k, \psi_{k+1}) = \mathcal{H}_k(\bar{x}_k, \bar{u}_k, \bar{v}_k, \psi_{k+1}) = \max\limits_{v \in V_k} \mathcal{H}_k(\bar{x}_k, \bar{u}_k, v, \psi_{k+1});$

(ii) *transversality condition*

$$\lim_{k \to +\infty} \psi_k Z_k = 0,$$

where Z_k is defined by (9).

Proof. The triple $(\bar{\mathbf{x}}, \bar{\mathbf{u}}, \bar{\mathbf{v}})$ being an optimal process means that $(\bar{\mathbf{u}}, \bar{\mathbf{v}})$ is a saddle point of the criterion, i.e., $(\bar{\mathbf{u}}, \bar{\mathbf{v}}) \in \mathcal{U} \times \mathcal{V}$ satisfies the inequalities

$$J(x_0, \bar{\mathbf{u}}, \mathbf{v}) \leq J(x_0, \bar{\mathbf{u}}, \bar{\mathbf{v}}) \leq J(x_0, \mathbf{u}, \bar{\mathbf{v}}) \text{ for every } (\mathbf{u}, \mathbf{v}) \in \mathcal{U} \times \mathcal{V}. \quad (15)$$

Let us consider the following optimal control problem (P_v):

$$\sum_{k=0}^{\infty} g_k(x_k, \bar{u}_k, v_k) \longrightarrow \max$$

subject to

$$x_{k+1} = f_k(x_k, \bar{u}_k, v_k), \ x_0 \in G, v_k \in V_k, \ k = 0, 1, \dots .$$

The first inequality in (15) implies that the sequence $\bar{\mathbf{v}} := \{\bar{v}_k\}_{k=0}^{\infty}$ is a solution of (P_v).

Define the adjoint sequence $\psi^v = \{\psi_k^v\}_{k=1}^{\infty}$ by the equality

$$\psi_k^v = \sum_{s=k}^{\infty} \frac{\partial}{\partial \xi} g_s(x_s^{k,\xi}, \bar{u}_s, \bar{v}_s)_{|\xi=\bar{x}_k}, \quad k = 1, 2, \ldots$$

as well as the Hamiltonian

$$\mathcal{H}_k^+(x, u, v, \psi^v) = g_k(x, u, v) + \psi_{k+1}^v f_k(x, u, v), \quad k \in \mathbb{N}.$$

Corollary 1 implies that for each $k \in \mathbb{N}$,

$$\frac{\partial}{\partial v} \mathcal{H}_k^+(\bar{x}_k, \bar{u}_k, \bar{v}_k, \psi_{k+1}^v) w^v \leq 0 \text{ for every } w^v \in T_{V_k}(\bar{v}_k) \tag{16}$$

($T_{V_k}(\bar{v}_k)$ is the Bouligand tangent cone to the set V_k at the point $\bar{v}_k \in V_k$) and that for each $k = 1, 2, \ldots$, the sequence ψ^v satisfies the adjoint equation

$$\psi_k^v = \frac{\partial}{\partial x} \mathcal{H}_k^+(\bar{x}_k, \bar{u}_k, \bar{v}_k, \psi_{k+1}^v) = \frac{\partial}{\partial x} g_k(\bar{x}_k, \bar{u}_k, \bar{v}_k) + \psi_{k+1}^v \frac{\partial}{\partial x} f_k(\bar{x}_k, \bar{u}_k, \bar{v}_k)$$

together with the transversality condition

$$\lim_{k \to +\infty} \psi_k^v Z_k = 0.$$

Let us fix an arbitrary element $v_k \in V_k$. Consider the sequence of positive reals $\{\theta_\mu\}_{\mu=1}^{\infty}$ from the interval $(0, 1)$ that tends to 0 as μ tends to $+\infty$. The convexity of V_k implies that the inclusion $\bar{v}_k + \theta_\mu(v_k - \bar{v}_k) = \theta_\mu v_k + (1 - \theta_\mu)\bar{v}_k \in V_k$ is fulfilled for every $\mu = 1, 2, \ldots$. Using the definition of the Bouligand tangent cone, we have that $(v_k - \bar{v}_k) \subset T_{V_k}(\bar{v}_k)$.

According to Remark 3, the function $\mathcal{H}_k^+(\bar{x}_k, \bar{u}_k, \cdot, \psi_{k+1}^v) : V_k \to \mathbb{R}^n$ is concave. Because $\frac{\partial}{\partial v} \mathcal{H}_k^+(\bar{x}_k, \bar{u}_k, \bar{v}_k, \psi_{k+1}^v)$ belongs to the subdifferential of the Hamiltonian at $(\bar{x}_k, \bar{u}_k, \bar{v}_k, \psi_{k+1}^v)$, the definition of a subgradient of a concave function implies that

$$\mathcal{H}_k^+(\bar{x}_k, \bar{u}_k, v_k, \psi_{k+1}^v) - \mathcal{H}_k^+(\bar{x}_k, \bar{u}_k, \bar{v}_k, \psi_{k+1}^v) \leq \frac{\partial}{\partial v} \mathcal{H}_k^+(\bar{x}_k, \bar{u}_k, \bar{v}_k, \psi_{k+1}^v)(v_k - \bar{v}_k).$$

Taking into account (16), we obtain that

$$\mathcal{H}_k^+(\bar{x}_k, \bar{u}_k, v_k, \psi_{k+1}^v) - \mathcal{H}_k^+(\bar{x}_k, \bar{u}_k, \bar{v}_k, \psi_{k+1}^v) \leq 0.$$

Because v_k is an arbitrary element of V_k, we obtain that

$$\mathcal{H}_k^+(\bar{x}_k, \bar{u}_k, \bar{v}_k, \psi_{k+1}^v) = \max_{v_k \in V_k} \mathcal{H}_k^+(\bar{x}_k, \bar{u}_k, v_k, \psi_{k+1}^v).$$

Similarly, we fix the control $\bar{\mathbf{v}}$ and consider the following optimal control problem (P_u):

$$-\sum_{k=0}^{\infty} g_k(x_k, u_k, \bar{v}_k) \longrightarrow \max$$

subject to:

$$x_{k+1} = f_k(x_k, u_k, \bar{v}_k), \ x_0 \in G, u_k \in U_k, \ k \in \mathbb{N}.$$

The second inequality in (15) implies that the sequence $\bar{\mathbf{u}} := \{\bar{u}_k\}_{k=0}^{\infty}$ is a solution of (P_u).

Define the adjoint sequence $\psi^u = \{\psi_k^u\}_{k=1}^{\infty}$ as:

$$\psi_k^u = -\sum_{s=k}^{\infty} \frac{\partial}{\partial \xi} g_s(x_s^{k,\xi}, \bar{u}_s, \bar{v}_s)_{|\xi=\bar{x}_k}, \quad k \in \mathbb{N}$$

and the Hamiltonian as:

$$\mathcal{H}_k^-(x, u, v, \psi^u) = -g_k(x, u, v) + \psi_{k+1}^u f_k(x, u, v), \ k \in \mathbb{N}.$$

Clearly, $\psi_k^u = -\psi_k^v$ for each $k = 1, 2, \ldots$, and thus

$$\mathcal{H}_k^-(x, u, v, \psi_{k+1}^u) = -g_k(x, u, v) + \psi_{k+1}^u f_k(x, u, v)$$
$$= -\Big(g_k(x, u, v) - \psi_{k+1}^u f_k(x, u, v)\Big) \qquad (17)$$
$$= -\Big(g_k(x, u, v) + \psi_{k+1}^v f_k(x, u, v)\Big) = -\mathcal{H}_k^+(x, u, v, \psi_{k+1}^v).$$

Corollary 1 implies that for each $k \in \mathbb{N}$,

$$\frac{\partial}{\partial u} \mathcal{H}_k^-(\bar{x}_k, \bar{u}_k, \bar{v}_k, \psi_{k+1}^u) w^u \leq 0 \quad \text{for every } w^u \in T_{U_k}(\bar{u}_k) \qquad (18)$$

$(T_{U_k}(\bar{u}_k)$ is the Bouligand tangent cone to the set U_k at the point $\bar{u}_k \in U_k$) and that for each $k = 1, 2, \ldots$, the sequence ψ^u satisfies the adjoint equation

$$\psi_k^u = \frac{\partial}{\partial x} \mathcal{H}_k^-(\bar{x}_k, \bar{u}_k, \bar{v}_k, \psi_{k+1}^u) = -\frac{\partial}{\partial x} g_k(\bar{x}_k, \bar{u}_k, \bar{v}_k) + \psi_{k+1}^u \frac{\partial}{\partial x} f_k(\bar{x}_k, \bar{u}_k, \bar{v}_k)$$

together with the transversality condition

$$\lim_{k \to +\infty} \psi_k^u Z_k = 0.$$

Let us fix an arbitrary element $u_k \in U_k$. As before, one can prove that the vector $(u_k - \bar{u}_k)$ belongs to $T_{U_k}(\bar{u}_k)$, hence (18) implies that

$$\frac{\partial}{\partial u} \mathcal{H}_k^-(\bar{x}_k, \bar{u}_k, \bar{v}_k, \psi_{k+1}^u)(u_k - \bar{u}_k) \leq 0 \text{ for every } u_k \in U_k. \qquad (19)$$

According to Remark 3, the function $\mathcal{H}_k^+(\bar{x}_k, \cdot, \bar{v}_k, \psi_{k+1}^v) : U_k \to \mathbb{R}^n$ is convex on U_k. Considering (17), we obtain that

$$\mathcal{H}_k^-(\bar{x}_k, \cdot, \bar{v}_k, \psi_{k+1}^u) = -\mathcal{H}_k^+(\bar{x}_k, \cdot, \bar{v}_k, \psi_{k+1}^v),$$

hence the function $\mathcal{H}_k^-(\bar{x}_k, \cdot, \bar{v}_k, \psi_{k+1}^u) : U_k \to \mathbb{R}^n$ is concave on U_k. From here, taking into account (18), analogously, we obtain that

$$\mathcal{H}_k^-(\bar{x}_k, \bar{u}_k, \bar{v}_k, \psi_{k+1}^u) = \max_{u_k \in U_k} \mathcal{H}_k^-(\bar{x}_k, u_k, \bar{v}_k, \psi_{k+1}^u).$$

Furthermore, one can check directly that

$$
\begin{aligned}
\max_{v_k \in V_k} \mathcal{H}_k^+(\bar{x}_k, \bar{u}_k, v_k, \psi_{k+1}^v) &= \mathcal{H}_k^+(\bar{x}_k, \bar{u}_k, \bar{v}_k, \psi_{k+1}^v) \\
&= g_k(\bar{x}_k, \bar{u}_k, \bar{v}_k) + \psi_{k+1}^v f_k(\bar{x}_k, \bar{u}_k, \bar{v}_k) \\
&= g_k(\bar{x}_k, \bar{u}_k, \bar{v}_k) - \psi_{k+1}^u f_k(\bar{x}_k, \bar{u}_k, \bar{v}_k) \\
&= -\left(-g_k(\bar{x}_k, \bar{u}_k, \bar{v}_k) + \psi_{k+1}^u f_k(\bar{x}_k, \bar{u}_k, \bar{v}_k)\right) \\
&= -\mathcal{H}_k^-(\bar{x}_k, \bar{u}_k, \bar{v}_k, \psi_{k+1}^u) \\
&= -\max_{u_k \in U_k} \mathcal{H}_k^-(\bar{x}_k, u_k, \bar{v}_k, \psi_{k+1}^u) \\
&= -\max_{u_k \in U_k} \left\{-g_k(\bar{x}_k, u_k, \bar{v}_k) + \psi_{k+1}^u f_k(\bar{x}_k, u_k, \bar{v}_k)\right\} \\
&= \min_{u_k \in U_k} \left\{g_k(\bar{x}_k, u_k, \bar{v}_k) - \psi_{k+1}^u f_k(\bar{x}_k, u_k, \bar{v}_k)\right\} \\
&= \min_{u_k \in U_k} \left\{g_k(\bar{x}_k, u_k, \bar{v}_k) + \psi_{k+1}^v f_k(\bar{x}_k, u_k, \bar{v}_k)\right\} \\
&= \min_{u_k \in U_k} \mathcal{H}_k^+(\bar{x}_k, u_k, \bar{v}_k, \psi_{k+1}^v).
\end{aligned}
$$

We finish the proof by setting $\psi := \psi^v$ and $\mathcal{H} := \mathcal{H}^+$.   □

4   Example

The well-known linear quadratic regulators are usually used for modelling and solving problems in engineering, economics, and other fields (see, for example, [4], Chap. 1, Example 1.3.3; [6], Chap. 4.2, Problem 11; [7]; Chap. 5.9.2, Example 2; [5,8]).

Below, we present a discrete version of a simple linear quadratic regulator. The aim is only to demonstrate the applicability of our result. We examine a deterministic game whose dynamics is determined by the following system:

$$x_{k+1} = x_k + u_k + 2v_k, \quad x_0 \in \mathbb{R}, \ u_k \in \mathbb{R}, \ v_k \in \mathbb{R}, \ k \in \mathbb{N}. \tag{20}$$

The players influence the state variable x_k, $k \in \mathbb{N}$, by selecting the controls u_k and v_k (for player 1 and player 2, respectively). Given (20), we consider the following infinite-time horizon linear quadratic game:

$$\min_u J(x_0, \mathbf{u}, \mathbf{v}), \ \max_v J(x_0, \mathbf{u}, \mathbf{v}), \tag{21}$$

where the objective function of the game is defined by the following quadratic functional:

$$J(x_0, \mathbf{u}, \mathbf{v}) = \sum_{k=0}^{\infty} \left(\frac{1}{6}x_k^2 + u_k^2 - 5v_k^2 \right).$$

Let the control pair $(\bar{\mathbf{u}}, \bar{\mathbf{v}}) \in \mathcal{U} \times \mathcal{V}$ determine a Nash equilibrium. The corresponding Hamiltonian functions are

$$\mathcal{H}_k(x_k, u_k, v_k, \psi_{k+1}) = \left(\frac{1}{6}x_k^2 + u_k^2 - 5v_k^2 \right) + \psi_{k+1}\left(x_k + u_k + 2v_k \right), k \in \mathbb{N},$$

where the adjoint sequence ψ is defined by (7).

Using Theorem 1, for every $k \in \mathbb{N}$, we obtain:

– the adjoint equation:

$$\psi_k = \frac{1}{3}\bar{x}_k + \psi_{k+1} \tag{22}$$

– an equation derived from the Hamiltonian minimization with respect to u_k:

$$0 = 2\bar{u}_k + \psi_{k+1}, \tag{23}$$

– an equation derived from the Hamiltonian maximization with respect to v_k:

$$0 = -10\bar{v}_k + 2\psi_{k+1}, \tag{24}$$

We can deduce from the last two equations that

$$\bar{u}_k = -\frac{5}{2}\bar{v}_k \text{ for every } k \in \mathbb{N}. \tag{25}$$

From (20), (22), (24), and (25) we obtain that

$$\frac{1}{3}\bar{x}_k = \psi_k^v - \psi_{k+1}^v = 5\bar{v}_{k-1} - 5\bar{v}_k = 10(\bar{x}_{k-1} - \bar{x}_k) - 10(\bar{x}_k - \bar{x}_{k+1}),$$

which is equivalent to the following difference equation:

$$\bar{x}_{k+1} - \frac{61}{30}\bar{x}_k + \bar{x}_{k-1} = 0.$$

The general solution of this equation is given by:

$$\bar{x}_k = c_1 r_1^k + c_2 r_2^k, \tag{26}$$

where $r_1 = 6/5$ and $r_2 = 5/6$ are the roots of the following quadratic equation:

$$x^2 - \frac{61}{30}x + 1 = 0. \tag{27}$$

From the adjoint equation (22) and the transversality condition $\lim_{k\to\infty}\psi_k = 0$, we obtain that

$$\bar{x}_k \to 0. \tag{28}$$

From here and taking into account (26), we get the equality

$$\bar{x}_k = c_1(r_1)^k + c_2(r_2)^k. \tag{29}$$

Finally, using (28) and (29), we conclude that $c_1 = 0$. Using that the starting point is x_0, we obtain that $c_2 = x_0$. Hence

$$\bar{x}_k = r_2^k x_0.$$

Furthermore, from the dynamics (20), we obtain the state feed-back control laws

$$\bar{u}_k = 5\Big(r_2^{k+1}x_0 - r_2^k x_0\Big) = 5\Big(\frac{5}{6} - 1\Big)r_2^k x_0 = -\frac{5}{6}\bar{x}_k,$$

$$\bar{v}_k = -2\Big(r_2^{k+1}x_0 - r_2^k x_0\Big) = -2\Big(\frac{5}{6} - 1\Big)r_2^k x_0 = \frac{1}{3}\bar{x}_k.$$

Using (7), one can directly calculate the adjoint sequence $\{\psi_k\}_{k=1}^{\infty}$. Indeed, the following equalities hold true:

$$\psi_k = \sum_{s=k}^{\infty} \frac{\partial}{\partial\xi} g_s(x_s^{k,\xi}, \bar{u}_s, \bar{v}_s)_{|\xi=\bar{x}_k}$$

$$= \sum_{s=k}^{\infty} \frac{\partial}{\partial x} g_s(x_s^{k,\xi}, \bar{u}_s, \bar{v}_s) \prod_{i=k}^{s-1} \frac{\partial}{\partial x} f_i(x_i^{k,\xi}, \bar{u}_i, \bar{v}_i)_{|\xi=\bar{x}_k}$$

$$= \sum_{s=k}^{\infty} \frac{\bar{x}_s}{3} = \frac{x_0}{3}\sum_{s=k}^{\infty} r_2^s = \frac{x_0 r_2^k}{3(1-r_2)}$$

for every $k = 1, 2, \dots$.

To ensure that the obtained triple $(\bar{\mathbf{x}}, \bar{\mathbf{u}}, \bar{\mathbf{v}})$ is an optimal process, one can apply Bellman's optimality principle (cf., for example, [3], Chap. 3), and prove that the control pair $(\bar{\mathbf{u}}, \bar{\mathbf{v}})$ determines a Nash equilibrium for this example.

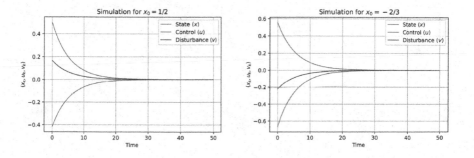

Fig. 1. State trajectory and control functions.

Finally, the simulations of the state trajectories and the control functions for the initial values $x_0 = 1/2$ and $x_0 = -2/3$ are shown in Fig. 1. As illustrated in the figure, the state and closed-loop trajectories converge in both cases. These values are obtained for $k \to 50$.

References

1. Aseev, S.M., Krastanov, M.I., Veliov, V.M.: Optimality conditions for discrete-time optimal control on infinite horizon. Pure Appl. Funct. Anal. **2**, 395–409 (2017)
2. Aubin, J.P., Frankowska, H.: Set-valued analysis. Birkhauser, Boston (1990)
3. Başar, T., Bernhard, P.: H^∞-Optimal Control and Related Minimax Design Problems. MBC, Birkhäuser Boston, Boston, MA (2008). https://doi.org/10.1007/978-0-8176-4757-5
4. Bertsekas, D.P.: Dynamic Programming and Optimal Control 4th Edition, Athena Scientific, Belmont, Massachusetts (2017)
5. Cochrane, J.: The New Keynsian Liquidity Trap, NBER Working Paper 19476. Cambridge, MA (2013)
6. Geering, H.P.: Optimal Control with Engineering Applications. Springer-Verlag, Berlin Heidelberg (2007). https://doi.org/10.1007/978-3-540-69438-0
7. Glizer, V.Y., Kelis, O.: Singular linear-quadratic zero-sum differential games and H_∞ control problems: regularization approach, Birkhäuser (2022)
8. Werning, I.: Managing a liquidity trap: monetary and fiscal policy. MIT Working Paper, Cambridge, MA (2012)

An Inequality for Polynomials
on the Standard Simplex

Lozko Milev$^{(\boxtimes)}$ and Nikola Naidenov

Faculty of Mathematics and Informatics, Sofia University "St. Kliment Ohridski", 5,
James Bourchier Blvd., 1164 Sofia, Bulgaria
{milev,nikola}@fmi.uni-sofia.bg

Abstract. Let $\Delta := \{(x,y) \in \mathbb{R}^2 : x \geq 0,\ y \geq 0,\ x + y \leq 1\}$ be the standard simplex in $\mathbb{R}^2$ and $\partial\Delta$ be the boundary of Δ. We use the notations $\|f\|_\Delta$ and $\|f\|_{\partial\Delta}$ for the uniform norm of a continuous function f on Δ and $\partial\Delta$, respectively.

Denote by π_n the set of all real algebraic polynomials of two variables and of total degree not exceeding n. Let $B_\Delta := \{p \in \pi_2 : \|p\|_\Delta \leq 1\}$ and $B_{\partial\Delta} := \{p \in \pi_2 : \|p\|_{\partial\Delta} \leq 1\}$.

Recently we described the set of all extreme points of B_Δ. In the present paper we give a full description of the strictly definite extreme points of $B_{\partial\Delta}$. As an application we prove the following sharp inequality:

$$\|p\|_\Delta \leq \frac{5}{3}\|p\|_{\partial\Delta}, \quad \text{for every } p \in \pi_2.$$

We also establish two generalizations of the above inequality, for any triangle in $\mathbb{R}^2$ and for any dimension $d \geq 2$.

We hope that our results can be useful in studying numerical problems related with estimates for uniform norms of polynomials and splines.

Keywords: Polynomial inequalities · Extreme points · Convexity

1 Introduction and Statement of the Results

Denote by π_n the set of all real algebraic polynomials of two variables and of total degree not exceeding n. Let

$$\Delta := \{(x,y) \in \mathbb{R}^2 : x \geq 0,\ y \geq 0,\ x + y \leq 1\}$$

be the standard simplex in $\mathbb{R}^2$ and $\partial\Delta$ be the boundary of Δ.

We use the notations $\|f\|_\Delta$ and $\|f\|_{\partial\Delta}$ for the uniform norm of f on Δ and $\partial\Delta$, respectively. Let $B_\Delta := \{p \in \pi_2 : \|p\|_\Delta \leq 1\}$ and $B_{\partial\Delta} := \{p \in \pi_2 : \|p\|_{\partial\Delta} \leq 1\}$ be the unit balls corresponding to these norms.

Recall that a point p of a convex set B in a linear space is *extreme* if the equality $p = \lambda p_1 + (1 - \lambda)p_2$, for some $p_1, p_2 \in B$ and $\lambda \in (0,1)$, implies $p_1 = p_2 = p$.

Supported by the Sofia University Research Fund under Contract 80–10-111/ 13.05.2022.

I. Georgiev et al. (Eds.): NMA 2022, LNCS 13858, pp. 221–232, 2023.
https://doi.org/10.1007/978-3-031-32412-3_20

According to the Krein-Milman theorem, every nonempty, compact and convex set B in a Banach space has an extreme point. Moreover, if E denotes the set of all extreme points of B, then B can be represented as the closure of the convex hull of E. This result motivates the study of the extreme points of the unit ball of various polynomial spaces. We refer to the papers [3,5–13]. The description of the extreme points can be useful in finding exact constants in polynomial inequalities involving convex functionals, see [1,2,4,17,18].

We described in [14–16] the set of all extreme points of B_Δ. In the present paper we give a full description of the strictly definite extreme points of $B_{\partial\Delta}$.

In order to formulate our results we need some additional notations. Let $E_{\partial\Delta}$ be the set of all extreme points of $B_{\partial\Delta}$ and $E_{\partial\Delta}^-$ be the subset of $E_{\partial\Delta}$ which consists of the strictly concave polynomials.

We denote the vertices of Δ by $O(0,0)$, $A(1,0)$ and $B(0,1)$. The sides OA, OB and AB of Δ will be denoted by l_1, l_2 and l_3, respectively. Let $\vec{l_1} = \overrightarrow{OA}$, $\vec{l_2} = \overrightarrow{OB}$, $\vec{l_3} = \overrightarrow{AB}$.

First we give a qualitative characterization of the elements of $E_{\partial\Delta}^-$.

Theorem 1. *Let $p \in \pi_2$ be a strictly concave function. Then $p \in E_{\partial\Delta}^-$ if and only if the following conditions hold true:*

(i) *For every $i = 1, 2, 3$ there exists a unique point $X_i \in \operatorname{int} l_i$ such that*
$$p(X_i) = 1 \text{ and } \frac{\partial p}{\partial \vec{l_i}}(X_i) = 0;$$
(ii) $\min\{p(O), p(A), p(B)\} = -1$.

Note that all strictly convex elements of $E_{\partial\Delta}$ have the form $q = -p$, where $p \in E_{\partial\Delta}^-$.

The next result provides an explicit form of the elements of $E_{\partial\Delta}^-$.

Theorem 2. *A polynomial $p \in \pi_2$ is a strictly concave element of $E_{\partial\Delta}$ if and only if*
$$p(x,y) = a + bx + cy + dx^2 + 2exy + fy^2, \tag{1}$$

where

$$
\begin{aligned}
a &= \gamma, \\
b &= 2\sqrt{1-\gamma}\left(\sqrt{1-\alpha} + \sqrt{1-\gamma}\right), \\
c &= 2\sqrt{1-\gamma}\left(\sqrt{1-\beta} + \sqrt{1-\gamma}\right), \\
d &= -\left(\sqrt{1-\alpha} + \sqrt{1-\gamma}\right)^2, \\
e &= \sqrt{1-\alpha}\sqrt{1-\beta} - \sqrt{1-\gamma}\left(\sqrt{1-\alpha} + \sqrt{1-\beta} + \sqrt{1-\gamma}\right), \\
f &= -\left(\sqrt{1-\beta} + \sqrt{1-\gamma}\right)^2,
\end{aligned}
\tag{2}
$$

and

$$\alpha, \beta, \gamma \in [-1, 1), \quad \min\{\alpha, \beta, \gamma\} = -1. \tag{3}$$

The following theorem provides a sharp inequality connecting the norms of a polynomial $f \in \pi_2$ on the standard triangle and its boundary.

Theorem 3. *The inequality*

$$\|f\|_\Delta \le \frac{5}{3}\,\|f\|_{\partial\Delta} \tag{4}$$

holds true for every $f \in \pi_2$. The equality is attained if and only if $f = cf^*$, where c is a constant and

$$f^*(x,y) = -1 + 8(x+y) - 8(x^2 + xy + y^2). \tag{5}$$

Note that the extremal polynomial f^* has the following geometrical properties: $\min_{X \in \Delta} f^*(X) = -1$ is attained only at the vertices of Δ and $\max_{X \in \Delta} f^*(X) = \frac{5}{3}$ is attained only at the centroid of Δ. In addition, $\|f^*\|_{\partial\Delta} = 1$ is attained only at the vertices and the midpoints of the sides of Δ.

The inequality (4) can be generalized for any triangle in $\mathbb{R}^2$. More precisely, the following result holds true.

Theorem 4. *Let T be a given triangle. For every $f \in \pi_2$ we have the inequality*

$$\max_{X \in T} |f(X)| \le \frac{5}{3} \max_{X \in \partial T} |f(X)|. \tag{6}$$

The equality is attained if and only if $f(X) = cf^*(L(X))$, where f^* is defined by (5), c is a constant and L is an affine one-to-one transformation from T to Δ.

Another extension of inequality (4) is for any dimension $d \ge 3$. In order to formulate the result, we need some additional notations.

Let π_n^d be the set of all algebraic polynomials of d variables and of total degree not exceeding n and $\Delta_r^d := \{(x_1, \ldots, x_d) \in \mathbb{R}^d : x_i \ge 0,\ i = 1, \ldots, d,\ \sum_{i=1}^d x_i \le r\}$, $r > 0$.

Theorem 5. *The inequality*

$$\max_{X \in \Delta_r^d} |f(X)| \le \frac{5}{3} \max_{X \in \partial\Delta_r^d} |f(X)| \tag{7}$$

holds true for every $f \in \pi_2^d$, $d \ge 3$.

2   Proofs of Theorems 1 and 2

Lemma 1. *Suppose that $p \in E_{\partial\Delta}$ and p is not a constant. Then p attains the values ± 1 on $\partial\Delta$.*

Proof. For definiteness we shall prove that p attains the value 1 at some point of $\partial\Delta$. Let us assume that, on the contrary $A := \max_{X \in \partial\Delta} p(X) < 1$. We consider the polynomials $f_{\pm\epsilon} := p \pm \epsilon q$, where $q := 1 + p$ and $\epsilon > 0$. Note that $q \not\equiv 0$. We shall prove that $f_{\pm\epsilon} \in B_{\partial\Delta}$ for every sufficiently small ϵ, which will be a contradiction with the definition of an extreme point.

Since $q \ge 0$ we have $f_{+\epsilon}(X) \ge p(X) \ge -1$ for every $X \in \partial\Delta$. The assumption $A < 1$ implies $f_{+\epsilon}(X) \le A + \epsilon q < 1$ ($X \in \partial\Delta$), provided ϵ is sufficiently small. Therefore $f_{+\epsilon} \in B_{\partial\Delta}$.

The fact that $f_{-\epsilon} \in B_{\partial\Delta}$ follows from the representation $f_{-\epsilon} = (1-\epsilon)p + \epsilon r$, where $r := -1$. Lemma 1 is proved. □

We shall say that $X \in \partial\Delta$ is a *critical* point of a function $f \in C(\partial\Delta)$ if $|f(X)| = \|f\|_{C(\partial\Delta)}$.

Recall that if $\vec{v}$ is a nonzero vector in $\mathbb{R}^n$ and $X \in \mathbb{R}^n$ then the derivative of a function f at the point X in direction $\vec{v}$ is defined by

$$\frac{\partial f}{\partial \vec{v}}(X) = \lim_{t \to 0+} \frac{f(X + t\vec{v}) - f(X)}{t}.$$

The next lemma gives a sufficient condition for a $p \in B_{\partial\Delta}$ to be an extreme point.

Lemma 2. *Let $p \in \pi_2$ and $\|p\|_{\partial\Delta} = 1$. Suppose that $q \equiv 0$ is the only polynomial from π_2, which satisfies the conditions:*

(i) *$q(X) = 0$ for every critical point X of p;*

(ii) *If $X \in \partial\Delta$ is a critical point of p and $\dfrac{\partial p}{\partial \overrightarrow{XM}}(X) = 0$ for a point $M \neq X$,*

$[XM] \subset \partial\Delta$ then $\dfrac{\partial q}{\partial \overrightarrow{XM}}(X) = 0$.

Then p is an extreme point of $B_{\partial\Delta}$.

Proof. We assume on the contrary that p is not an extreme point of $B_{\partial\Delta}$, i.e. there exist polynomials $p_1, p_2 \in B_{\partial\Delta}$, $p_1 \neq p_2$ and a number $\lambda \in (0,1)$ such that $p = \lambda p_1 + (1 - \lambda)p_2$. We shall show that $q := p_1 - p_2$ satisfies (i) and (ii), hence $q = 0$, which is a contradiction.

Let us consider the polynomials $f_{\pm\epsilon} := p \pm \epsilon q$. It is easily seen that $f_{\pm\epsilon} \in B_{\partial\Delta}$ for every sufficiently small $\epsilon > 0$.

Let X be a critical point of p. Without loss of generality we may assume that $p(X) = 1$. It follows from $f_{\pm\epsilon}(X) \leq 1$ that $q(X) = 0$. It remains to prove (ii). We fix a point $M \in \partial\Delta$ such that $M \neq X$, $[XM] \subset \partial\Delta$ and $\dfrac{\partial p}{\partial \overrightarrow{XM}}(X) = 0$. Set $g_{\pm\epsilon}(t) := f_{\pm\epsilon}(X + t(M - X))$ for $t \in [0,1]$. By the Taylor's formula we have $g(t) = g(0) + g'(0)t + \frac{1}{2}g''(0)t^2$. Since $g(0) = f_{\pm\epsilon}(X) = p(X) = 1$ and $g'(0) = \dfrac{\partial f_{\pm\epsilon}}{\partial \overrightarrow{XM}}(X) = \pm\epsilon \dfrac{\partial q}{\partial \overrightarrow{XM}}(X)$, we obtain

$$g(t) = 1 \pm \epsilon \frac{\partial q}{\partial \overrightarrow{XM}}(X)t + O(t^2). \tag{8}$$

Suppose that $\dfrac{\partial q}{\partial \overrightarrow{XM}}(X) \neq 0$ and, for definiteness, let it be positive. We fix an $\epsilon_0 > 0$ such that $f_{+\epsilon_0} \in B_{\partial\Delta}$. It follows from (8) that there exists a $t_0 \in (0,1)$ such that $g_{+\epsilon_0}(t_0) > 1$ hence $\|f_{+\epsilon_0}\|_{\partial\Delta} > 1$, which is a contradiction. This completes the proof of Lemma 2. $\square$

Lemma 3. *Let p be a strictly concave extreme point of $B_{\partial\Delta}$. Then for every $i = 1, 2, 3$ there exists a unique point $X_i \in \text{int}\, l_i$ such that $p(X_i) = 1$ and consequently $\dfrac{\partial p}{\partial \overrightarrow{l_i}}(X_i) = 0$.*

Proof. Step 1. We claim that there exists $X_i \in l_i$ such that $p(X_i) = 1$, $i = 1, 2, 3$.

For definiteness, we shall consider the side $l_1 = [OA]$. On the contrary, we assume that $p(X) < 1$ for every $X \in l_1$. We set $q(x, y) := x(x + y - 1)$. We shall prove that $f_{\pm\epsilon} = p \pm \epsilon q \in B_{\partial\Delta}$ for every sufficiently small positive number ϵ, which leads to a contradiction. By the definition of q, $f_{\pm\epsilon}(X) = p(X)$ for $X \in l_2 \cup l_3$.

Suppose now that $X \in l_1$. The strict concavity of p implies that $f_{\pm\epsilon}$ is strictly concave, provided ϵ is sufficiently small. Consequently the restriction of $f_{\pm\epsilon}$ on l_1 is strictly concave, too. This implies

$$\min_{X \in l_1} f_{\pm\epsilon}(X) = \min\{f_{\pm\epsilon}(O), f_{\pm\epsilon}(A)\} = \min\{p(O), p(A)\} \geq -1.$$

By the assumption for p, we have $c := \max_{X \in l_1} p(X) < 1$. Then $f_{\pm\epsilon}(X) \leq c + \epsilon\|q\|_{C(\partial\Delta)} < 1$ for every $X \in l_1$ provided ϵ is sufficiently small. This finishes the proof of Step 1.

Step 2. It remains to prove that $p(O) < 1$, $p(A) < 1$, $p(B) < 1$.

We shall consider only the first inequality, the proof of the remaining ones is similar. Suppose that $p(O) = 1$. Since $p \in B_{\partial\Delta}$ we have

$$\frac{\partial p}{\partial x}(O) \leq 0, \quad \frac{\partial p}{\partial y}(O) \leq 0. \tag{9}$$

By Step 1, there exists a point $X_3 \in l_3$ such that $p(X_3) = 1$. Let us consider the restriction

$$F(t) = p(O + tX_3) = p(\alpha t, \beta t),$$

where $X_3 = (\alpha, \beta)$. By Taylor's formula $F(t) = F(0) + F'(0) + \frac{1}{2}F''(0)t^2$. Note that $F(0) = p(O) = 1$, $F'(0) = \alpha\frac{\partial p}{\partial x}(O) + \beta\frac{\partial p}{\partial y}(O) \leq 0$, in view of $\alpha, \beta \geq 0$ and (9), and $F''(0) < 0$ by the strict concavity of p. Hence $F(t) < 1$ for every $t \in (0, 1]$. In particular $p(X_3) = F(1) < 1$, which is a contradiction. The uniqueness of X_i follows from the strict concavity of p. Lemma 3 is proved. □

Proof of Theorem 1. Necessity. The conditions (i) are established in Lemma 3. Lemma 1 provides the existence of a point $X \in \partial\Delta$ such that $p(X) = -1$. In addition, the strict concavity of p implies $p(X) > -1$ for every $X \in \partial\Delta \setminus \{O, A, B\}$. This gives (ii) and completes the proof of the necessity.

Sufficiency. Let $p \in \pi_2$ be strictly concave and satisfies (i) and (ii). We shall prove that $p \in E_{\partial\Delta}$.

We claim that $\|p\|_{\partial\Delta} = 1$. Indeed, the strict concavity of p and the condition (ii) imply

$$\min_{X \in \partial\Delta} p(X) = \min\{p(O), p(A), p(B)\} = -1.$$

It remains to prove that $p(X) \leq 1$ for every $X \in \partial\Delta$. Without loss of generality we suppose that $X \in [OA]$. Let $\tilde{p}(x) := p(x, 0)$ be the corresponding restriction of p. Clearly $\tilde{p}$ is a strictly concave univariate polynomial hence there exists a unique $x^* \in \mathbb{R}$ such that $\tilde{p}(x^*) = \max_{x \in \mathbb{R}} \tilde{p}(x)$. As a consequence $\tilde{p}'(x^*) = 0$. This is equivalent to $\dfrac{\partial p}{\overrightarrow{\partial l_1}}(X^*) = 0$, where $X^* := (x^*, 0)$. Noticing that $\dfrac{\partial p}{\overrightarrow{\partial l_1}}(X_1) = 0$ we conclude that $X^* = X_1$ and $\max_{X \in [OA]} p(X) = 1$. The claim is proved.

In order to prove that $p \in E_{\partial \Delta}$ we shall use Lemma 2. Suppose that $q \in \pi_2$ satisfies (i) and (ii) from Lemma 2. The points $X \in \partial \Delta$ where p attains the value 1 are X_i, $i = 1, 2, 3$. (Note that by the strict concavity of p we have $p(X) < 1$ for $X \in \{O, A, B\}$.) Furthermore, $p(X) > -1$ for every $X \in \partial \Delta \setminus \{O, A, B\}$. Indeed if $p(X) = -1$ for a point $X \in \operatorname{int} l_i$ then the derivative of the restriction $p|_{l_i}$ would have two different zeros in l_i, which is a contradiction. Therefore the points $X \in \partial \Delta$ where p attains the value -1 are amongst the vertices $\{O, A, B\}$. For definiteness we assume that $p(O) = -1$. The condition (i) for q implies

$$q(O) = 0, \ q(X_i) = 0, \text{ for } i = 1, 2, 3. \tag{10}$$

Since $\dfrac{\partial p}{\overrightarrow{\partial l_i}}(X_i) = 0$ for $i = 1, 2, 3$ we get by (ii)

$$\dfrac{\partial q}{\overrightarrow{\partial l_i}}(X_i) = 0, \text{ for } i = 1, 2, 3. \tag{11}$$

The conditions (10) and (11) lead to $q(X) = 0$ for every $X \in \partial \Delta$, hence $q \equiv 0$. Theorem 1 is proved. □

Lemma 4. *Let p and q be real numbers which are smaller than 1. Then there exist a unique polynomial $f(t) = At^2 + Bt + C$ and $t_1 \in (0, 1)$ such that*

$$f(0) = p, \ f(1) = q, \ f(t_1) = 1, \ f'(t_1) = 0. \tag{12}$$

Moreover,

$$t_1 = \dfrac{\sqrt{1 - p}}{\sqrt{1 - p} + \sqrt{1 - q}}.$$

Proof. Conditions (12) are equivalent to the following nonlinear system:

$$\begin{aligned} C &= p \\ A + B + C &= q \\ At_1^2 + Bt_1 + C &= 1 \\ 2At_1 + B &= 0 \, . \end{aligned} \tag{13}$$

It can be derived from (13) that $t_1 = \dfrac{2(1 - p)}{B}$, where B satisfies the equation $B^2 - 4(1 - p)B + 4(1 - p)(q - p) = 0$. Hence $B = 2\sqrt{1 - p}(\sqrt{1 - p} \pm \sqrt{1 - q})$ and the minus sign is excluded by $t_1 \in (0, 1)$. The relations $A = -\dfrac{B^2}{4(1 - p)}$ and $C = p$ show that A and C are also uniquely determined. Lemma 4 is proved. □

Proof of Theorem 2. Necessity. Let p be a strictly concave extreme point of $B_{\partial \Delta}$. We set $\alpha := p(A)$, $\beta := p(B)$ and $\gamma := p(O)$. Theorem 1 and the strict concavity of p imply that α, β, γ satisfy (3). It remains to prove that the coefficients in the representation (1) are given by the formulas (2).

By Theorem 1, for every $i = 1, 2, 3$ there exists a unique point $X_i \in \operatorname{int} l_i$ such that $p(X_i) = 1$ and $\dfrac{\partial p}{\partial \overrightarrow{l_i}}(X_i) = 0$. Let $X_1 = (\lambda, 0)$, $X_2 = (0, \mu)$ and $X_3 = (1 - \nu, \nu)$, where $0 < \lambda, \mu, \nu < 1$.

Applying Lemma 4 to the restriction $p|_{l_i}$, $i = 1, 2, 3$ we get

$$\lambda = \frac{\sqrt{1 - \gamma}}{\sqrt{1 - \gamma} + \sqrt{1 - \alpha}}, \ \mu = \frac{\sqrt{1 - \gamma}}{\sqrt{1 - \gamma} + \sqrt{1 - \beta}}, \ \nu = \frac{\sqrt{1 - \alpha}}{\sqrt{1 - \alpha} + \sqrt{1 - \beta}}.$$

The conditions

$$p(X_i) = 1, \ i = 1, 2, 3, \ p(A) = \alpha, \ p(B) = \beta, \ p(O) = \gamma \tag{14}$$

form a linear system for the coefficients of p. Solving (14) we obtain the formulas (2).

Sufficiency. Suppose that the polynomial p has the form (1) where the coefficients are given by (2) and the parameters α, β, γ satisfy (3). We shall prove that p is a strictly concave extreme point of $B_{\partial \Delta}$. By the Sylvester's criterion, p is strictly concave if and only if $d < 0$ and $\det M > 0$, where $M := \begin{pmatrix} d & e \\ e & f \end{pmatrix}$. It is clear from (2) and (3) that $d < 0$. Furthermore, a computation gives

$$\det M = 4\sqrt{1 - \alpha}\sqrt{1 - \beta}\sqrt{1 - \gamma}\,(\sqrt{1 - \alpha} + \sqrt{1 - \beta} + \sqrt{1 - \gamma}) > 0,$$

hence p is a strictly concave polynomial.

In order to prove that $p \in E_{\partial \Delta}$ we shall use Theorem 1. We set

$$\lambda := \frac{\sqrt{1 - \gamma}}{\sqrt{1 - \gamma} + \sqrt{1 - \alpha}}, \ \mu := \frac{\sqrt{1 - \gamma}}{\sqrt{1 - \gamma} + \sqrt{1 - \beta}}, \ \nu := \frac{\sqrt{1 - \alpha}}{\sqrt{1 - \alpha} + \sqrt{1 - \beta}},$$

and

$$X_1 := (\lambda, 0), \ X_2 := (0, \mu), \ X_3 := (1 - \nu, \nu).$$

The way the coefficients in the necessity part were derived shows that p satisfies conditions (14). We have

$$\min\{p(O), p(A), p(B)\} = \min\{\gamma, \alpha, \beta\} = -1,$$

hence condition (ii) from Theorem 1 is fulfilled.

In view of (14) and the strict concavity of p it remains to show that $\dfrac{\partial p}{\partial \overrightarrow{l_i}}(X_i) = 0$ for $i = 1, 2, 3$. Without loss of generality we can assume that $i = 1$.

Let f be the solution of the problem (12) with $p := \gamma$ and $q := \alpha$. There exists a unique $t_1 \in (0, 1)$ such that $f(t_1) = 1$ and $f'(t_1) = 0$. In addition

$$t_1 = \frac{\sqrt{1 - \gamma}}{\sqrt{1 - \gamma} + \sqrt{1 - \alpha}} = \lambda.$$

The restriction $\tilde{p} = p|_{l_1}$ satisfies the conditions: $\tilde{p}(0) = \gamma$, $\tilde{p}(t_1) = 1$, and $\tilde{p}(1) = \alpha$. The uniqueness in the Lagrange interpolation problem implies $\tilde{p} = f$. In particular, $\tilde{p}'(t_1) = f'(t_1) = 0$, i.e. $\dfrac{\partial p}{\partial \overrightarrow{l_1}}(X_1) = 0$ as desired. Thus p satisfies the conditions (i) and (ii) from Theorem 1, consequently $p \in E_{\partial \Delta}$. Theorem 2 is proved.   □

3   Proof of Theorem 3

Let us consider the extremal problem

$$S = \sup_{f \in B_{\partial \Delta}} \|f\|_{\Delta}. \tag{15}$$

We shall prove that f^* is the only solution of (15) which implies the statement of Theorem 3.

Since $\|f\|_{\Delta}$ is a convex functional on the convex set $B_{\partial \Delta}$, by the Krein-Milman's theorem we get

$$S = \sup_{f \in E_{\partial \Delta}} \|f\|_{\Delta}.$$

Clearly $S \geq 1$. Note that if f is an indefinite or semidefinite polynomial then $\|f\|_{\Delta} = \|f\|_{\partial \Delta}$. Therefore

$$S = \sup_{f \in E_{\partial \Delta}^-} \|f\|_{\Delta}.$$

Our next goal is to find an explicit formula for $\|f\|_{\Delta}$ provided $f \in E_{\partial \Delta}^-$. The gradient system for f is equivalent to

$$M \begin{pmatrix} x \\ y \end{pmatrix} = -\frac{1}{2} \begin{pmatrix} b \\ c \end{pmatrix}, \tag{16}$$

where $M = \begin{pmatrix} d & e \\ e & f \end{pmatrix}$.

Set $A := \sqrt{1 - \alpha}$, $B := \sqrt{1 - \beta}$ and $C := \sqrt{1 - \gamma}$. As we mentioned in the proof of Theorem 2, $\det M = 4ABC(A + B + C) > 0$. Hence the system (16) has a unique solution which is

$$x_0 = \frac{B + C}{2(A + B + C)}, \quad y_0 = \frac{A + C}{2(A + B + C)}.$$

The strict convexity of f ensures $f(x_0, y_0) > f(x, y)$ for every $(x, y) \neq (x_0, y_0)$. On the other hand, it is easily seen that $(x_0, y_0) \in \text{int } \Delta$. Thus, using the explicit form of f established in Theorem 2, we obtain

$$\max_{(x,y) \in \Delta} f(x, y) = f(x_0, y_0) = g(A, B, C),$$

where

$$g(A, B, C) := 1 + \frac{ABC}{A + B + C}.$$

Since $\|f\|_{\partial\Delta} = 1$ we conclude that $\|f\|_{\Delta} = g(A, B, C)$.

Consequently,

$$S = \sup_{(A,B,C)\in\mathcal{D}} g(A, B, C),$$

where

$$\mathcal{D} := \left\{(A, B, C) \in \mathbb{R}^3 : A, B, C \in (0, \sqrt{2}], \max\{A, B, C\} = \sqrt{2}\right\}.$$

Since $\dfrac{\partial g}{\partial A}, \dfrac{\partial g}{\partial B}, \dfrac{\partial g}{\partial C} > 0$ provided $A, B, C > 0$ we have

$$S = g(\sqrt{2}, \sqrt{2}, \sqrt{2}) = \frac{5}{3}.$$

Equivalently, this proves the inequality (4).

Next we shall find the extremal polynomials in (15). The function g attains its maximal value only at $(A, B, C) = (\sqrt{2}, \sqrt{2}, \sqrt{2})$ which implies that the only polynomial from $E_{\partial\Delta}^-$ extremizing (15) is f^* which corresponds to $\alpha = \beta = \gamma = -1$. Clearly, $-f^*$ also solves (15). Note that

$$\sup_{f\in E_{\partial\Delta}\setminus\{E_{\partial\Delta}^-\cup E_{\partial\Delta}^+\}} \|f\|_{\Delta} = 1,$$

where $E_{\partial\Delta}^+ := \{f \in E_{\partial\Delta} : f \text{ is strictly convex}\}$. Therefore $\pm f^*$ are the only solutions of (15) which are extreme points of $B_{\partial\Delta}$.

It remains to prove that every extremal polynomial belongs to $E_{\partial\Delta}$. We assume on the contrary that there exists a polynomial $f_0 \in B_{\partial\Delta} \setminus E_{\partial\Delta}$ such that $\|f_0\|_{\Delta} = S$. It is easily seen that $\|f_0\|_{\partial\Delta} = 1$. Applying the Caratheodory's theorem to the convex set $B_{\partial\Delta}$ we get the representation

$$f_0 = \sum_{i=1}^{m} \alpha_i f_i,$$

where $0 < \alpha_i \le 1$, $\sum_{j=1}^{m} \alpha_j = 1$, and $f_i \in E_{\partial\Delta}$, for $i = 1, \ldots, m$. (In fact, $m \le \dim \pi_2 + 1 = 7$.) The assumption $f_0 \notin E_{\partial\Delta}$ implies $m \ge 2$ and $0 < \alpha_i < 1$ for every $i = 1, \ldots, m$. By the extremality of f_0 we obtain

$$S = \|f_0\|_{\Delta} \le \sum_{i=1}^{m} \alpha_i \|f_i\|_{\Delta} \le S \sum_{i=1}^{m} \alpha_i = S.$$

Consequently all f_i, $i = 1, \ldots, m$ are extremal for (15). The previous analysis leads to the conclusion that $m = 2$. Let us suppose for definiteness that $f_1 = f^*$ and $f_2 = -f^*$. This gives $f_0 = \alpha_1 f^* - (1 - \alpha_1)f^* = (2\alpha_1 - 1)f^*$. Hence

$$\|f_0\|_{\partial\Delta} = |2\alpha_1 - 1|\,\|f^*\|_{\partial\Delta} = |2\alpha_1 - 1| < 1,$$

which contradicts to $\|f_0\|_{\partial\Delta} = 1$. The proof of Theorem 3 is completed.   □

4   Proofs of Theorems 4 and 5

Proof of Theorem 4. Let us fix the affine transformation L and $f \in \pi_2$. Applying (4) for the polynomial $f(L^{-1}(U))$, $U \in \Delta$ and using the fact that L maps the boundary of T onto the boundary of Δ, we get

$$
\begin{aligned}
\max_{X \in T} |f(X)| &= \max_{U \in \Delta} |f(L^{-1}(U))| \\
&\leq \frac{5}{3} \max_{U \in \partial \Delta} |f(L^{-1}(U))| \\
&= \frac{5}{3} \max_{X \in \partial T} |f(X)|.
\end{aligned}
$$

The inequality (6) is proved.

The equality is attained if and only if $f(L^{-1}(U)) = cf^*(U)$, i.e. $f(X) = cf^*(L(X))$, where c is a constant.

Note that the extremal polynomials do not depend on the choice of L. (There are six possibilities for L that correspond to the mapping of the vertices of T to those of Δ.) Indeed, let us fix L and set $p^*(X) := f^*(L(X))$. Let p be a polynomial from π_2, such that $p \neq \pm p^*$ and $\max_{X \in \partial T} |p(X)| = 1$. Let $f(U) := p(L^{-1}(U))$. Clearly,

$$
\max_{U \in \partial \Delta} |f(U)| = \max_{X \in \partial T} |p(X)| = 1.
$$

The condition $p \neq \pm p^*$ implies $f \neq \pm f^*$. The uniqueness in Theorem 3 ensures that $\max_{U \in \Delta} |f(U)| < \frac{5}{3}$, hence $\max_{X \in T} |p(X)| < \frac{5}{3}$. Therefore p is not extremal in (6). Theorem 4 is proved.   □

Proof of Theorem 5. We proceed by induction on d for every $r > 0$. The base case $d = 2$ is contained in Theorem 4.

Assume that (7) is proved for dimension $d - 1 \geq 2$ and we shall prove it for d. Let us define

$$
\Sigma_{r,h}^d := \left\{ (x_1, \ldots, x_{d-1}, h) : x_i \geq 0, \ i = 1, \ldots, d-1, \ \sum_{i=1}^{d-1} x_i \leq r - h \right\},
$$

for $h \in [0, r]$. Note that

$$
\Delta_r^d = \bigcup \left\{ \Sigma_{r,h}^d : h \in [0, r] \right\}
$$

and

$$
\partial \Delta_r^d = \bigcup \left\{ \partial \Sigma_{r,h}^d : h \in [0, r] \right\}.
$$

Using the representation

$$
\Sigma_{r,h}^d = \left\{ (X', h) : X' \in \Delta_{r-h}^{d-1} \right\}
$$

and the induction hypothesis, we get

$$
\begin{aligned}
\max_{X \in \Delta_r^d} |p(X)| &= \max_{h \in [0,r]} \max_{X \in \Sigma_{r,h}^d} |p(X)| \\
&= \max_{h \in [0,r]} \max_{X' \in \Delta_{r-h}^{d-1}} |p(X', h)| \\
&\leq \frac{5}{3} \max_{h \in [0,r]} \max_{X' \in \partial \Delta_{r-h}^{d-1}} |p(X', h)| \\
&= \frac{5}{3} \max_{h \in [0,r]} \max_{X \in \partial \Sigma_{r,h}^d} |p(X)| \\
&= \frac{5}{3} \max_{X \in \partial \Delta_r^d} |p(X)|.
\end{aligned}
$$

Theorem 5 is proved.   □

References

1. Araújo, G., Jiménez-Rodríguez, P., Muñoz-Fernández, G.A., Seoane-Sepúlveda, J.B.: Polynomial inequalities on the $\pi/4$-circle sector. J. Convex Anal. **24**, 927–953 (2017)
2. Araújo, G., Muñoz-Fernández, G.A., Rodríguez-Vidanes, D.L., Seoane-Sepúlveda, J.B.: Sharp Bernstein inequalities using convex analysis techniques. Math. Inequalities Appl. **23**, 725–750 (2020)
3. Bernal-González, L., Muñoz-Fernández, G.A., Rodríguez-Vidanes, D.L., Seoane-Sepúlveda, J.B.: A complete study of the geometry of 2-homogeneous polynomials on circle sectors. Revista de la Real Academia de Ciencias Exactas, Fisicas y Naturales - Serie A: Matematicas **114**, Article number 160 (2020)
4. Jiménez-Rodríguez, P., Muñoz-Fernández, G.A., Pellegrino, D., Seoane-Sepúlveda, J.B.: Bernstein-Markov type inequalities and other interesting estimates for polynomials on circle sectors. Math. Inequalities Appl **20**, 285–300 (2017)
5. Jiménez-Rodríguez, P., Muñoz-Fernández, G.A., Rodríguez-Vidanes, D.L.: Geometry of spaces of homogeneous trinomials on $\mathbb{R}^2$. Banach J. Math. Anal. **15**, Article number 61 (2021)
6. Kim, S.G.: Extreme 2-homogeneous polynomials on the plane with a hexagonal norm and applications to the polarization and unconditional constants. Studia Scientiarum Mathematicarum Hung. **54**, 362–393 (2017)
7. Kim, S.G.: Extreme bilinear forms on $\mathbb{R}^n$ with the supremum norm. Periodica Mathematica Hung. **77**, 274–290 (2018)
8. Kim, S.G.: Extreme points of the space $\mathcal{L}(^2 l_\infty)$. Commun. Korean Math. Soc. **35**, 799–807 (2020)
9. Kim, S.G.: The unit balls of $L(^n l_\infty^m)$ and $L_s(^n l_\infty^m)$. Studia Scientiarum Mathematicarum Hung. **57**, 267–283 (2020)
10. Kim, S.G.: Extreme and exposed symmetric bilinear forms on the space $\mathcal{L}_s(^2 l_\infty^2)$. Carpatian Math. Publ. **12**, 340–352 (2020)
11. Kim, S.G.: Extreme points of $\mathcal{L}_s(^2 l_\infty)$ and $\mathcal{P}(^2 l_\infty)$. Carpatian Math. Publ. **13**, 289–297 (2021)
12. Kim, S.G.: Geometry of bilinear forms on the plane with the octagonal norm. Bull. Transilvania Univ. Braşov, Ser. III: Math. Comput. Sci. **63**(1), 161–190 (2021)

13. Kim, S.G.: Geometry of multilinear forms on $\mathbb{R}^m$ with a certain norm. Acta Scientiarum Mathematicarum **87**, 233–245 (2021)
14. Milev, L., Naidenov, N.: Strictly definite extreme points of the unit ball in a polynomial space. Comptes Rendus de l'Académie bulgare des Sciences **61**, 1393–1400 (2008)
15. Milev, L., Naidenov, N.: Indefinite extreme points of the unit ball in a polynomial space. Acta Scientiarum Mathematicarum (Szeged) **77**, 409–424 (2011)
16. Milev, L., Naidenov, N.: Semidefinite extreme points of the unit ball in a polynomial space. J. Math. Anal. Appl. **405**, 631–641 (2013)
17. Milev, L., Naidenov, N.: Some exact Bernstein-Szegő inequalities on the standard triangle. Math. Inequalities Appl. **20**, 815–824 (2017)
18. Milev, L., Naidenov, N.: An Exact Pointwise Markov Inequality on the Standard Triangle. In: Ivanov, K., Nikolov, G., Uluchev, R. (eds.) CONSTRUCTIVE THEORY OF FUNCTIONS, Sozopol 2016, pp. 187–205. Prof. Marin Drinov Academic Publishing House, Sofia (2018)

A Theoretical Analysis on the Bound Violation Probability in Differential Evolution Algorithm

Mădălina-Andreea Mitran$^{(\boxtimes)}$ (iD)

Department of Computer Science, West University of Timisoara, Timisoara, Romania
madalina.mitran96@e-uvt.ro

Abstract. This study is focused on Differential Evolution (DE) algorithm in the context of solving continuous bound-constrained optimization problems. The mutation operator involved in DE might lead to infeasible elements, i.e. one or all of their components exceed the lower or upper bound. The infeasible components become the subject of a correction method, that deflects the algorithm from its canonical behavior. The repairing strategy considered in this work is a stochastic variant of the projection to bounds strategy, known as "exponentially confined". The main aim of this study is to determine the analytical expression of the bound violation probability of components generated by mutation operator in conjunction with "exponentially confined".

Keywords: Differential Evolution · Bound Constraint Handling Methods · Exponentially Confined · Bound Violation Probability

1 Preliminaries

Formally, the optimization problem addressed in this work can be enunciated as finding the minimum of a real-valued objective function $f : \Omega \to \mathbf{R}$, where $\Omega = [a_1, b_1] \times [a_2, b_2] \times \cdots \times [a_D, b_D]$. Here, Ω is the feasible domain defined by the lower (a_i) and upper (b_i) bounds of each dimension i, with $i = \overline{1, D}$. The practical relevance of this study is given by the fact that in many real-world problems, the design variables are limited to physical ranges and since values exceeding the bounds cannot be used to compute the objective function they must be dismissed or repaired.

This study is focused on *Differential Evolution (DE)* algorithm [10], a population based metaheuristic that is one of the most effective search methods for solving continuous optimization problems. The search is performed in a stochastic manner, relying on *mutation, crossover* and *selection* operators. The mutation operator involved in DE might lead to *infeasible individuals* that become the subject of a repairing or substitution strategy referred to in this paper as *Bound Constraint Handling Method (BCHM)*. The experimental analysis in [8] illustrates that many elements generated by DE mutation operator contain at least one infeasible component and the number of such elements increases with the problem size.

I. Georgiev et al. (Eds.): NMA 2022, LNCS 13858, pp. 233–245, 2023.
https://doi.org/10.1007/978-3-031-32412-3_21

The first systematic study on different BCHMs [2] highlighted that the efficiency of the Differential Evolution algorithm on the CEC2005 benchmark suite is significantly impacted by the chosen BCHM strategy. The results obtained by authors of [6] for the CEC2017 benchmark also suggest that the BCHMs have a relevant influence on the results obtained by the DE algorithm.

The impact of the repairing method on the algorithm performance is related to the number of components that violate the bound constraints and require correction [8]. The expected value of this number is $D \cdot p_v$, with p_v denoting the probability of generating infeasible components within the vector obtained using the mutation operator (*bound violation probability*). Most of the studies on the bound violation probability in the case of DE are empirical and previous works on theoretical results [11] refer only to BCHM based on uniform distribution.

The main aim of this study is to determine the analytical expression of the bound violation probability of components generated by the mutation operator in conjunction with a BCHM that is suitable for objective functions having the global optimum located near the bounds of the feasible domain. *Exponentially Confined* (exp_c) correction method was proposed in [9] and consists in bringing back the infeasible component in a region located between its old position and the exceeded bound, according to an exponential probability distribution that favors locations near the bound. The method was introduced under the motivation that a newly created infeasible point exceeds a bound because the optimum solution lies close to it, therefore enabling finding it. The analysis is conducted under the assumption that the current population corresponding to the first generations is uniformly distributed. This is a simplifying assumption that allows the analytical computation of the bound violation probability.

The rest of this paper is organized as follows: after briefly introducing the DE algorithm with a focus on the mutation operator and setting the notations, Sect. 2 presents a selection of constraint-handling methods. Section 3 is divided into four parts: (i) first, related theoretical results are reviewed; (ii) the problem definition and assumptions are presented; (iii) and (iv) describe the computation of the *probability density function (pdf)* of the elements generated by mutation operator. Section 4 presents the computation of the bound violation probability, under some simplifying assumptions, and compares the theoretical results with empirical estimations. Section 5 concludes the paper.

2   Differential Evolution

The population $P = (x_1, x_2, \ldots, x_N)$, processed by the DE algorithm incorporates N individuals that are encoded as D-dimensional real number vectors. After the random *initialization*, the population iteratively undergoes *mutation*, *crossover* and *selection* until a stopping condition (such as exhaustion of the maximal number of objective function evaluations) is satisfied. Storn and Price [10] introduced the notation $DE/X/Y/Z$ to represent different DE variants with X denoting the mutation method, Y - the number of difference vectors and Z - the crossover operator.

The differential *mutation* operator is responsible for generating a *mutant vector* for each target element in the population. The original DE algorithm, i.e. *"DE/rand/1"* strategy (also considered for this study) consists of randomly selecting three elements from the population and adding to the first of them (also known as base vector) the scaled difference between the other two, as exemplified in Eq. 1, with a scale factor $F \in (0, 2)$.

$$z = x_{r_1} + F \cdot (x_{r_2} - x_{r_3}) \tag{1}$$

The mutation operator involved in DE can lead to infeasible components that should be corrected. Components of the mutant vector are selected to be included in the trial vector based on a probability related to the crossover rate, $C_r \in (0, 1]$.

A mutant z is considered *feasible* if all its components are in the feasible domain, i.e. $z^j \in [a_j, b_j], j = \overline{1, D}$. Elements having at least one component that exceeds the lower or the upper bound are referred to as *infeasible elements*. The *bound violation probability*, p_v, is defined as $p_v(z^j) = Prob(z^j \notin [a_j, b_j])$

2.1 Bound Constraints Handling Methods

Bound constraint handling methods (BCHMs) can be divided into two main groups: (i) strategies that perform the correction *component-wise*, where only the components exceeding their corresponding bounds are altered; (ii) strategies that perform the correction *vector-wise* where the vector with component(s) exceeding the variable bounds is brought back into the search space along a vector direction; in this case, components that explicitly do not violate their bounds may also get modified [9].

Regarding component-wise BCHMs, they can be further classified as: (i) *deterministic:* the same feasible value is returned for the same infeasible value corrected; (ii) *stochastic:* the correction is based on a random value, sampled from a specific random variable. Some frequently used deterministic BCHMs are presented in the following.

$$c(z^j) = \begin{cases} a_j, & \text{if } z^j < a_j \\ b_j, & \text{if } z^j > b_j \end{cases} \qquad \text{(Saturation)} \tag{2}$$

$$c(z^j) = \begin{cases} \frac{1}{2} \cdot (a_j + x^j), & \text{if } z^j < a_j \\ \frac{1}{2} \cdot (b_j + x^j), & \text{if } z^j > b_j \end{cases} \qquad \text{(Midpoint Target)} \tag{3}$$

$$c(z^j) = \begin{cases} \frac{1}{2} \cdot (a_j + x_{r_1}^j), & \text{if } z^j < a_j \\ \frac{1}{2} \cdot (x_{r_1}^j + b_j), & \text{if } z^j > b_j \end{cases} \qquad \text{(Midpoint Base)} \tag{4}$$

A general description of deterministic BCHMs is specified in Eq. (5) where $\alpha \in [0, 1]$ and R is a feasible reference vector. For $\alpha = 1$ one obtains `Saturation`, while $\alpha = 1/2$ corresponds to a midpoint strategy (when R is the target vector then one obtains the Midpoint Target and when R is the base vector one obtains Midpoint Base).

$$c(z^j) = \begin{cases} \alpha a_j + (1 - \alpha)R^j, & \text{if } z^j < a_j \\ \alpha b_j + (1 - \alpha)R^j, & \text{if } z^j > b_j \end{cases} \quad \text{, where } \alpha \in (0, 1] \tag{5}$$

Stochastic BCHMs can be described as in Eq. (6) where ξ_j is a realization of a random variable ξ.

$$c(z^j) = \begin{cases} a_j + \xi_j & \text{if } z^j < a_j \\ b_j - \xi_j & \text{if } z^j > b_j \end{cases} \quad (6)$$

Depending on the distribution of the random variable ξ, one can obtain different stochastic BCHMs. Exponentially Confined (exp_c) was proposed in [9] as a stochastic variant of Saturation, where the infeasible component is brought back inside the feasible space in a region located between the reference point R and the violated bound, according to an exponential distribution. This approach will probabilistically create more solutions closer to the exceeded boundary, making this method suitable for objective functions with the global optimum situated close to the variable boundary. In [5] and [3] it is investigated empirically the impact of this correction on interior search algorithms.

The Exponentially Confined strategy can be defined as in Eq. 7, where $\mathcal{U}$ denotes a uniform distribution.

$$c(z^j) = \begin{cases} a_j + \ln(1/r_j^L), \ z^j < a_j & r_j^L \sim \mathcal{U}[\exp(a_j - R_j), 1] \\ b_j - \ln(1/r_j^U), \ z^j > b_j & r_j^U \sim \mathcal{U}[\exp(R_j - b_j), 1] \end{cases} \quad (7)$$

The probability density functions of the distribution of corrected elements (f_L for lower bound violation and f_U for upper bound violation), as derived from [9] are described in Eq. 8.

$$f_L(c) = \frac{e^{R-c}}{e^{(R-a)} - 1}, \quad c \in [a, R]; \qquad f_U(c) = \frac{e^{c-R}}{e^{(b-R)} - 1}, \quad c \in [R, b]. \quad (8)$$

3   Analysis for the Exponentially Confined Strategy

As mentioned above, mutation in DE is a difference-based perturbation scheme. Regarding the construction of a mutant vector z_i, $i = \overline{1, N}$, described in Sect. 2, notations X_1, X_2, X_3, Y are used to denote the continuous random variables which take the values $x_{r_1}, x_{r_2}, x_{r_3}, y$ respectively during the evolutionary process. The theoretical analysis presented further is performed at component level and considers $F = 1$ as scale factor in Eq. (1). Without losing generality, let us consider the bounds $a = 0$ and $b = 1$ (therefore, $R \in [0, 1]$).

The main aim of this study is to compute the bound violation probability in the case of DE/rand/1 mutation in conjunction with exp_c correction. This computation requires the determination of the probability density function (pdf) of $Z = X_1 + F \cdot (X_2 - X_3)$ for $F = 1$ and under some assumptions which correspond to the case when the analysis is conducted for the first generations.

3.1   Related Work

The context for this study was provided by the analysis performed in study [1], where the authors derived the expressions corresponding to the distribution of elements generated by the $DE/rand/1$ mutation operator. The authors assumed that mutation is applied in the first generation when the population is uniformly distributed inside the feasible domain. This result allowed the computation of the probability for a mutant to violate the bounds as $F/3$ in paper [11].

The bound violation probability was theoretically estimated in the case of the *first generation* of a particle swarm optimization algorithm by Helwig et al. [4]. The authors concluded that many particles leave the feasible domain and that the correction strategy plays a critical role.

Motivated by the very few papers available in the literature that concern the theoretical grounding for using optimization algorithms in bounded search spaces, this study is a one-step analysis at one component level using a generalized variant of exp_c correction method as an attempt to provide a theoretical grounding for empirically observed behavior.

3.2   Method and Assumptions

The steps performed in the theoretical analysis are detailed as follows:
1. *Computation of the probability density function f_Z of $Z = X_1 + (X_2 - X_3)$.*
First, we proceed by determining the pdf of $Y = X_2 - X_3$:

$$f_Y(y) = \int_{-\infty}^{\infty} f_{X_2}(x_2) f_{X_3}(x_2 - y) dx_2 \tag{9}$$

Then the pdf of Z can be computed using the convolution product:

$$f_Z(z) = \int_{-\infty}^{\infty} f_Y(y) f_{X_1}(z - y) dy \tag{10}$$

2. *Computation of the lower and upper bound violation probabilities* using f_Z, as:

$$Prob(Z < a) = \int_{-\infty}^{a} f_Z(z) dz \qquad Prob(Z > b) = 1 - \int_{-\infty}^{b} f_Z(z) dz \tag{11}$$

As stated in [1], during DE algorithm runs, a contraction process drives the population elements in the vicinity of the global optimum solution. The authors considered three stages of the contraction process of the population: (i) in the *initial stage*, the distribution of the population elements is uniform or near uniform in Ω; (ii) in the *intermediate stage*, the elements are distributed in the regions of attraction of different local optima; (iii) in the *final stage*, the elements are distributed in the region of attraction of the global optimum.

Let us consider that the population corresponding to generation g, denoted as $P(g)$ is in an initial search stage. $P(g)$ consists of two subpopulations: $P(g) = P_w(g) \cup P_c(g)$, where $P_w(g)$ is the subpopulation corresponding to feasible mutants that were not the subject of any BCHMs in the previous generation and $P_c(g)$ is the subpopulation corresponding to mutants whose components were corrected using exp_c in the previous generation. It is assumed that the elements of $P_w(g)$ are uniformly distributed in $[0, 1]$, while elements of $P_w(g)$ follow the distributions f_L and f_U described in Eq. (8), depending on which bound has been violated.

3.3   Probability Distribution Function of the Difference Term

In a one-step analysis, one can consider that each of the elements involved in the difference involved in the construction of the mutant can be selected, with a specific probability, either from P_w or from P_c. The selection probabilities are specified in Table 1 where p_v denotes the bound violation probability corresponding to the previous generation, meaning that the expected number of elements in the subpopulation P_w is $N \cdot (1 - p_v)$.

Table 1. Combinations of elements involved in the difference and their corresponding probabilities.

Case	X_2	X_3	Probability
1.	$\in P_c$	$\in P_w$	$p_v(1 - p_v)$
2.	$\in P_w$	$\in P_c$	$p_v(1 - p_v)$
3.	$\in P_w$	$\in P_w$	$(1 - p_v)^2$
4.	$\in P_c$	$\in P_c$	p_v^2

Each case is discussed as follows, considering that elements extracted from P_c have violated the lower bound and are distributed according to f_L (Eq. (8)). A similar analysis has been conducted in the case of elements violating the upper bound (see the supplementary material[1]).

Cases 1–2. $X_2 \in P_c$, $X_3 \in P_w$ with the corresponding probability density functions:

$$f_{X_2}(x_2) = \frac{e^{R-x_2}}{e^R - 1}, \qquad\qquad x_2 \in [0, R] \qquad\qquad (12)$$

$$f_{X_3}(x_3) = 1, \qquad\qquad x_3 \in [0, 1] \qquad\qquad (13)$$

The probability density function of $Y = X_2 - X_3$ is computed using the convolution product as specified in Eq. (9) which, for case **1**, leads to:

$$f_Y(y) = \begin{cases} k \cdot (\sinh(y+1) - \cosh(y+1) + 1), & y \in [-1, R-1] \\ k \cdot (\sinh(R) - \cosh(R) + 1), & y \in [R-1, 0] \\ k \cdot (\sinh(R) - \cosh(R) - \sinh(y) + \cosh(y)), & y \in [0, R] \end{cases} \qquad (14)$$

with $k = e^R/(e^R - 1)$. The computations for the second case ($X_2 \in P_w$, $X_3 \in P_c$) follows the same line of reasoning and are presented in the supplementary material[1].

Case 3. $X_2 \in P_w$, $X_3 \in P_w$, with the probability density function of Y being determined following the same procedure as in the previous case:

$$f_Y(y) = \begin{cases} 1 + y, & y \in [-1, 0] \\ 1 - y, & y \in [0, 1] \end{cases} \qquad\qquad (15)$$

---

[1] https://github.com/madalinami/NMA22.

Case 4. $X_2 \in P_c$, $X_3 \in P_c$, with the probability density function of Y being determined as:

$$f_Y(y) = \begin{cases} k_1 \cdot \sinh(R+y), & y \in [-R,0] \\ k_1 \cdot \sinh(R-y), & y \in [0,R] \end{cases} \qquad k_1 = e^R/(e^R - 1)^2 \qquad (16)$$

3.4   Probability Distribution Function of the Mutant

In the next step, the pdf of $Z = X_1 + Y$, with $Y = X_2 - X_3$, is computed using the formula indicated in Eq. (10), applied to the cases specified in Table 2. In this section is presented the computation of the pdf corresponding to **Case I**, under the assumption that the infeasible elements violated only the upper bound. The other cases can be addressed similarly (see supplementary material).

Table 2. Combinations of elements involved in mutation and their corresponding probabilities.

Case	X_1	X_2	X_3	Probability
I	$\in P_w$	$\in P_c$	$\in P_w$	$p_1 = p_v(1-p_v)^2$
II	$\in P_c$	$\in P_c$	$\in P_w$	$p_2 = p_v^2(1-p_v)$
III	$\in P_w$	$\in P_c$	$\in P_c$	$p_3 = p_v^2(1-p_v)$
IV	$\in P_c$	$\in P_c$	$\in P_c$	$p_4 = p_v^3$
V	$\in P_w$	$\in P_w$	$\in P_w$	$p_5 = (1-p_v)^3$
VI	$\in P_c$	$\in P_w$	$\in P_w$	$p_6 = p_v(1-p_v)^2$
VII	$\in P_w$	$\in P_w$	$\in P_c$	$p_7 = p_v(1-p_v)^2$
VIII	$\in P_c$	$\in P_w$	$\in P_c$	$p_8 = p_v^2(1-p_v)$

Essentially, the convolution product allows the factorization of the joint distribution of X_1 and Y into their individual probability density functions contained in Eq. (13), respectively Eq. (14) and illustrated in Fig. 2. Considering that we can obtain a non-zero value for the pdf of Z only when the red and the blue distributions overlap, z is used in order to translate f_{X_1} to the right. The first overlapping pattern is obtained for $z \in [-1, R-1)$. Thus, the convolution product is computed as:

$$f_Z(z) = k \cdot \int_{-1}^{z} (\sinh(y+1) - \cosh(y+1) + 1)dy \qquad (17)$$

The next overlapping pattern is obtained for $z \in [R-1, 0)$, and the convolution product of the pdfs is computed accordingly as:

$$f_Z(z) = k \cdot \int_{-1}^{R-1} (\sinh(y+1) - \cosh(y+1) + 1)dy + \int_{R-1}^{z} 1 dy \qquad (18)$$

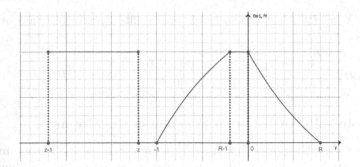

Fig. 1. Probability distribution functions for X_1 (red) and Y (blue). (Color figure online)

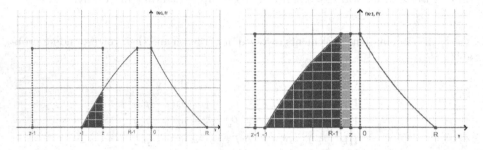

Fig. 2. Illustrating the overlap between distributions of X_1 (red) and Y (blue). (Color figure online)

Following the same reasoning, one can obtain the pdf for the remaining cases, as presented in Eq. (19).

$$
f_{Z_{(1)}}(z) = \begin{cases} A, & z \in [-1, R-1) \\ B, & z \in [R-1, 0) \\ C, & z \in [0, R) \\ D, & z \in [R, 1) \\ E, & z \in [1, R+1] \end{cases} \tag{19}
$$

In Eq. (19) the following notations have been used:
$A = k \cdot (z - \sinh(z+1) - \cosh(z+1))$
$B = k \cdot R \cdot e^{-R} + z$
$C = k \cdot (R - z - \sinh(R) + \cosh(R) + \sinh(z) - \cosh(z) - R + 1$
$D = k \cdot (-R \cdot e^{-R} + \sinh(R) - \cosh(R) + 1) - z + 1$
$E = k \cdot [(-z + R + 2) \cdot \sinh(R) + (z - R - 2) \cdot \cosh(R) - e \cdot (\sinh(z) - \cosh(z)]$
In **Case V**, where $X_1, X_2, X_3 \in P_w$, the pdf $f_{Z_{(5)}}(z)$ was computed in [1]. The analysis of the remaining cases is included in the supplementary material[2].

---

[2] https://github.com/madalinami/NMA22.

4   Estimation of the Bound Violation Probability

The *bound violation probability* is defined as the probability of generating infeasible components within the vector obtained using the mutation operator. In the previous section, the probability density function of elements generated using the mutation operator was computed considering the scenarios depending on the origin of elements, as presented in Table 2. Considering Eq. (11) and the analytic expression of $f_Z(z)$, the bound violation probability was computed for each scenario. Let us consider **Case I** for which the analytical expression of the bound violation probability corresponding is presented in Eq. (20) and the dependence on the positions of the reference point, R, is illustrated in Fig. 3.

$$p_{v(1)} = \frac{3e^R - 2R^2 - 2R - 3}{2(e^R - 1)} \tag{20}$$

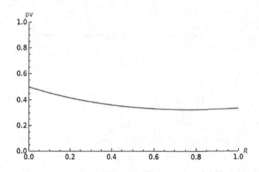

Fig. 3. Theoretical probability of bounds violation for $X_1 \in P_w, X_2 \in P_c, X_3 \in P_w$.

From Fig. 3 one can easily observe that $p_{v(1)} < 0.5$ and it is maximal when R is 0. As R approaches 1, the domain in which corrected elements are generated is larger and the bound violation probability is close to the probability of violation when elements are uniformly distributed, which is $1/3$ [11].

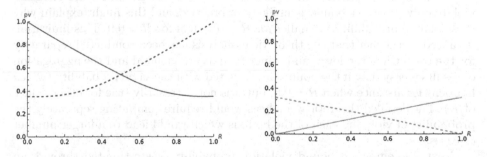

Fig. 4. Theoretical probability of bounds violation for $X_1, X_2 \in P_c, X_3 \in P_w$ (blue) and $X_1 \in P_w, X_2, X_3 \in P_c$ (green), lower bound violation (solid) and upper bound violation (dashed). (Color figure online)

As it is illustrated in Fig. 4 the dependence of the violation probability on R depends both on the analyzed case and on the violated bound. In **Case II**, for $R \in \{0,1\}$ the violation probability can be 1. For instance, $R = 0$ implies $X_1 = X_2 = 0$ and the generated mutant will be $-X_3 \in [-1,0]$, thus always infeasible. Similarly, in **Case III** (right graph in Fig. 4) when $R \in \{0,1\}$ the violation probability is 0.

Considering the associated probability of each case, the expected violation probability can be computed as $p_v = \sum_{i=1}^{8} p_{v(i)} p_i$, where $p_{v(i)}$ are the violation probabilities computed for the previous cases and p_i is the probability associated with each case, $i = \overline{1,8}$, as described in Table 2.

The influence of R on the expected bound violation probability for the next mutation steps (under the assumption that the population of feasible elements, P_w, has a uniform distribution and the corrected elements have been obtained from infeasible elements which exceeded the lower bound) is illustrated in Fig. 5.

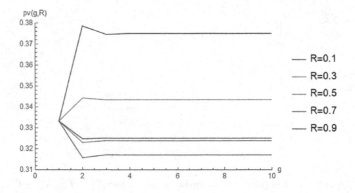

Fig. 5. The impact of R on the evolution of the bound violation probability.

The bound violation probability for generation g depends on the bound violation probability corresponding to the previous generation. For the analysis presented so far, we considered $p_v = 1/3$ in the first generation.

It should be emphasized that the scenario in which the infeasible elements violate only the lower bound is unlikely in practice, and this might explain why the violation probability is smaller for $R = 0.7$ than for $R = 0.9$. This limitation is caused by the fact that the theoretical analysis has been conducted separately for the cases when the lower and upper bounds are violated and the aggregation of results is easy only if the bounds are violated with the same probability (which happens for instance when $R = 0.5$, but it is not necessarily true for other values of R). The detailed theoretical analysis would require computing separately the probability of violating each of the bounds which would lead to a larger number of cases.

Next, the empirical bound violation probability was estimated through an experimental analysis for a population consisting of 100 elements of size 30 processed by "DE/rand/1/bin" combined with exp_c, in 50 independent runs, for

100 generations, with $F = 1$ and $C_r = 1$. One must consider that as the search moves forward, under the selection pressure exercised by the objective function, the individuals in the population are driven in the region of attraction of the global optimum. Aiming not to introduce any bias on the search process, we considered as objective functions: f_0, where $\forall x \in \mathbf{R}^D : f_0(x) \sim U(0, 1)$ and f_1, where $(\forall x) \in \mathbf{R}^D : f_1(x) = 1$. Here, "bias" is used in order to denote individuals being attracted to or by avoiding certain regions of the search space, as discussed in [7]. The objective functions were introduced motivated by the fact that the theoretical analysis was conducted under the assumption that elements in P_w are uniformly distributed in the search space, so in the absence of selection pressure, we expect elements to preserve their initial uniform distribution.

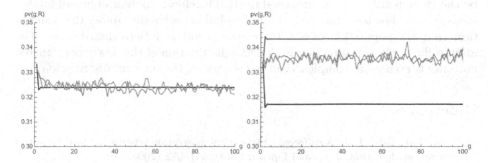

Fig. 6. Bound violation probability estimation: theoretical (blue for lower bound violation, red for upper bound violation), empirical estimation on f_0 (orange) and f_1 (green); left: $R = 0.5$, right: $R = 0.7$. (Color figure online)

Figure 6 illustrates that when the reference point used in exp_c correction is in the middle of the search space ($R = 0.5$), the theoretical estimation of the bound violation probability is close to that obtained empirically under the assumption of no selection pressure (functions f_0 and f_1). On the other hand, for other values of R, the theoretical estimations (obtained under the simplifying assumption that either the lower or the upper bound is always violated) of the violation probability provide only some (rather large) margins for the empirical values.

5   Conclusions

The analysis conducted in this paper is a first step toward obtaining theoretical estimations of the bound violation probability in the case when the exp_c correction strategy is applied in the context of DE/rand/1 mutation. Even if the analysis is based on several simplifying assumptions, i.e. uniform distribution of the current population, scale factor equal to 1, separate analysis of the cases when the upper and the lower bounds are violated, the obtained results are in line with experimental results (in the case of $R = 0.5$) and might provide useful guidelines in the choice of the reference point, R.

It should be remarked that the exp_c correction strategy analyzed in this paper is a generalization of that used in [9], as the reference point is not necessarily the target or the base element. Chosing appropriate values for R might reduce the number if infeasible mutants in the next generation, and consequently reducing the additional cost induced by applying correction methods.

The assumption that the population of feasible elements is uniformly distributed is not realistic for all stages of the population evolution. However, as it is illustrated in Fig. 6, it seems that it captures the overall behavior of the population during the first generations, in the absence of selection pressure.

A more realistic distribution of the population of feasible mutants seems to be the Beta distribution, as proposed in [1]. Therefore, a potential line of further research would be to estimate the bound violation probability under the assumption that the population of feasible elements follow a Beta distribution. The main difficulty in this direction is that the distribution of the newly constructed mutants is even more complex than in the case of the uniform distribution.

References

1. Ali, M.M., Fatti, L.P.: A differential free point generation scheme in the differential evolution algorithm. J. Global Optim. **35**(4), 551–572 (2006)
2. Arabas, J., Szczepankiewicz, A., Wroniak, T.: Experimental comparison of methods to handle boundary constraints in differential evolution. In: Schaefer, R., Cotta, C., Kołodziej, J., Rudolph, G. (eds.) PPSN 2010. LNCS, vol. 6239, pp. 411–420. Springer, Heidelberg (2010). https://doi.org/10.1007/978-3-642-15871-1_42
3. Gandomi, A.H., Kashani, A.R., Zeighami, F.: Retaining wall optimization using interior search algorithm with different bound constraint handling. Int. J. Numer. Anal. Meth. Geomech. **41**, 1304–1331 (2017)
4. Helwig, S., Wanka, R.: Particle swarm optimization in high-dimensional bounded search spaces. In: IEEE Swarm Intelligence Symposium, pp. 198–205 (2007)
5. Kashani, A.R., Chiong, R., Dhakal, S., Gandomi, A.H.: Investigating bound handling schemes and parameter settings for the interior search algorithm to solve truss problems. In: Engineering Reports, 3(10), p. e12405 (2021)
6. Kreischer, V., Magalhaes, T.T., Barbosa, H.J., Krempser, E.: Evaluation of bound constraints handling methods in differential evolution using the cec2017 benchmark. In: XIII Brazilian Congress on Computational Intelligence (2017)
7. van Stein, B., Caraffini, F., Kononova, A.V.: Emergence of structural bias in differential evolution. In: Proceedings of the Genetic and Evolutionary Computation Conference Companion, pp. 1234–1242 (2021)
8. Kononova, A.V., Caraffini, F., Bäck, T.: Differential evolution outside the box. Inf. Sci. **581**, 587–604 (2021)
9. Padhye, N., Mittal, P., Deb, K.: Feasibility preserving constraint-handling strategies for real parameter evolutionary optimization. Comput. Optim. Appl. **62**(3), 851–890 (2015)

10. Storn, R., Price, K.: Differential evolution-a simple and efficient heuristic for global optimization over continuous spaces. J. Global Optim. **11**(4), 341–359 (1997)
11. Zaharie, D., Micota, F.: Revisiting the analysis of population variance in Differential Evolution algorithms. In: 2017 IEEE Congress on Evolutionary Computation (CEC), pp. 1811–1818 (2017)

On the Extreme Zeros of Jacobi Polynomials

Geno Nikolov[✉][iD]

Faculty of Mathematics and Informatics, Sofia University "St. Kliment Ohridski",
5, James Bourchier blvd., 1164 Sofia, Bulgaria
geno@fmi.uni-sofia.bg

Abstract. By applying the Euler–Rayleigh method to a specific representation of the Jacobi polynomials as hypergeometric functions, we obtain new bounds for their largest zeros. In particular, we derive upper and lower bound for $1 - x_{nn}^2(\lambda)$, with $x_{nn}(\lambda)$ being the largest zero of the n-th ultraspherical polynomial $P_n^{(\lambda)}$. For every fixed $\lambda > -1/2$, the limit of the ratio of our upper and lower bounds for $1 - x_{nn}^2(\lambda)$ does not exceed 1.6. This paper is a continuation of [16].

Keywords: Jacobi polynomials · Gegenbauer polynomials · Laguerre polynomials · Euler-Rayleigh method

1 Introduction and Statement of the Results

The extreme zeros of the classical orthogonal polynomials of Jacobi, Laguerre and Hermite have been a subject of intensive study. We refer to Szegő's monograph [22] for earlier results, and to [2–4,6,7,9–14,16,21] for some more recent developments.

Throughout this paper we use the notation

$$x_{1n}(\alpha, \beta) < x_{2n}(\alpha, \beta) < \cdots < x_{nn}(\alpha, \beta)$$

for the zeros of the n-th Jacobi polynomial $P_n^{(\alpha,\beta)}$, $\alpha, \beta > -1$, and the zeros of the n-th Gegenbauer polynomial $P_n^{(\lambda)}$, $\lambda > -1/2$ are denoted by

$$x_{1n}(\lambda) < x_{2n}(\lambda) < \cdots < x_{nn}(\lambda).$$

In the recent paper [16] we applied the Euler–Rayleigh method to the Jacobi and, in particular, the Gegenbauer polynomials, represented as hypergeometric functions, to derive new bounds for their extreme zeros. Below we state some of the bounds obtained in [16], which improve upon some results of Driver and Jordaan [6].

Research supported by the Bulgarian National Research Fund under Contract KP-06-N62/4.

Theorem A. *([16, Theorem 1.4]) For every $n \geq 3$ and α, $\beta > -1$, the largest zero $x_{nn}(\alpha, \beta)$ of the Jacobi polynomial $P_n^{(\alpha, \beta)}$ satisfies*

$$1 - x_{nn}(\alpha, \beta) < \frac{2(\alpha + 1)(\alpha + 3)}{(n + \alpha + 1)(n + \alpha + \beta + 1)\left[2 - \frac{(\alpha+1)(2n+\beta-1)}{(n+\alpha+1)(n+\alpha+\beta+1)-(\alpha+1)(\alpha+2)}\right]}.$$

Corollary A. *([16, Corollary 1.6]) For every $n \geq 3$ and $\lambda > -1/2$, the largest zero $x_{nn}(\lambda)$ of the Gegenbauer polynomial $P_n^{(\lambda)}$ satisfies*

$$1 - x_{nn}(\lambda) < \frac{(2\lambda + 1)(2\lambda + 5)}{(n + 2\lambda)(2n + 2\lambda + 1)\left[2 - \frac{(2\lambda+1)(4n+2\lambda-3)}{2(n+2\lambda)(2n+2\lambda+1)-(2\lambda+1)(2\lambda+3)}\right]}.$$

Theorem B. *([16, Theorem 1.1]) For every $n \geq 3$ and $\lambda > -1/2$, the largest zero $x_{nn}(\lambda)$ of the Gegenbauer polynomial $P_n^{(\lambda)}$ satisfies*

$$1 - x_{nn}^2(\lambda) < \frac{(2\lambda + 1)(2\lambda + 5)}{2n(n + 2\lambda) + 2\lambda + 1 + \frac{2(\lambda+1)(2\lambda+1)^2(2\lambda+3)}{n(n+2\lambda)+2(2\lambda+1)(2\lambda+3)}}.$$

The above results provide lower bounds for the largest zeros of the Jacobi and Gegenbauer polynomials. It is instructive to compare Theorem B with the following upper bound for the largest zeros of the Gegenbauer polynomials:

Theorem C. *([15, Lemma 3.5]) For every $\lambda > -1/2$, the largest zero $x_{nn}(\lambda)$ of the Gegenbauer polynomial $P_n^{(\lambda)}$ satisfies*

$$1 - x_{nn}^2(\lambda) > \frac{(2\lambda + 1)(2\lambda + 9)}{4n(n + 2\lambda) + (2\lambda + 1)(2\lambda + 5)}.$$

We observe that, for any fixed $\lambda > -1/2$ and large n, the ratio of the upper and the lower bound for $1 - x_{nn}^2(\lambda)$, given by Theorems B and C, does not exceed 2. With Corollary 1 below this ratio is reduced to 1.6.

In the present paper we apply the Euler–Rayleigh method to the Jacobi polynomial $P_n^{(\alpha, \beta)}$, represented as a hypergeometric function, to obtain further bounds for the largest zeros of the Jacobi and Gegenbauer polynomials. As at some points the calculations become unwieldy, we have used the assistance of *Wolfram Mathematica*.

The following is the main result in this paper.

Theorem 1. *For every $n \geq 4$ and α, $\beta > -1$, the largest zero $x_{nn}(\alpha, \beta)$ of the Jacobi polynomial $P_n^{(\alpha, \beta)}$ satisfies*

$$1 - x_{nn}(\alpha, \beta) < \frac{4(\alpha + 1)(\alpha + 2)(\alpha + 4)}{(5\alpha + 11)\left[n(n + \alpha + \beta + 1) + \frac{1}{3}(\alpha + 1)(\beta + 1)\right]}. \tag{1}$$

Moreover, if either $n \geq \max\{4, \alpha + \beta + 3\}$ or $\beta \leq 4\alpha + 7$, then

$$1 - x_{nn}(\alpha, \beta) < \frac{4(\alpha + 1)(\alpha + 2)(\alpha + 4)}{(5\alpha + 11)\left[n(n + \alpha + \beta + 1) + \frac{1}{2}(\alpha + 1)(\beta + 1)\right]}. \tag{2}$$

Since $P_n^{(\alpha,\beta)}(x) = (-1)^n P_n^{(\beta,\alpha)}(-x)$, Theorem 1 can be equivalently formulated as

Theorem 2. *For every $n \geq 4$ and $\alpha, \beta > -1$, the smallest zero $x_{1n}(\alpha, \beta)$ of the Jacobi polynomial $P_n^{(\alpha,\beta)}$ satisfies*

$$1 + x_{1n}(\alpha, \beta) < \frac{4(\beta+1)(\beta+2)(\beta+4)}{(5\beta+11)\left[n(n+\alpha+\beta+1) + \frac{1}{3}(\alpha+1)(\beta+1)\right]}.$$

Moreover, if either $n \geq \max\{4, \alpha+\beta+3\}$ or $\alpha \leq 4\beta+7$, then

$$1 + x_{1n}(\alpha, \beta) < \frac{4(\beta+1)(\beta+2)(\beta+4)}{(5\beta+11)\left[n(n+\alpha+\beta+1) + \frac{1}{2}(\alpha+1)(\beta+1)\right]}.$$

The assumption $\beta \leq 4\alpha + 7$ is satisfied, in particular, when $\beta = \alpha > -1$. Therefore, as a consequence of Theorem 1, we obtain a bound for the largest zero of the ultraspherical polynomial $P_n^{(\lambda)} = c\, P_n^{(\alpha,\alpha)}$, $\alpha = \lambda - \frac{1}{2}$.

Theorem 3. *For every $n \geq 4$ and $\lambda > -1/2$, the largest zero $x_{nn}(\lambda)$ of the Gegenbauer polynomial $P_n^{(\lambda)}$ satisfies*

$$1 - x_{nn}(\lambda) < \frac{(2\lambda+1)(2\lambda+3)(2\lambda+7)}{(10\lambda+17)\left[n(n+2\lambda) + \frac{1}{8}(2\lambda+1)^2\right]}. \tag{3}$$

Theorem 3 and the trivial inequality $1 - x_{nn}^2(\lambda) < 2(1 - x_{nn}(\lambda))$ imply immediately the following:

Corollary 1. *For every $n \geq 4$ and $\lambda > -1/2$, the largest zero $x_{nn}(\lambda)$ of the Gegenbauer polynomial $P_n^{(\lambda)}$ satisfies*

$$1 - x_{nn}^2(\lambda) < \frac{2(2\lambda+1)(2\lambda+3)(2\lambda+7)}{(10\lambda+17)\left[n(n+2\lambda) + \frac{1}{8}(2\lambda+1)^2\right]}. \tag{4}$$

Usually, comparison of the various bounds for the extreme zeros of the classical orthogonal polynomials is not an easy task due to the parameters involved. We show that, at least for large n, the bounds provided by Theorem 1, Theorem 3 and Corollary 1 are sharper than those in Theorem A, Corollary A and Theorem B, respectively. By a limit passage from Theorem 1 we reproduce a result of Gupta and Muldoon [9] concerning the smallest zero of the Laguerre polynomial. These and some other observations are given in Sect. 4 of the paper. Although our bounds are less precise when n is fixed and parameters (α, β or λ) grow, they are easy to work with because to their simple forms. We refer to [12–14] for bounds for the zeros of classical orthogonal polynomials which are uniform with respect to parameters in the weight functions.

The rest of the paper is organized as follows. In Sect. 2 we present the necessary facts about the Euler–Rayleigh method and the Newton identities. The proof of Theorem 1 is given in Sect. 3.1. For the reader's convenience, in Sect. 3.2 we include a short proof of Theorem C.

2   The Euler–Rayleigh Method

As was already mentioned, the proof of our results exploits the so-called Euler–Rayleigh method (see [11]). Here (and also in [16]) the Euler–Rayleigh method is applied to real-root polynomials, and for the reader's convenience we provide some details from [16].

Let P be a monic polynomial of degree n with zeros $(x_i)_1^n$,

$$P(x) = x^n - b_1 x^{n-1} + b_2 x^{n-2} - \cdots + (-1)^n b_n = \prod_{i=1}^{n} (x - x_i). \qquad (5)$$

For $k \in \mathbb{N}_0$, the power sums

$$p_k = p_k(P) := \sum_{i=1}^{n} x_i^k, \qquad p_0 = n = \deg P,$$

and the coefficients $(b_i)_1^n$ of P are connected by the Newton identities (cf. [23])

$$p_r + \sum_{i=1}^{\min\{r-1,n\}} (-1)^i p_{r-i} b_i + (-1)^r r\, b_r = 0.$$

From Newton's identities one easily obtains:

Lemma 1. *Assuming $n \geq r$, the following formulae hold for p_r, $1 \leq r \leq 4$:*

$$p_1(P) = b_1;$$
$$p_2(P) = b_1^2 - 2b_2;$$
$$p_3(P) = b_1^3 - 3b_1 b_2 + 3b_3;$$
$$p_4(P) = b_1^4 - 4b_1^2 b_2 + 2b_2^2 + 4b_1 b_3 - 4b_4.$$

Let us set

$$\ell_k(P) := \frac{p_k(P)}{p_{k-1}(P)}, \qquad u_k(P) := \left[p_k(P) \right]^{1/k}, \qquad k \in \mathbb{N}.$$

The following statement is Proposition 2.2 in [16]; it is a slight modification of Lemma 3.2 in [11].

Proposition 1. *Let P be as in (5) with positive zeros $x_1 < x_2 < \cdots < x_n$. Then the largest zero x_n of P satisfies the inequalities*

$$\ell_k(P) < x_n < u_k(P), \qquad k \in \mathbb{N}.$$

Moreover, $\{\ell_k(P)\}_{k=1}^{\infty}$ is monotonically increasing, $\{u_k(P)\}_{k=1}^{\infty}$ is monotonically decreasing, and

$$\lim_{k \to \infty} \ell_k(P) = \lim_{k \to \infty} u_k(P) = x_n.$$

3 Proof of the Results

3.1 Proof of Theorem 1

The starting point for the proof of Theorem 1 is the following representation of $P_n^{(\alpha,\beta)}$ (cf. [22, Eq. (4.21.2)]):

$$P_n^{(\alpha,\beta)}(x) = \frac{(\alpha+1)_n}{n!} \, {}_2F_1\left(-n, n+\alpha+\beta+1; \alpha+1; \frac{1-x}{2}\right) \qquad (6)$$

(for the proof of Theorem A we have used another representation of $P_n^{(\alpha,\beta)}$ as a hypergeometric function, namely, [22, Eq. (4.3.2)]). Here we use Szegő's notation for the hypergeometric ${}_2F_1$ function,

$$F(a,b;c;z) = 1 + \sum_{k=1}^{\infty} \frac{(a)_k}{k!} \frac{(b)_k}{(c)_k} z^k, \qquad (a)_k := a(a+1)\cdots(a+k-1).$$

It follows from (6) that the monic polynomial

$$P(z) = z^n + \sum_{i=1}^{n} (-1)^i b_i \, z^{n-i}$$

with coefficients

$$b_i = b_i(P) = \binom{n}{i} \frac{(n+\alpha+\beta+1)_i}{(\alpha+1)_i}, \qquad i=1,\ldots,n, \qquad (7)$$

has n positive zeros $z_1 < z_2 < \cdots < z_n$, connected with the zeros of $P_n^{(\alpha,\beta)}$ by the relation

$$z_i = \frac{2}{1-x_{in}(\alpha,\beta)}, \qquad i=1,\ldots,n.$$

According to Proposition 1, $p_{k+1}(P)/p_k(P) < z_n < \left[p_k(P)\right]^{1/k}$, $k \in \mathbb{N}$, and consequently

$$\frac{2}{\left[p_k(P)\right]^{1/k}} < 1 - x_{nn}(\alpha,\beta) < \frac{2p_k(P)}{p_{k+1}(P)}, \qquad k \in \mathbb{N}. \qquad (8)$$

At this point, we find it suitable to substitute

$$a := \alpha+1, \; b := \beta+1,$$
$$t := n(n+\alpha+\beta+1),$$

thus $a, b > 0$ and $t = n(n+a+b-1)$. With this notation, the first four coefficients $b_i(P)$ in (7) are given by

$$b_1(P) = \frac{t}{a}, \quad b_2(P) = \frac{t(t-a-b)}{2a(a+1)}, \quad b_3(P) = \frac{t(t-a-b)\left[t-2(a+b+1)\right]}{6a(a+1)(a+2)},$$

$$b_4(P) = \frac{t(t-a-b)\left[t-2(a+b+1)\right]\left[t-3(a+b+2)\right]}{24a(a+1)(a+2)(a+3)}.$$

Using Lemma 1, we find $p_1(P) = b_1(P) = t/a$,

$$p_2(P) = \frac{t[t + a(a + b)]}{a^2(a + 1)}, \tag{9}$$

$$p_3(P) = \frac{t \, q_2(t)}{a^3(a + 1)(a + 2)}, \tag{10}$$
$$q_2(t) = 2t^2 + a(2a + 3b)t + a^2(a + b)(a + b + 1),$$

$$p_4(P) = \frac{t \, q_3(t)}{a^4(a + 1)^2(a + 2)(a + 3)},$$
$$q_3(t) = (5a + 6)t^3 + 2a(3a^2 + 5ab + 4a + 6b)t^2 \tag{11}$$
$$+ a^2(3a^3 + 9a^2b + 6ab^2 + 6a^2 + 15ab + 7b^2 + 2a + 4b)t$$
$$+ a^3(a + 1)(a + b)(a + b + 1)(a + b + 2).$$

Theorem 1 follows from the right-hand inequality in (8) with $k = 3$. In order to show this, we observe that, according to (10) and (11),

$$1 - x_{nn}(\alpha, \beta) < \frac{2p_3(P)}{p_4(P)} = \frac{2a(a + 1)(a + 3)q_2(t)}{q_3(t)} = \frac{4(\alpha + 1)(\alpha + 2)(\alpha + 4)}{\frac{2q_3(t)}{q_2(t)}}.$$

Hence, to prove the first part of Theorem 1, it suffices to show that if a, b and t are positive, then

$$\frac{2q_3(t)}{q_2(t)} \geq (5a + 6)\left(t + \frac{ab}{3}\right) = (5\alpha + 11)\left[n(n + \alpha + \beta + 1) + \frac{1}{3}(\alpha + 1)(\beta + 1)\right].$$

With the help of *Wolfram Mathematica* we find

$$2q_3(t) - (5a + 6)\left(t + \frac{ab}{3}\right)q_2(t) = \frac{a}{3}r_2(t),$$

where

$$r_2(t) = (6a^2 + 5ab + 12a + 6b)t^2$$
$$+ a(3a^3 + 14a^2b + 6ab^2 + 3a^2 + 27ab + 6b^2 - 6a + 6b)t$$
$$+ a^2(a + b)(a + b + 1)(6a^2 + ab + 18a + 12).$$

It is clear now that $r_2(t) > 0$: the single negative summand in the right-hand side, $-6a^2 t$, is neutralized by $6a^2 t^2$, since $t = n(n + a + b - 1) > n(n - 1) \geq 12$ for $n \geq 4$. Consequently,

$$\frac{2q_3(t)}{q_2(t)} \geq (5a + 6)\left(t + \frac{ab}{3}\right),$$

and the first claim of Theorem 1 is proved.

For the proof of the second claim of Theorem 1 we need to show that

$$2q_3(t) - (5a + 6)\left(t + \frac{ab}{2}\right)q_2(t) \geq 0 \tag{12}$$

provided either $\beta \leq 4\alpha + 7$ or $n \geq \max\{4, \alpha + \beta + 3\} = \max\{4, a + b + 1\}$.
With the assistance of *Mathematica* we find

$$2q_3(t) - (5a + 6)\left(t + \frac{ab}{2}\right)q_2(t) = \frac{1}{2}a^2(a + 2)\, s_2(a, b; t),$$

where

$$s_2(a, b; t) = 4t^2 + (2a^2 + 6ab - b^2 - 2a + 2b)t + a(a + b)(a + b + 1)(4a - b + 4).$$

Firstly, assume that $\beta \leq 4\alpha + 7$, which is equivalent to $b \leq 4a + 4$. Then obviously the constant term in the quadratic $s_2(a, b; \cdot)$ is non-negative. We shall prove that the sum of the other two terms is positive. Indeed, since for $n \geq 4$ we have

$$t = n(n + a + b - 1) \geq 4(a + b + 3) > 0,$$

we need to show that $4t + 2a^2 + 6ab - b^2 - 2a + 2b > 0$. This inequality follows from

$$4t + 2a^2 + 6ab - b^2 - 2a + 2b \geq 16(a + b + 3) + 2a^2 + 6ab - b^2 - 2a + 2b$$
$$= 2a^2 + b(6a + 18 - b) + 14a + 48$$
$$\geq 2a^2 + b(2a + 14) + 14a + 48 > 0.$$

Secondly, assume that $n \geq \max\{4, \alpha + \beta + 3\} = \max\{4, a + b + 1\}$. We observe that

$$t = n(n + a + b - 1) \geq 2(a + b)(a + b + 1). \tag{13}$$

Therefore,

$$4t^2 + (2a^2 + 6ab - b^2 - 2a + 2b)t \geq 8(a + b)(a + b + 1)t + (2a^2 + 6ab - b^2 - 2a + 2b)t$$
$$= (10a^2 + 22ab + 7b^2 + 6a + 10b)t > 0.$$

Using this last inequality and applying (13) once again, we conclude that

$$s_2(a, b; t) \geq (10a^2 + 22ab + 7b^2 + 6a + 10b)t + a(a + b)(a + b + 1)(4a - b + 4)$$
$$\geq (a + b)(a + b + 1)\left[20a^2 + 44ab + 14b^2 + 12a + 20b + a(4a - b + 4)\right]$$
$$= (a + b)(a + b + 1)(24a^2 + 43ab + 14b^2 + 16a + 20b) > 0.$$

Thus, (12) holds true in the case $n \geq \max\{4, \alpha + \beta + 3\}$, which completes the proof of the second claim of Theorem 1.

3.2   Proof of Theorem C

The original proof of Theorem C in [15] makes use of an idea from [22, Paragraph 6.2], based on the following observation of Laguerre: if f is a real-valued polynomial of degree n having only real and distinct zeros, and $f(x_0) = 0$, then

$$3(n-2)\left[f''(x_0)\right]^2 - 4(n-1)f'(x_0)f'''(x_0) \geq 0.   \qquad (14)$$

In [19] Uluchev and the author proved a conjecture of Foster and Krasikov [8], stating that if f is a real-valued and real-root polynomial of degree n, then for every integer m satisfying $0 \leq 2m \leq n$ the following inequalities hold true:

$$\sum_{j=0}^{2m}(-1)^{m+j}\binom{2m}{j}\frac{(n-j)!(n-2m+j)!}{(n-m)!(n-2m)!}f^{(j)}(x)f^{(2m-j)}(x) \geq 0, \qquad x \in \mathbb{R}.$$

It was shown in [19] that these inequalities provide a refinement of the Jensen inequalities for functions from the Laguerre-Pólya class, specialized to the subclass of real-root polynomials. In [4], (14) was deduced from the above inequalities in the special case $m = 2$, and then applied for the derivation of certain bounds for the zeros of classical orthogonal polynomials.

Let us substitute in (14) $f = P_n^{(\lambda)}$ and $x_0 = x_{nn}(\lambda)$. We make use of $f(x_0) = 0$ and the second order differential equations for f and f',

$$(1-x^2)\,f'' - (2\lambda+1)x\,f'(x) + n(n+2\lambda)\,f = 0\,,$$
$$(1-x^2)f''' - (2\lambda+3)x\,f''(x) + (n-1)(n+2\lambda+1)\,f' = 0\,,$$

to express $f'(x_0)$ and $f'''(x_0)$ in terms of $f''(x_0)$ as follows:

$$f'(x_0) = \frac{1-x_0^2}{(2\lambda+1)x_0}\,f''(x_0)\,,$$

$$f'''(x_0) = \left[\frac{2\lambda+3)x_0}{1-x_0^2} - \frac{(n-1)(n+2\lambda+1)}{(2\lambda+1)x_0}\right]f''(x_0)\,.$$

Putting these expressions in (14), canceling out the positive factor $\left[f''(x_0)\right]^2$ and solving the resulting inequality with respect to x_0^2, we arrive at the condition

$$x_0^2 \leq \frac{(n-1)(n+2\lambda+1)}{(n+\lambda)^2 + 3\lambda + \frac{5}{4} + 3\frac{(\lambda+1/2)^2}{n-1}}\,.$$

Hence,

$$x_0^2 < \frac{(n-1)(n+2\lambda+1)}{(n+\lambda)^2 + 3\lambda + \frac{5}{4}} = 1 - \frac{(2\lambda+1)(2\lambda+9)}{4n(n+2\lambda) + (2\lambda+1)(2\lambda+5)}\,.$$

This accomplishes the proof of Theorem C.

4   Remarks

1. As was mentioned in the introduction, at least for large n, the bounds given in Theorem 1, Theorem 3 and Corollary 1 are sharper than those in Theorem A, Corollary A and Theorem B, respectively. For instance, for fixed α, $\beta > -1$ the upper bounds for $1 - x_{nn}(\alpha, \beta)$ in Theorem A and Theorem 1 are respectively

$$\frac{(\alpha + 1)(\alpha + 3)}{n^2} + o(n^{-2}), \qquad \frac{4(\alpha + 1)(\alpha + 2)(\alpha + 4)}{(5\alpha + 11)n^2} + o(n^{-2}), \qquad n \to \infty,$$

and

$$(\alpha + 1)(\alpha + 3) - \frac{4(\alpha + 1)(\alpha + 2)(\alpha + 4)}{5\alpha + 11} = \frac{(\alpha + 1)^3}{5\alpha + 11} > 0, \qquad \alpha > -1.$$

The same conclusion is drawn for the other two pairs of bounds when $\lambda > -1/2$ is fixed and n is large (it follows from the above consideration with $\lambda = \alpha - 1/2$).

2. Theorem 1 is deduced from the second inequality in (8) with $k = 3$. Note that (8) with $k = 2$ together with (9) and (10) implies the estimate

$$1 - x_{nn}(\alpha, \beta) < \frac{2(\alpha + 1)(\alpha + 3)}{2n(n + \alpha + \beta + 1) + (\alpha + 1)(\beta + 1)},$$

which however is less precise than the estimate in Theorem A, and also than the estimate of Driver and Jordaan from [6],

$$1 - x_{nn}(\alpha, \beta) < \frac{2(\alpha + 1)(\alpha + 3)}{2n(n + \alpha + \beta + 1) + (\alpha + 1)(\alpha + \beta + 2)}.$$

Of course, having found the power sums $p_i(P)$, $1 \le i \le 4$, one could apply Proposition 1 for derivation of lower bounds for $1 - x_{n,n}(\alpha, \beta)$ as well. For instance, the first inequality in (8) with $k = 4$ yields

$$1 - x_{nn}(\alpha, \beta) > \frac{2}{[p_4(P)]^{1/4}}$$

with $p_4(P)$ given by (11) and $a = \alpha + 1$, $b = \beta + 1$, $t = n(n + \alpha + \beta + 1)$. However, the expression on the right-hand side looks rather complicated to be of any use.

3. In [9] Gupta and Muldoon proved the following upper bound for the smallest zero $x_{1n}(\alpha)$ of the n-th Laguerre polynomial $L_n^{(\alpha)}$:

$$x_{1n}(\alpha) < \frac{(\alpha + 1)(\alpha + 2)(\alpha + 4)(2n + \alpha + 1)}{(5\alpha + 11)n(n + \alpha + 1) + (\alpha + 1)^2(\alpha + 2)}. \qquad (15)$$

Let us demonstrate how this result can be deduced from the proof of Theorem 1 and the well-known limit relation

$$x_{1n}(\alpha) = \lim_{\beta \to \infty} \frac{\beta}{2} \left(1 - x_{nn}(\alpha, \beta) \right).$$

Since
$$\frac{1}{2}\left(1 - x_{nn}(\alpha, \beta)\right) \leq \frac{p_3(P)}{p_4(P)} = \frac{a(a+1)(a+3)q_2(t)}{q_3(t)}$$
with $a = \alpha + 1$, $b = \beta + 1$, $t = n(n + \alpha + \beta + 1)$, and $q_2(t)$, $q_3(t)$ given in (10) - (11), we have

$$x_{1n}(\alpha) = \lim_{\beta \to \infty} \frac{\beta}{2}\left(1 - x_{nn}(\alpha, \beta)\right) \leq a(a+1)(a+3) \lim_{b \to \infty} \frac{b\,q_2(t)}{q_3(t)}. \qquad (16)$$

From
$$\lim_{b \to \infty} \frac{t}{b} = n$$
and the explicit form of $q_2(t)$ and $q_3(t)$ we find

$$\lim_{b \to \infty} \frac{b\,q_2(t)}{q_3(t)} = \lim_{b \to \infty} \frac{q_2(t)/b^2}{q_3(t)/b^3}$$
$$= \frac{2n^2 + 3a\,n + a^2}{(5a+6)n^3 + 2a(5a+6)n^2 + a^2(6a+7)n + a^3(a+1)}$$
$$= \frac{2n + a + 1}{(5a+6)n(n+a) + a^2(a+1)}.$$

Substituting the last expression in (16) and setting $a = \alpha + 1$, we arrive at (15).

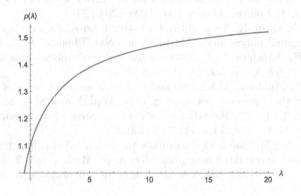

Fig. 1. The graph of $\varrho(\lambda)$.

4. We already mentioned in the introduction that, for every fixed $\lambda > -1/2$, the ratio $r(\lambda, n)$ of the upper and the lower bound for $1 - x_{nn}^2(\lambda)$, given by Theorem 1 and Theorem C, respectively, tends to a limit which does not exceed 1.6. More precisely,
$$r(\lambda, n) = \varrho(\lambda)\psi(\lambda, n),$$
where
$$\varrho(\lambda) = \frac{8(2\lambda + 3)(2\lambda + 7)}{(2\lambda + 9)(10\lambda + 17)}, \qquad \varphi(\lambda, n) = \frac{n(n + 2\lambda) + (2\lambda + 1)(2\lambda + 5)/4}{n(n + 2\lambda) + (2\lambda + 1)^2/8}.$$

The function $\varrho(\lambda)$ is monotonically increasing in the interval $(-1/2, \infty)$ assuming values between 1 and 1.6 (see Fig. 1) while, for a fixed $\lambda > -1/2$, $\lim_{n \to \infty} \varphi(\lambda, n) = 1$.

5. The Euler–Rayleigh approach assisted with computer algebra has been applied in [21] to the derivation of bounds for the extreme zeros of the Laguerre polynomials, and in [1,5,17,18,20] to the estimation of the extreme zeros of some non-classical orthogonal polynomials, which are related to the sharp constants in some Markov-type inequalities in weighted L_2 norms.

References

1. Aleksov, D., Nikolov, G.: Markov L_2-inequality with the Gegenbauer weight. J. Approx. Theory **225**, 224–241 (2018)
2. Area, I., Dimitrov, D.K., Godoy, E., Rafaeli, F.: Inequalities for zeros of Jacobi polynomials via Obreshkoff's theorem. Math. Comp. **81**, 991–1012 (2012)
3. Area, I., Dimitrov, D.K., Godoy, E., Ronveaux, A.: Zeros of Gegenbauer and Hermite polynomials and connection coefficients. Math. Comp. **73**, 1937–1951 (2004)
4. Dimitrov, D.K., Nikolov, G.P.: Sharp bounds for the extreme zeros of classical orthogonal polynomials. J. Approx. Theory **162**, 1793–1804 (2010)
5. Dimitrov, D.K., Nikolov, G.: A discrete Markov-Bernstein inequality for sequences and polynomials. J. Math. Anal. Appl. **493**, 124522 (2021)
6. Driver, K., Jordaan, K.: Bounds for extreme zeros of some classical orthogonal polynomials. J. Approx. Theory **164**, 1200–1204 (2012)
7. Driver, K., Jordaan, K.: Inequalities for extreme zeros of some classical orthogonal and q-orthogonal polynomials. Math. Model. Nat. Phenom. **8**(1), 48–59 (2013)
8. Foster, W.H., Krasikov, I.: Inequalities for real-root polynomials and entire functions. Adv. Appl. Math. **29**, 102–114 (2002)
9. Gupta, D.P., Muldoon, M.E.: Inequalities for the smallest zeros of Laguerre polynomials and their q-analogues. J. Ineq. Pure Appl. Math. **8**(1), 24 (2007)
10. Ismail, M.E.H., Li, X.: Bounds on the extreme zeros of orthogonal polynomials. Proc. Amer. Math. Soc. **115**, 131–140 (1992)
11. Ismail, M.E.H., Muldoon, M.E.: Bounds for the small real and purely imaginary zeros of Bessel and related functions. Met. Appl. Math. Appl. **2**, 1–21 (1995)
12. Krasikov, I.: Bounds for zeros of the Laguerre polynomials. J. Approx. Theory **121**, 287–291 (2003)
13. Krasikov, I.: On zeros of polynomials and allied functions satisfying second order differential equation. East J. Approx. **9**, 51–65 (2003)
14. Krasikov, I.: On extreme zeros of classical orthogonal polynomials. J. Comp. Appl. Math. **193**, 168–182 (2006)
15. Nikolov, G.: Inequalities of Duffin-Schaeffer type. II. East J. Approx. **11**, 147–168 (2005)
16. Nikolov, G.: New bounds for the extreme zeros of Jacobi polynomials. Proc. Amer. Math. Soc. **147**, 1541–1550 (2019)
17. Nikolov, G., Shadrin, A.: On the L_2 Markov inequality with Laguerre weight. In: Govil, N.K., Mohapatra, R., Qazi, M.A., Schmeisser, G. (eds.) Progress in Approximation Theory and Applicable Complex Analysis. SOIA, vol. 117, pp. 1–17. Springer, Cham (2017). https://doi.org/10.1007/978-3-319-49242-1_1

18. Nikolov, G., Shadrin, A.: Markov-type inequalities and extreme zeros of orthogonal polynomials. J. Approx. Theory **271**, 105644 (2021)
19. Nikolov, G., Uluchev, R.: Inequalities for real-root polynomials. proof of a conjecture of Foster and Krasikov. In: Dimitrov, D.K., Nikolov, G., Uluchev, R. (eds.) Approximation Theory: a volume dedicated to Borislav Bojanov, pp. 201–216. Prof. Marin. Drinov Academic Publishing House, Sofia (2004)
20. Nikolov, G., Uluchev, R.: Estimates for the best constant in a Markov L_2-inequality with the assistance of computer algebra. Ann. Univ. Sofia, Ser. Math. Inf. **104**, 55–75 (2017)
21. Nikolov, G., Uluchev, R.: Bounds for the extreme zeros of Laguerre polynomials. In: Nikolov, G., Kolkovska, N., Georgiev, K. (eds.) NMA 2018. LNCS, vol. 11189, pp. 243–250. Springer, Cham (2019). https://doi.org/10.1007/978-3-030-10692-8_27
22. Szegő, G.: Orthogonal Polynomials, 4th edn. American Mathematical Society Colloquium Publications, Providence, RI (1975)
23. Van der Waerden, B.L.: Modern Algebra, vol. 1. Frederick Ungar Publishing Co, New York (1949)

Sensitivity Analysis of an Air Pollution Model with Using Innovative Monte Carlo Methods in Calculating Multidimensional Integrals

Tzvetan Ostromsky[1]([✉]), Venelin Todorov[1,2][iD], Ivan Dimov[1], and Rayna Georgieva[1]

[1] Department of Parallel Algorithms, Institute of Information and Communication Technologies, Bulgarian Academy of Sciences, Acad. G. Bonchev Str., Block 25 A, 1113 Sofia, Bulgaria
{ceco,venelin,rayna}@parallel.bas.bg, ivdimov@bas.bg
[2] Department of Information Modeling, Institute of Mathematics and Informatics, Bulgarian Academy of Sciences, Acad. Georgi Bonchev Str., Block 8, 1113 Sofia, Bulgaria
vtodorov@math.bas.bg

Abstract. Large-scale models are mathematical models with a lot of natural uncertainties in their input data sets and parameters. Sensitivity analysis (SA) is a powerful tool for studying the impact of these uncertainties on the output results and helps to improve the reliability of these models. In this article we present some results of a global sensitivity study of the Unified Danish Eulerian Model (UNI-DEM). A large number of heavy numerical experiments must be carried out in order to collect the necessary data for such comprehensive sensitivity study. One of the largest supercomputers in Europe and the most powerful in Bulgaria, the petascale EuroHPC supercomputer Discoverer is used to perform efficiently this huge amount of computations.

One of the most important features of UNI-DEM is its advanced chemical scheme, called Condensed CBM IV, which considers a large number of chemical species and all significant reactions between them. The ozone is one of the most harmful pollutants, that is why it is important for many practical applications to study it precisely. Stochastic methods based on Adaptive approach and Sobol sequences are used for computing the corresponding sensitivity measures. We show by experiments that the stochastic algorithms for calculating the multidimensional integrals under consideration are one of the best stochastic techniques for computing the small in value sensitivity indices.

The presented work was supported by the Bulgarian National Science Fund under the Bilateral Project KP-06-Russia/17 "New Highly Efficient Stochastic Simulation Methods and Applications" and by the Bulgarian National Science Fund under Project KP-06-N52/5 "Efficient methods for modeling, optimization and decision making".

Keywords: Sensitivity analysis · Sobol global sensitivity indices ·
Monte Carlo methods · Environmental modeling · Air pollution ·
Supercomputer

1 Introduction

Sensitivity analysis (SA) is a procedure for studying how sensitive are the output results of large-scale mathematical models to some uncertainties of the input data. Algorithms based on analysis of variances technique (ANOVA) for calculating numerical indicators of sensitivity and computationally efficient Monte Carlo integration techniques have recently been developed [4,5]. A careful SA is needed in order to decide where and how simplifications of the model can be done. A good candidate for reliable SA of models containing nonlinearity is the variance based method [11]. The idea of this approach is to estimate how the variation of an input parameter or a group of inputs contributes to the variance of the model output. As a measure of this analysis we use the *total sensitivity indices (TSI)*, described as multidimensional integrals:

$$I = \int_{\Omega} g(\mathbf{x}) p(\mathbf{x}) \, d\mathbf{x}, \quad \Omega \subset \mathbf{R}^d, \tag{1}$$

where $g(\mathbf{x})$ is a square integrable function in Ω and $p(\mathbf{x}) \geq 0$ is a probability density function, such that $\int_{\Omega} p(\mathbf{x}) \, d\mathbf{x} = 1$.

Clearly, the progress in the area of sensitivity analysis is closely related to the advance in finding reliable and efficient algorithms for multidimensional integration.

2 Advanced Stochastic Algorithms

For computing the above sensitivity parameters we'll use the standard Sobol sequence [1,7], the modified Sobol $\Lambda\Pi_\tau$ sequences [6] and a modified optimal lattice rule [8].

The first algorithm (MCA-MSS-1), based on shaking of the original random point on a certain distance ρ (parameter of the algorithm), was proposed in [2]. For the second Monte Carlo algorithm (MCA-MSS-2) [6] two pseudo random points are chosen - the first one is selected following the concept of the first algorithm MCA-MSS-1, while the second pseudo random point is chosen to be symmetric to the first point according to the central point. In a previous work of the authors, it has been proven that the Monte Carlo algorithm MCA-MSS-2 has an optimal rate of convergence for the class of continuous functions with continuous first derivatives and bounded second derivatives. Since the procedure of shaking is a computationally expensive procedure, especially for a large number of points, a modification of MCA-MSS-2 algorithm has been developed – the algorithm MCA-MSS-2-S [6]. The algorithm MCA-MSS-2-S has an optimal rate of convergence for the class of continuous functions with continuous first derivatives and bounded second derivatives [6].

Now we will consider a special lattice rule with generating vector based on the component by component construction method [9]. The lattice rule applies a polynomial transformation function to a nonperiodic integrand to make it suitable for applying a lattice rule. For this method we will use the notation LR-CC. At the first step of the algorithm d dimensional optimal generating vector $\mathbf{z} = (z_1, z_2, \ldots z_d)$ is generated [8]. On the second step of the algorithm we generate the points of lattice rule by formula $\mathbf{x}_k = \left\{\frac{k}{N}\mathbf{z}\right\}$, $k = 1, \ldots, N$. And at the third and last step of the algorithm an approximate value I_N of the multidimensional integral is evaluated by the formula: $I_N = \frac{1}{N} \sum_{k=1}^{N} f\left(\left\{\frac{k}{N}\mathbf{z}\right\}\right)$.

3   Short Review, Parallelization and Scalability of UNI-DEM

The Unified Danish Eulerian Model (UNI-DEM) [13–16] is a complex and powerful tool for modeling the atmosphere physics and chemistry and calculating the concentrations of a large number of pollutants and other chemical species in the air and their variation along certain (long) time period. Its large spatial domain (4800 × 4800 km) covers the whole Europe, the Mediterranean and some parts of Asia and Africa. It takes into account the main physical, chemical and photochemical processes in the atmosphere. Large input data sets are needed for the emissions and the quickly changing meteorological conditions, which must also be taken into account. The model is described mathematically by a complex system of partial differential equations. In order to simplify it splitting procedure is applied first. As a result the initial system is replaced by three simpler systems (submodels), connected with the main physical and chemical processes. Discretization of the domain is the next step towards the numerical solution of these systems. The dynamics of the atmospheric processes require small time-step to be used (at least, for the chemistry submodel) in order to get a stable numerical solution of the corresponding system. All this makes the treatment of large-scale air pollution models a tuff and heavy computational task. It has been always a serious challenge, even for the fastest and most powerful state-of-the-art supercomputers [10,14,15].

3.1   Parallel Implementation and Optimizations

The parallelization strategy is based on the space domain partitioning [10,13,14, 16]. The MPI standard library is used to create parallel tasks. The space domain is divided into sub-domains (the number of the sub-domains is equal to the number of MPI tasks). Each MPI task works on its own sub-domain. On each time step there is no data dependency between the MPI tasks on both the chemistry and the vertical exchange stages. This is not so with the advection-diffusion stage. Spatial grid partitioning between the MPI tasks requires overlapping of the inner boundaries and exchange of boundary values between the neighboring subgrids (to be used for boundary conditions). As a result, there is always certain overhead on this stage (in terms of additional computations and communications), increasing proportionally with the nunber of MPI tasks. The effect of this overhead on the speed-up of the advection-diffusion stage (in comparison with the chemistry stage) can be observed comparing the corresponding columns in Tables 1 or 2 below.

In order to achieve good data locality, the relatively small calculation tasks are grouped in chunks (if appropriate) for more efficient utilization of the fastest cache memory of the corresponding hardware unit. An input parameter CHUNK-SIZE is provided, which controls the amount of short-term reusable data in order to reduce the transfer between the cache and the main (slower access) memory. It should be tuned with respect to the cache size of the target machine.

Another important parallel optimization issue is the load-balance. MPI barriers, used to force synchronization between the processes in data transfer commands, often do not allow good load-balance. This obstacle can be avoided to some extent by using non-blocking communication routines from the MPI standard library.

3.2 Scalability Experiments with UNI-DEM on the Bulgarian EuroHPC Petascale Supercomputer DISCOVERER

The most powerful Bulgarian supercomputer Discoverer, part of the European network of high-performance machines of the EuroHPC Joint Undertaking, has been installed in Sofia Tech Park in the summer of 2021 by Atos company with the support of the consortium "Petascale Supercomputer – Bulgaria". Petascale-Bulgaria Supercomputer is a consortium that combines the knowledge of the member organisations with over 15 years of experience in applied sciences. The consortium is the executor of the project for creation of this new supercomputing system, part of the EuroHPC JU network in Bulgaria. The project is funded by the budgets of the Bulgarian state and EuroHPC JU. The system is based on the BullSequana XH2000 platform (CPU type – AMD EPYC 7H12/2.6 GIIz, 280 W) and has 1128 nodes with 144384 cores in total (128 cores per node, 128 GB RAM per node). The total RAM of the system is 300 TB, the total disk storage – about 2 PB.

Results of scalability experiments with UNI-DEM on various Discoverer configurations are shown in this section. The coarse and the fine grid versions of UNI-DEM has been run to produce modelling results for one-year time period. The computing times, speed-up and efficiency obtained from these experiments are shown in Tables 1 and 2.

4 Sensitivity Analysis of UNI-DEM with Respect to Emission Levels

The first task of our long-term research on the sensitivity of UNI-DEM is to study the sensitivity of its output (in terms of mean monthly concentrations of several important pollutants) with respect to variation of the emissions of anthropogenic pollution (quite uncertain and critical input data for the model). The anthropogenic emissions are determined as vectors with 4 separate components $E = (E_1, E_2, E_3, E_4)$, which correspond to 4 different groups of pollutants as follows:

Table 1. Time (T) in seconds and speed-up (Sp) (with respect to the number of MPI tasks, given in the first column) for running UNI-DEM (the coarse grid version) on the EuroHPC supercomputer DISCOVERER (in Sofia Tech Park, Bulgaria).

Time (T) in seconds and speed-up (Sp) of UNI-DEM on DISCOVERER (96 × 96 × 1) grid, 35 species, CHUNKSIZE=32								
NP	#	Advection		Chemistry		TOTAL		
(MPI)	nodes	T [s]	(Sp)	T [s]	(Sp)	T [s]	(Sp)	E [%]
1	1	439	(1.0)	902	(1.0)	1454	(1.0)	100%
2	1	212	(2.1)	448	(2.0)	747	(1.9)	97%
4	1	108	(4.1)	223	(4.0)	394	(3.7)	92%
8	1	55	(8.0)	113	(8.0)	206	(7.1)	88%
12	1	38	(11.6)	76	(11.9)	164	(8.9)	74%
16	1	31	(14.2)	57	(15.8)	127	(11.4)	72%
24	2	23	(19.1)	38	(23.7)	101	(14.4)	60%
32	2	19	(23.1)	29	(31.1)	92	(15.8)	49%
48	3	15	(29.3)	19	(47.5)	84	(17.3)	36%

Table 2. Time (T) in seconds and speed-up (Sp) (with respect to the number of MPI tasks, given in the first column) for running UNI-DEM (the fine grid version) on the EuroHPC supercomputer DISCOVERER (in Sofia Tech Park, Bulgaria).

Time (T) in seconds and speed-up (Sp) of UNI-DEM (fine grid version) on DISCOVERER (480 × 480 × 1) grid, 35 species, CHUNKSIZE=32								
NP	#	Advection		Chemistry		TOTAL		
(MPI)	nodes	T [s]	(Sp)	T [s]	(Sp)	T [s]	(Sp)	E [%]
1	1	93022	(1.0)	81850	(1.0)	195036	(1.0)	100%
4	1	23408	(4.0)	21615	(3.9)	48604	(4.0)	100%
8	1	11830	(7.9)	11072	(7.4)	25045	(7.7)	96%
16	1	6023	(15.4)	5438	(15.1)	13061	(14.8)	92%
24	2	4075	(22.8)	3630	(22.5)	9150	(21.1)	88%
48	3	2216	(42.0)	1845	(44.4)	4805	(40.1)	84%
96	6	1243	(74.8)	824	(99.3)	2701	(71.4)	74%
120	8	1072	(86.8)	662	(123.6)	2394	(80.6)	67%
160	10	895	(103.9)	498	(164.4)	2052	(94.0)	59%

$E_1 = E^{(A)}$ – ammonia (NH_3)
$E_2 = E^{(N)}$ – nitrogen oxides $(NO + NO_2)$
$E_3 = E^{(S)}$ – sulphur dioxide (SO_2)
$E_4 = E^{(HC)}$ – anthropogenic hydrocarbons

The most commonly used output of UNI-DEM are the mean monthly concentrations of a set of dangerous chemical species (or groups of such), calculated in the grid points of the computational domain. In this work we consider the results for the following 3 species/groups: the ammonia (NH_3), the ozone (O_3), as well as the ammonium sulphate and ammonium nitrate together $(NH_4SO_4 + NH_4NO_3)$, which UNI-DEM evaluate in common. With respect to the spatial grid of the computational domain we have chosen three European cities with different climate and level of pollution: (i) Milan, (ii) Manchester and (iii) Edinburgh.

Table 3. f_0, D, first-order, second-order and total sensitivity indices of the concentration of 3 pollutants in 3 different cities with respect to the input emissions

Estimated value	pollutant / city								
	NH_3			O_3			$NH_4SO_4 + NH_4NO_3$		
	Milan	Manchester	Edinburgh	Milan	Manchester	Edinburgh	Milan	Manchester	Edinburgh
f_0	0.048	0.049	0.049	0.059	0.068	0.062	0.044	0.045	0.044
D	2e-04	4e-04	3e-04	1e-05	5e-05	4e-06	4e-04	4e-05	4e-05
S_1	0.889	0.812	0.845	2e-06	6e-06	3e-05	0.152	0.393	0.295
S_2	2e-04	1e-04	4e-04	0.156	0.791	0.387	0.017	0.006	0.012
S_3	0.109	0.181	0.148	2e-06	6e-06	3e-05	0.818	0.575	0.647
S_4	4e-05	8e-05	3e-04	0.826	0.209	0.589	0.002	0.002	0.008
$\sum_{i=1}^{4} S_i$	0.999	0.994	0.994	0.983	0.999	0.976	0.991	0.976	0.962
S_{12}	2e-05	5e-06	1e-05	7e-07	1e-07	8e-06	0.001	2e-04	4e-04
S_{13}	0.001	0.006	0.006	7e-07	1e-07	8e-06	0.007	0.024	0.033
S_{14}	2e-06	7e-06	5e-05	7e-07	1e-07	8e-06	1e-04	1e-04	0.001
S_{23}	5e-06	7e-08	5e-06	7e-07	1e-07	8e-06	3e-04	3e-05	6e-05
S_{24}	5e-06	2e-06	1e-04	0.017	3e-04	0.024	4e-04	1e-04	0.004
S_{34}	7e-07	3e-06	2e-05	7e-07	1e-07	8e-06	2e-05	5e-05	3e-04
$\sum_{i<j;\ i,j=1}^{4} S_{ij}$	0.001	0.006	0.006	0.017	3e-04	0.024	0.009	0.024	0.038
S_1^{tot}	0.891	0.818	0.851	4e-06	6e-06	6e-05	0.161	0.417	0.329
S_2^{tot}	2e-04	1e-04	6e-04	0.174	0.791	0.411	0.019	0.006	0.016
S_3^{tot}	0.110	0.188	0.154	4e-06	6e-06	6e-05	0.826	0.598	0.679
S_4^{tot}	5e-05	9e-05	5e-04	0.844	0.209	0.613	0.003	0.003	0.013

The mean value (f_0), the variance (D), first- and second-order as well as total sensitivity indices in each case study are given in Table 3. It should be noted that all third- and forth-order sensitivity indices are very small, that is why they are not given in the table. Another common feature for all case studies is that the sum of main effects is close to 1. It means that the model is additive according to input parameters under consideration - the most important air pollutants [5].

Consider, for example, the sensitivity indices of the ammonia concentrations in the air for the city of Milan (the first data column of Table 3). The main input emissions which contribute to the concentration of ammonia are the ammonia emissions −89% and the sulphur dioxide emissions −11%. A detailed analysis is available in [5].

The results for the relative errors of the evaluation of the quantities f_0, total variance and first-order and total sensitivity indices using various stochastic approaches for numerical integration are presented in Tables 4, 5, and 6. The quantity f_0 is represented by a 4-dimensional integral and the rest of quantities under consideration are represented by double integrals. The algorithm MCA-MSS-2-S is the most efficient in terms of relative error, where the advantage is in the range of 1–3 orders. The algorithm has a significant advantage in the computation of the 4-dimensional integral for f_0 (Table 4). For some of the sensitivity

Table 4. Relative error for the approximate evaluation of $f_0 \approx 0.048$.

# of points	Sobol	MSS-1		MSS-2		MSS-2-S	LR-CC
n	Rel. error	ρ	Rel. error	ρ	Rel. error	Rel. error	Rel. error
2^8	2e-03	2e-02	5e-04				2e-02
(2×2^8)				8e-05	3e-03	1e-05	
2^{10}	4e-04	7e-04	3e-05				8e-04
(2×2^{10})				7e-04	0.06	4e-06	
2^{14}	3e-05	1e-04	3e-06				3e-06
(2×2^{14})				1e-05	4e-05	2e-07	
2^{16}	7e-06	1e-04	6e-07				4e-07
(2×2^{16})				9e-05	8e-04	2e-08	

Table 5. Relative error for the approximate evaluation of the total variance $D \approx 0.0002$.

# of points	Sobol	MSS-1		MMS-2		MMS-2-S	LR-CC
	Rel. error	ρ	Rel. error	ρ	Rel. error	Rel. error	Rel. error
2^8	2e-02	2e-02	2e-02				3e-01
(2×2^8)				3e-03	9e-03	9e-04	
2^{10}	4e-03	7e-04	9e-03				2e-02
(2×2^{10})				7e-04	3e-04	3e-04	
2^{14}	4e-04	1e-04	3e-04				1e-03
(2×2^{14})				1e-05	2e-04	2e-05	
2^{16}	4e-05	1e-04	2e-05				2e-03
(2×2^{16})				9e-05	7e-04	2e-06	

Table 6. Relative error for estimation of sensitivity indices of input parameters using different Monte Carlo and quasi-Monte Carlo methods ($n \approx 2^{16}$).

Sens. Index	Ref. value	Sobol	MSS-1	MSS-2	MSS-2-S	LR-CC
S_1	9e-01	8e-05	8e-05	6e-06	5e-04	7e-04
S_2	2e-04	3e-02	3e-02	4e-03	7e-02	3e-02
S_3	1e-01	8e-04	8e-04	7e-05	1e-02	4e-03
S_4	4e-05	7e-02	7e-02	1e-02	6e-01	2e-02
S_1^{tot}	9e-01	8e-05	8e-05	1e-05	1e-03	5e-04
S_2^{tot}	2e-04	5e-03	2e-03	1e-03	3e-03	2e-01
S_3^{tot}	1e-01	7e-04	7e-04	4e-05	4e-03	6e-03
S_4^{tot}	5e-05	6e-02	6e-02	1e-02	1e-01	2e-01

indices LR-CC gives very close results to the MCA-MSS-2-S, but for small in value sensitivity indices the difference is more pronounced. The numerical results in Tables 4, 5 and 6 show that all stochastic approaches under consideration give reliable relative errors for sufficiently large sample sizes, where the most efficient in terms of computational complexity is the algorithm MCA-MSS-2-S.

5   Sensitivity Analysis of UNI-DEM with Respect to Chemical Reaction Rates

Our next aim here is to study the sensitivity of the ozone concentration according to variation of some chemical reactions rates, in particular ## 1, 3, 7, 22 (time-dependent) and ##27, 28 (time independent), as they are denoted in the condensed CBM-IV scheme [13,15]. For the purpose of our SA study these parameters will be considered as components of the 6-dimensional vector $\mathcal{R} = (\mathcal{R}^1, \mathcal{R}^2, \mathcal{R}^3, \mathcal{R}^4, \mathcal{R}^5, \mathcal{R}^6)$. Their original numbers in the condensed CBM-IV scheme and the simplified chemical equations of the corresponding reactions are as follows:

index in $\mathcal{R}$	# in CBM-IV	chemical reaction equation
$\mathcal{R}^1$	# 1	$NO_2 + h\nu \Longrightarrow NO + O$
$\mathcal{R}^2$	# 3	$NO_2 + h\nu \Longrightarrow NO + O$
$\mathcal{R}^3$	# 7	$O_3 + NO \Longrightarrow NO_2$
$\mathcal{R}^4$	# 22	$NO_2 + O_3 \Longrightarrow NO_3$
$\mathcal{R}^5$	# 27	$HO_2 + NO \Longrightarrow OH + NO_2$
$\mathcal{R}^6$	# 28	$HO_2 + HO_2 \Longrightarrow H_2O_2$

The most frequently used output of UNI-DEM are the mean monthly concentrations of a set of dangerous chemical species (or groups of species) in dependence with the particular chemical scheme, calculated in the grid points of the computational domain.

Table 7 contains first-, second-order and total sensitivity indices of model inputs under consideration. Analyzing the values of TSI for the reactions under consideration one can conclude that the influence of rates of the reactions $\mathcal{R}^1$, $\mathcal{R}^2$ and $\mathcal{R}^4$, on the ozone concentrations is very important. The impact of the rates of the reactions $\mathcal{R}^3$, and $\mathcal{R}^5$ is smaller but significant, while the influence of the rate of the reaction $\mathcal{R}^6$ can be neglected. As expected, the values of higher-order sensitivity indices are relatively small and close to zero (taking into account that the values of both first-order and total sensitivity indices are close to each other). It means that the mathematical model is additive according to the chosen input parameters. At the same time one can observe that the rate of the chemical reaction $\mathcal{R}^1$ is of primary importance.

We should specify that the quantity f_0 is represented by a 6-dimensional integral, while for the rest of sensitivity measures 12-dimensional integrals must be calculated. The experimental results for the relative errors and computational

Table 7. Sensitivity indices for the ozone concentrations with respect to 6 chemical rate coefficients.

Estimated	city			
value	Genova	Milan	Manchester	Edinburgh
f_0	0.26588	0.26566	0.26526	0.26616
D	0.00249	0.00256	0.00245	0.00136
S_1	0.35858	0.36281	0.37165	0.33487
S_2	0.29485	0.29936	0.26509	0.23399
S_3	0.04652	0.04129	0.00997	0.05559
S_4	0.26462	0.26276	0.32358	0.30133
S_5	4.34e-07	1.8e-07	0.00023	0.00009
S_6	0.01904	0.01703	0.00857	0.04653
$\sum_{i=1}^{6} S_i$	0.98361	0.98325	0.97909	0.97241
S_{12}	0.00556	0.00574	0.00568	0.00457
S_{13}	0.00048	0.00049	0.00024	0.00106
S_{14}	0.00516	0.00563	0.00809	0.00837
S_{16}	0.00031	0.00025	0.00018	0.00104
S_{23}	0.00038	0.00033	0.00005	0.00075
S_{24}	0.00349	0.00343	0.00516	0.00457
S_{34}	0.00045	0.00040	0.00015	0.00068
S_{36}	0.00016	0.00014	0.00039	0.00435
$\sum_{i<j;\ i,j=1}^{6} S_{ij}$	0.01639	0.01675	0.02092	0.02759
S_1^{tot}	0.37009	0.37493	0.38599	0.34993
S_2^{tot}	0.30442	0.30897	0.27625	0.24471
S_3^{tot}	0.04799	0.04267	0.01098	0.06274
S_4^{tot}	0.27391	0.27239	0.33719	0.31559
S_5^{tot}	0.00015	0.00013	0.00089	0.00091
S_6^{tot}	0.01983	0.01766	0.00963	0.05371

times for the computation of f_0, total variance and first-order and total sensitivity indices using different stochastic methods for numerical integration are presented in Tables 8, 9, and 10. Here LR-CC gives very close results to the MCA-MSS-2-S but for small in value sensitivity indices the difference is much more pronounced. The algorithm MCA-MSS-2 is the most computationally expensive, but it is also the most robust, stable and reliable – Tables 8, 9, 10.

Table 8. Relative error and computational time for the approximate evaluation of $f_0 \approx 0.27$.

# of points n	Sobol Rel. error	MSS-1 ρ	MSS-1 Rel. error	MSS-2 ρ	MSS-2 Rel. error	MSS-2-S Rel. error	LR-CC Rel. error
2^6	2e-03	7e-04	2e-03				2e-01
2×2^6				7e-03	3e-04	9e-05	
2^{12}	5e-05	3e-02	6e-06				7e-03
2×2^{12}				4e-04	7e-05	3e-06	
2^{14}	2e-05	2e-04	1e-05				4e-05
2×2^{14}				2e-04	9e-06	9e-08	
2^{16}	2e-06	2e-04	2e-07				1e-05
2×2^{16}				2e-05	4e-06	5e-07	

Table 9. Relative error for the approximate evaluation of the total variance $D \approx 0.0025$.

# of points n	Sobol Rel. error	MSS-1 ρ	MSS-1 Rel. error	MMS-2 ρ	MMS-2 Rel. error	MMS-2-S Rel. error	LR-CC Rel. error
2^6	1e-02	7e-04	1e-02				9e-01
(2×2^6)				7e-03	9e-03	8e-03	
2^{12}	6e-04	4e-03	3e-05				9e-02
(2×2^{12})				4e-04	4e-04	5e-04	
2^{14}	1e-04	2e-04	1e-04				8e-04
(2×2^{14})				2e-04	2e-04	6e-06	
2^{16}	2e-05	2e-04	7e-05				9e-04
(2×2^{16})				2e-05	2e-05	1e-04	

Table 10. Relative error for estimation of sensitivity indices with respect to 6 reaction rate coefficients, calculated by using different Monte Carlo and quasi-Monte Carlo methods $(n \approx 2^{16})$.

Sens. Index	Ref. value	Sobol	MSS-1	MSS-2	MSS-2-S	LR-CC
S_1	4e-01	1e-04	4e-04	2e-04	2e-02	1e-02
S_2	3e-01	3e-05	2e-04	3e-04	6e-02	2e-02
S_3	5e-02	2e-04	2e-03	9e-04	8e-02	8e-02
S_4	3e-01	3e-04	2e-05	2e-04	4e-03	7e-03
S_5	4e-07	3e-01	7e+00	7e-02	2e+02	3e+03
S_6	2e-02	3e-04	1e-03	3e-04	4e-02	1e-02
S_1^{tot}	4e-01	1e-04	4e-05	2e-04	5e-02	1e-02
S_2^{tot}	3e-01	4e-05	5e-04	2e-04	3e-02	2e-02
S_3^{tot}	5e-02	3e-04	2e-03	8e-04	4e-02	5e-02
S_4^{tot}	3e-01	2e-04	5e-04	2e-04	4e-02	2e-02
S_5^{tot}	2e-04	7e-03	1e-02	4e-03	1e+00	9e+01
S_6^{tot}	2e-02	4e-04	1e-03	3e-04	4e-02	8e-02
S_{12}	6e-03	2e-04	5e-03	1e-03	7e-01	2e-01
S_{14}	5e-03	2e-03	2e-02	2e-03	1e+00	1e+00
S_{15}	8e-06	3e-02	1e-01	5e-02	5e+00	5e+00
S_{24}	3e-03	1e-03	2e-02	6e-03	1e+00	6e-01
S_{45}	1e-05	4e-02	1e-01	2e-02	4e+00	4e+01

6   Conclusions

In this paper the resent developments and a high performance implementation of the Unified Danish Eulerian Model (UNI-DEM) on the new petascale EuroHPC supercomputer Discoverer were presented. The MPI parallel implementation proved to be efficient and scalable up to the level of granularity of the domain discretization (in dependence with the spatial grid size).

A large number of numerical experiments have been carried out to collect the necessary data for sensitivity analysis of the most important output data with respect to various critical input data and parameters. Several Monte Carlo algoritms for numerical integration were used in the process of computing Sobol global sensitivity indices. The algorithm MCA-MSS-2-S is the most efficient in terms of computational complexity for computing f_0 and the total variance D, but its efficiency decreases in the computation of the Sobol global sensitivity indices. The algorithm MCA-MSS-2-S overcomes the disadvantage of the algorithms MCA-MSS-1 and MCA-MSS-2 in terms of computational cost, but it is less efficient for computational problems where a loss of accuracy appears. The Monte Carlo algorithms MCA-MSS-1, MCA-MSS-2 and MCA-MSS-2-S have unimprovable rates of convergence and produce reliable numerical results. The development of efficient stochastic algorithms for multidimensional integration contributes to the progress in the area of sensitivity analysis and the study of the effects of the air pollution and the environmental changes.

Acknowledgements. The presented work was supported by the Bulgarian National Science Fund under the Bilateral Project KP-06-Russia/17 "New Highly Efficient Stochastic Simulation Methods and Applications" and by the Bulgarian National Science Fund under Project KP-06-N52/5 "Efficient methods for modeling, optimization and decision making".

References

1. Dimov, I.T.: Monte Carlo Methods for Applied Scientists. World Scientific, London, Singapore (2008)
2. Dimov, I., Georgieva, R.: Monte Carlo method for numerical integration based on Sobol's sequences. In: Dimov, I., Dimova, S., Kolkovska, N. (eds.) Monte Carlo method for numerical integration based on Sobol' sequences,. LNCS, vol. 6046, pp. 50–59. Springer, Heidelberg (2011). https://doi.org/10.1007/978-3-642-18466-6_5
3. Dimov, I.T., Georgieva, R., Ivanovska, S., Ostromsky, T., Zlatev, Z.: Studying the sensitivity of pollutants' concentrations caused by variations of chemical rates. J. Comput. Appl. Math. **235**, 391–402 (2010)
4. Dimov, I., Georgieva, R., Ostromsky, T., Zlatev, Z.: Variance-based sensitivity analysis of the unified Danish Eulerian model according to variations of chemical rates. In: Dimov, I., Faragó, I., Vulkov, L. (eds.) NAA 2012. LNCS, vol. 8236, pp. 247–254. Springer, Heidelberg (2013). https://doi.org/10.1007/978-3-642-41515-9_26
5. Dimov, I.T., Georgieva, R., Ostromsky, T., Zlatev, Z.: Sensitivity studies of pollutant concentrations calculated by UNI-DEM with respect to the input emissions. Centr. Eur. J. Math. **11**(8), 1531–1545 (2013). https://doi.org/10.2478/s11533-013-0256-2
6. Dimov, I.T., Georgieva, R., Ostromsky, T., Zlatev, Z.: Advanced algorithms for multidimensional sensitivity studies of large-scale air pollution models based on Sobol sequences. Spec. Issue Comput. Math. Appl. **65**(3), 338–351 (2013)
7. Joe, S., Kuo, F.: Constructing Sobol' sequences with better two-dimensional projections. SIAM J. Sci. Comput. **30**, 2635–2654 (2008)
8. Kuo, F.Y., Nuyens, D.: Application of quasi-Monte Carlo methods to elliptic PDEs with random diffusion coefficients–a survey of analysis and implementation. Found. Comput. Math. **16**, 1631–1696 (2016)
9. Nuyens, D., Cools, R.: Fast algorithms for component-by-component construction of rank-1 lattice rules in shift-invariant reproducing kernel Hilbert spaces. Math. Comp. **75**, 903–920 (2006)
10. Ostromsky, T., Dimov, I.T., Georgieva, R., Zlatev, Z.: Air pollution modelling, sensitivity analysis and parallel implementation. Int. J. Environ. Pollut. **46**(1/2), 83–96 (2011)
11. Saltelli, A., Tarantola, S., Campolongo, F., Ratto, M.: Sensitivity Analysis in Practice: A Guide to Assessing Scientific Models. Halsted Press, New York (2004)
12. Sobol, I.M.: Global sensitivity indices for nonlinear mathematical models and their Monte Carlo estimates. Math. Comput. Simul. **55**(1–3), 271–280 (2001)

13. Zlatev, Z.: Computer Treatment of Large Air Pollution Models. KLUWER Academic Publishers, Dorsrecht-Boston-London (1995)
14. Zlatev, Z., Dimov, I.T.: Computational and Numerical Challenges in Environmental Modelling. Elsevier, Amsterdam (2006)
15. Zlatev, Z., Dimov, I.T., Georgiev, K.: Modeling the long-range transport of air pollutants. IEEE Comput. Sci. Eng. **1**(3), 45–52 (1994)
16. Zlatev, Z., Dimov, I.T., Georgiev, K.: Three-dimensional version of the Danish Eulerian model. Z. Angew. Math. Mech. **76**(S4), 473–476 (1996)

Demanded Scale for Modern Numerical Optimisation

Kalin Penev[(✉)] [iD]

Faculty of Business, Law and Digital Technologies, Solent University, East Park Terrace, Southampton SO14 0YN, UK
Kalin.Penev@solent.ac.uk

Abstract. Modern online services facilitate many aspects of the human life such as goods and food shopping and delivery. The number of requested services is growing fast and at the same time can be seen growth of unattended constraints, unforeseen limitations and accidental events which affect the quality and feasibility of the service. Efficient management of online delivery requires optimisation at appropriate scale. This article discusses examples of online delivery services, which require optimization, points out advantages, and limitations, highlights their scale and the role of numerical methods for reducing the cost and time for delivery. Initial investigation illustrates algorithms capabilities to resolve such tasks and discusses current issues. Experimental results on high dimensional tests and consideration of further work conclude the article.

Keywords: Delivery optimisation · Numerical optimisation

1 Introduction

Numerical methods, Computational Intelligence and their application to optimisation are playing major role for delivery services. A classic example is well known Traveling Salesman Problem [33], which is one of the mostly studied problems in computational mathematics and operations research [5]. Efficient management in modern Industry 4 reality gradually increases the demanded scale of optimisation tasks, and introduces more sophisticated delivery problems, constraints, and limitations [5]. This article analyses examples of online delivery services, which require optimization, highlights their scale and the role of numerical methods for reducing the cost and time for delivery. It discusses requirements advantages, and limitations of optimisation methods and points out possible direction for research in this area. According to published investigations the number of internet users by July 2022 exceeds 5 billion [32]. Global number of internet users 2005–2021, shows world-wide growth up from 4.6 billion in 2020 to 4.9 billion on 2021 [14]. Perhaps one of the factors, which attracts and retains internet users are online

Supported by organization x.

services [3, 6], which facilitate various aspects of the human life such as goods and food internet shopping and home delivery, financial services, information, and media services etc. Percentage of internet sales are growing in line with growing number of internet users. Detailed information and time series can be seen in the publication [16]. The number of Internet users and the share of Internet retail sales are good indicators for growing scale of present delivery optimisation tasks. More detailed indicators for demanded scale of delivery optimisation are sales and parcel numbers per carrier, per country and globally. According to the [24] parcels volume reaches 159 billion in 2021, the highest on record. Consideration and optimisation of parcels deliveries at global level and country level could be a long-term strategy and could bring global outcomes. Based on the current rhythm of life delivery optimisation per second could be difficult and not feasible. Analysis and optimisation of delivery per day for inter country level, inter-city level and city level to person or household seems feasible tasks which could bring significant benefits. Evaluation of parcels and deliveries volume per country, city, person, or household could clarify and justify the demanded scale for modern optimisation [21]. Published [24] statistics amongst 13 countries presents shipping volume per country and analyses shipping per person and approximates parcels per household (Tables 1 and 2). For UK for 2021 delivered parcels are 5.4 billion, 80 per person and 192 per house-hold and average 15 million per day. Analysis based on Pitney Bowes statistic [24] state that UK has most parcels per person in the world [8]. According to [34] UK population is at the range of 68 million. Based on these figures for city with population of 250 thousand could be estimated to 55 thousand of parcels per day. Review of the statistics available for other countries suggests slightly lower but similarly growing range. Management of these deliveries requires optimisation and method capable of optimising task at this scale. According to the literature initial research on thousands dimensional task shows encouraging results [7, 11, 12, 17, 18, 27, 35].

2 Constraints and Limitations

Constraints and limitations increase the complexity of delivery optimisation. Next section discusses parameters with need to be included in the process of optimisation.

2.1 Time Frame Requirements

Pickup and delivery problems often require handling before or certain time or within a specific time window. This requirement adds to delivery routing problem scheduling problem. In contrast to classical scheduling optimisation where all information is static and is available in advance, delivery tasks are inherently dynamic [15]. Resolving time dependent problems requires very fast optimisation methods, which must find a solution before the task's time expire [13, 20, 29, 31]. Time for food delivery is also high priority requirement. For ready to eat food this time can be limited to less than one hour, which requires highly efficient

Table 1. Parcel shipping volume per country for 2021 [9].

Country	Parcels per year in billions
Australia	1.1
Brazil	1.5
Canada	1.5
China	108.000
France	1.7
Germany	4.5
India	2.7
Italy	1.4
Japan	9.2
Norway	000.114
Sweden	000.227
UK	5.4
US	21.60

Table 2. Parcel increase year-over year 2020–2021 [24].

Year	Parcels per second	Parcels per day in million	Parcels per person	Parcels per household
2020	4158	359	34	114
2021	5000	435	41	137

optimisation of time and route. Time scale for parcels delivery slightly differs from food delivery except from perishable goods. For parcels delivery optimal routing, items grouping and time window available for delivery seems with higher priority. This needs more research efforts and detailed investigation. Resolving time dependent tasks in general is hard problem and at present is difficult to find evidence for methods capable of resolving thousands dimensional time dependent tasks. More research efforts should be directed in this area.

2.2 Delivery Location

An attempt to mitigate obstacles caused buy time window limitations and add more flexibility is pick up location (delivery point or locker). This approach introduces new constraint such as delivery point capacity, time for delivery collection and utilisation of available capacity. Published in the literature investigations highlights encouraging results [10,26,31]. More research should be done towards enhancing scalability of the optimization methods on these tasks.

2.3 Zero Carbon Emissions and Vehicle Capacity

Delivery in urban areas with high density of orders and in the same time strict need for zero carbon emissions [4] add to the classical traveling salesman routing minimisation more parameters for optimisation such as minimisation of carbon emissions, maximisation of packages per vehicle, and mentioned above time window for delivery. According to the publications substantial efforts are conducted in this area such as electric vehicles, cargo bikes and carts for deliveries on foot through delivery service, all of which contribute to decarbonising the last-mile delivery process [4]. An attempt to mitigate obstacles caused buy time window for delivery limitation and add more flexibility is pick up location (delivery point or locker). This approach introduces new limitations such as delivery point capacity, time for delivery collection and utilisation of available capacity. Published in the literature investigations highlights encouraging approaches [26]. More research should be done towards enhancing scalability of the optimization methods on these tasks.

2.4   Accidental Events and Service Failure

In certain extent an excellent optimal solution when applied to the real-world practical task may lead to unexpected poor results [1]. Reason for this are usually accidental events not included in optimization process [25]. For example, published statistics for 2022 on Amazon delivery show that delays may exceed 10 − 15 percents from the total number of orders [28]. Publications shows that additional optimisation of delivery timing reduces almost twice delivery delays - from 11.4 percent of the online orders made in January 2020 and delivered later, and in May of the same year, with share increased to 15.9 percent, to 5 percent of the parcels not delivered in time in January 2021 [28]. Application of numerical optimization to real life tasks needs consideration off possible unforeseen and unpredictable factors such as traffic accidents and road-works. Issue with accidental delivery problems, service failure or delays need integration of heuristic risk analysis with the optimisation algorithms.

2.5   Packets Grouping and Packing Waste

Packing is undividable part off delivery. Each item or order must be packed appropriately. Optimisation and minimisation of the packet size add on an initial part or the process assessment and selection optimal packet size for each item. Minimisation of the packet size can significantly reduce the packing waste. It can contribute to more effective and more condensed packets grouping so in fact packing optimisation should be integrated with the whole process of delivery optimisation, routing and scheduling. Further, more detailed research could identify other parameters and limitations which should be subject of optimisation in planning and management of delivery process. Next section focuses on existing optimisation algorithm their ability to fulfil and cope with the required scale of the optimisation problems.

3   Optimisation Algorithms Applied to High Scale Tasks

Review of published in the literature articles on different aspects of optimisation and methods applied to thousands dimensional tasks, identifies the directions of current research efforts. A nature inspired meta-heuristic algorithm [18], applied to several high-dimensional benchmark functions (up to 10,000 dimensions), shows that the proposed approach is highly efficient for high-dimensional global optimization problems. Published results indicate that this algorithm could be applied to 100k dimensional tasks. It needs further evaluation. Interactive Cooperative Coevolutionary algorithm is applied to large scale black-box optimization [12]. This article identifies that large scale black-box optimization problems arise in many fields of science, engineering, and delivery services, and many of existing algorithms for these problems still suffer from the variations in high dimensionality. Presented results show that proposed method obtains competitive performance on high-dimensional optimization problems with different levels of dimensionality up to 10000. Evaluation on higher dimensionality should confirm the strength of this algorithm. A method which enhancers Whale optimization algorithm based on the prey-catching characteristics of the humpback whales is applied to twenty-five bench-mark functions using dimensions 100, 500, 1000, and 2000 and compared the results with the whale optimization algorithm and its variants and with seven basic metaheuristic algorithms [7]. Presented results suggests good performance of the proposed algorithm on higher-dimensional problems. Extending dimensionality of the tasks could further highlights advantages of this method. Published evaluation on dual Biogeography-based optimization based on sine cosine algorithm and dynamic hybrid mutation is applied to up to 10000-dimensions tasks. Results show an excellent performance [35]. Evaluation on tasks with unknown optimal solution with higher dimensionality could prove superiority on this method. A paper [17] presents an improved version of butterfly optimization algorithm applied to high-dimensional optimization problems and tested on 40 benchmarks with up to 10000 dimensions. Statistical results show better performance than compared methods. Perhaps this algorithm should demonstrate good performance on higher dimensions if tested. Based on the results of an investigation of genetic algorithm applied to large scale optimization problems with dimensions up to 100000 it can be inferred that the proposed genetic algorithm can find a good solution in a fairly short time [27]. This study needs further evaluation of tasks with unknown optimal solutions. Recently published study [30], which proposes algorithm that uses two unsupervised neural networks, restricted Boltzmann machine, denoising autoencoder and genetic operators is applied to large-scale multi-objective optimization problems. Presented results suggest that this method could be applicable to higher dimensions and need further evaluation. A novel black-box optimization algorithm is compared with 11 state-of-the-art optimization algorithms on 24 n-dimensional and 6 application-based benchmark problems [2]. Using identical computing resources, a compute cluster with 46,080 cores across 16 Graphical Processing Units, it is tested in up to 100000 dimensions, and demonstrates linear scalability in time with respect to dimensionality and computational resources. Superior performance of this algorithm could be further proved with evaluation of benchmarks with unknow optimal solutions.

4   Experimental Results

This study conducted experimental evaluation of 100 000-dimensional Michalewicz Test Function [19], well known global optimisation problem with unknown optimal solution for variety of dimensions. Achieved best optimal solutions for some particular numbers of dimensions are published in the literature. Selection of benchmark with unknow optimal solution intentional excludes attempts for adjusting optimisation method to test problems optimal values which could be widely used for testing benchmarks with known optimal solution. For experiments is used Free Search algorithm [22]. Initial achieved result for 100 000-dimensional Michalewicz Test Function is Fmax=99106.353407791452. All variables for which generate this solution are available at [23] and can be used for other methods as a benchmark for comparison and improvement.

5   Discussion

Evaluation of the existing publications suggests that optimization methods which are applied to high dimensions are using predominantly tests with known optimal values. This may be used for calibration of the optimization methods parameters but in practise where optimal solutions are unknown such calibration is not possible. Therefore, reliable evaluation of optimisation methods should be based on benchmarks with unknown solutions. Most of the existing publications on delivery recommend extensive approach such as increase of the number of vehicles, increase of the number of drivers, increase of the number of pickup locations for improvement. Extensive approach is naturally limited by available resources cannot continue indefinitely and does not seem acceptable solution. The direction which this study trying to recommend is future development of more efficient optimisation methods capable of resolving 100000 dimensional tasks and above, within an acceptable period of time and limited computational resources.

6   Conclusion

This article reviews current scale and complexity of delivery optimisation. It is identified that the demanded scale for delivery optimisation approaches 100 000 parameters with increasing complexity and need for integration of routing, packing, scheduling methods. Analysis of published optimisation algorithm applied to thousand dimensional tasks suggest initial efforts in this direction. More research should be done towards improvement of the optimisation algorithms.

References

1. ACSI: U.S. customer satisfaction with Amazon.com from 2000 to 2021 (index score) (2022). https://www.statista.com/statistics/185788/us-customer-satisfaction-with-amazon/. Accessed 17 Oct 2022
2. Albert, B.A., Zhang, A.Q.: SpartaPlex: a deterministic algorithm with linear scalability for massively parallel global optimization of very large-scale problems. Adv. Eng. Softw. **166**, 103090 (2022)
3. Amazon: annual net sales of Amazon in selected leading markets from 2014 to 2021 (in billion U.S. dollars) (2022). https://www.statista.com/statistics/672782/net-sales-of-amazon-leading-markets/. Accessed 17 Oct 2022
4. Amazon: In 2021, Amazon used zero emission vehicles, including more than 1,000 electric vans, cargo bikes, e-scooters, and walkers, to deliver more than 45 million packages to customers in the UK (2022). https://www.aboutamazon.co.uk/news/sustainability/amazon-delivered-45-million-packages-in-the-uk-with-zero-emission-vehicles-last-year. Accessed 22 Oct 2022
5. Bialic-Davendra, M., Davendra, D.: Novel Trends in the Traveling Salesman Problem. IntechOpen, London (2020)
6. Buck, A.: 57 Amazon statistics to know in 2022 (2022). LandingCube. https://landingcube.com/amazon-statistics/. Accessed 19 Oct 2022
7. Chakraborty, S., et al.: An enhanced whale optimization algorithm for large scale optimization problems. Knowl.-Based Syst. **233**, 107543 (2021)
8. Dawson, C.: 2021 UK has most Parcels per person in the world, ChannelX (2022). https://channelx.world/2021/10/uk-has-most-parcels-per-person-in-the-world. Accessed 19 Oct 2022
9. Dies, J.: Parcel shiping index, pitney bowes (2022). https://www.pitneybowes.com/content/dam/pitneybowes/us/en/shipping-index/22-pbcs-04529-2021-global-parcel-shipping-index-ebook-web-002.pdf. Accessed 19 Oct 2022
10. Karami, F., Vancroonenburg, W., Vanden Berghe, G.: A periodic optimization approach to dynamic pickup and delivery problems with time windows. J. Sched. **23**(6), 711–731 (2020). https://doi.org/10.1007/s10951-020-00650-x
11. Fujii, K., et al.: Solving challenging large scale QAPs (2021). https://arxiv.org/abs/2101.09629v1. Accessed 19 Oct 2022 (2022)
12. GE, H., et al.: Bi-space interactive cooperative coevolutionary algorithm for large scale black-box optimization. Appl. Soft Comput. **97**, 106798 (2020)
13. IANG, X., et al.: Time-dependent sequential optimization and possibility assessment method for time-dependent failure possibility-based design optimization. Aerosp. Sci. Technol. **110**, 106492 (2021)
14. ITU. Number of internet users worldwide from 2005 to 2021 (in millions) (2022). Statista Inc. https://www.statista.com/statistics/273018/number-of-internet-users-worldwide/. Accessed 19 Oct 2022
15. Karami, F., Vancroonenburg, W., Vanden, G.: BERGHE: a periodic optimization approach to dynamic pickup and delivery problems with time windows. J. Sched. **23**(6), 711–731 (2020)
16. Lewis, R.: Internet sales as a percentage of total retail sales (ratio), office of national statistics (2022). https://www.ons.gov.uk/businessindustryandtrade/retailindustry/timeseries/j4mc/drsi. Accessed 24 Oct 2022
17. LONG, W., et al.: A velocity-based butterfly optimization algorithm for high-dimensional optimization and feature selection. Expert Syst. Appl. **201**, 117217 (2022)

18. Miarnaeimi, F., Azizyan, G., Rashki, M.: Horse herd optimization algorithm: a nature-inspired algorithm for high-dimensional optimization problems. Knowl.-Based Syst. **213**, 106711 (2021)
19. Michalewicz, Z., Schoenauer, M.: Evolutionary algorithms for constrained parameter optimization problems. Evol. Comput. **4**(1), 1–32 (1996)
20. Nishida, K., Nishi, T.: Dynamic optimization of conflict-free routing of automated guided vehicles for just-in-time delivery. In: IEEE Transactions on Automation Science and Engineering (2022). https://doi.org/10.1109/ASE.2022.3194082
21. Ofcom: Annual volume of packages shipped in the United Kingdom (UK) from 2013 to 2021 (in millions) (2021). Statista Inc. https://www.statista.com/statistics/1010391/parcel-shipping-annual-volume-uk/. Accessed 22 Oct 2022 (2022)
22. Penev, K., Littlefair, G.: Free search - a novel heuristic method. In: Proceedings of PREP 2003, 14–16 April, Exeter, UK, pp. (133–134) (2003)
23. Penev, K.: An optimal value for 100 000-dimensional Michalewicz test (2022). https://pure.solent.ac.uk/ws/portalfiles/portal/33733992/100_000_dimensional_Michalewicz_test_2.pdf. Accessed 24 Oct 2022
24. Pitney Bowes Parcel Shipping Index (2022). https://www.pitneybowes.com/content/dam/pitneybowes/us/en/shipping-index/pb_globalshippingindexinfographic_2021stats_final.pdf. Accessed 20 Oct 2022
25. Rohden, S.F., de Matos, C.A.: Online service failure: how consumers from emerging countries react and complain. J. Cons. Market. **39**(1), 44–54 (2022)
26. Schnieder, M., Hinde, C., West, A.: Combining parcel lockers with staffed collection and delivery points: an optimization case study using real parcel delivery data (London, UK). J. Open Innov. **7**(3), 183 (2021)
27. SHAHAB, M.L., et al.: A genetic algorithm for solving large scale global optimization problems. J. Phys: Conf. Ser. **1821**(1), 12055 (2021)
28. Supply Chain Dive. (March 22, 2021). Share of Amazon orders arriving late from January 2020 to January 2021. In Statista. https://www.statista.com/statistics/1220033/share-of-amazon-orders-arriving-late/. Accessed 19 Oct 2022 (2022)
29. Teng, R., et al.: A dynamic routing optimization problem considering joint delivery of passengers and parcels. Neural Comput. Appl. **33**(16), 10323–10334 (2021)
30. TIAN, Y., et al.: Solving large-scale multiobjective optimization problems with sparse optimal solutions via unsupervised neural networks. IEEE Trans. Cybern. **51**(6), 3115–3128 (2021)
31. Vu, T.X., et al.: Dynamic bandwidth allocation and edge caching optimization for nonlinear content delivery through flexible multibeam satellites (2022)
32. We are social, datareportal, hootsuite (2022). Number of internet and social media users worldwide as of July 2022 (in billions). Statista Inc. https://www.statista.com/statistics/617136/digital-population-worldwide/. Accessed 19 Oct 2022 (2022)
33. Wolfe, J., et al.: The Traveling Salesman Problem. John Wiley Sons, Incorporated, Singapore (1985)
34. Worldometer 2022, UK Population. https://www.worldometers.info/world-population/uk-population/. Accessed 22 Oct 2022 (2022)
35. Zhang, Z., Gao, Y., Zuo, W.: A dual biogeography-based optimization algorithm for solving high-dimensional global optimization problems and engineering design problems. IEEE Access **10**, 55988–56016 (2022)

RMSD Calculations for Comparing Protein Three-Dimensional Structures

Fatima Sapundzhi[1], Metodi Popstoilov[1], and Meglena Lazarova[2(✉)]

[1] Department of Communication and Computer Engineering, Faculty of Engineering, South-West University "Neofit Rilski", 2700 Blagoevgrad, Bulgaria
sapundzhi@swu.bg, mpopstoilov@abv.bg
[2] Department of Mathematical Modelling and Numerical Methods, Faculty of Applied Mathematics and Informatics, Technical University of Sofia, 1000 Sofia, Bulgaria
meglena.laz@tu-sofia.bg

Abstract. The root Mean Square Deviation ($RMSD$) is a popular measure of structural similarity between protein structures in the field of bioinformatics. The $RMSD$ calculations involve alignment and optimal superposition between matched pairs of atoms, searching for the lowest $RMSD$ result for both structures. Among the popular methods for calculating the optimal rotation matrix that minimizes the $RMSD$ are the Kabsch algorithm and the Quaternion algorithm. The aim of this research is to present a simple tool for calculation of the $RMSD$ between pairs of aligned three dimensional structures by applying the methods mentioned above. As an implementation language the object-oriented $C\#$ programming language has been chosen.

Keywords: computer modelling · Root Mean Square Deviation ($RMSD$) · Protein sequences · C # programming language

1 Introduction

In the field of bioinformatics, the molecular structure alignment is the matching between two or more macromolecular structures based on the optimal superposition of their shape and three-dimensional (3D) conformation. A protein is usually compared to other proteins. It is useful for better understanding of their properties. Proteins with similar amino acid sequences or $3D$ structures are found to be evolutionary related and have similar functions, see [1–4]. An important topic in structural bioinformatics and computational chemistry is the analysis of protein sequences and their biological functions. Also important is the assessment of the protein structural similarities. These studies find wide application in both protein structure prediction and drug design, homology modeling approaches and even more. The difference between two conformations of molecules is most commonly measured by calculating the Root Mean Square Deviation ($RMSD$) which is the sum of the squares of the distances between the equivalent atom pairs. It is important to find the minimal value of $RMSD$. The main problem

I. Georgiev et al. (Eds.): NMA 2022, LNCS 13858, pp. 279–288, 2023.
https://doi.org/10.1007/978-3-031-32412-3_25

is that the conformation of a given molecule is randomly ordered in space, and calculating the $RMSD$ between two given conformations does not directly lead to appropriate results. The $RMSD$ calculations involve alignment (establishing a one to one correspondence between equivalent atoms in each structure) and optimal superposition (rotation and translation of the structure) so that the $RMSD$ between the equivalent atoms in both structures to be minimal. There are several approaches for finding the optimal alignment. Among these methods are the algorithms presented by Kabsch, see [5,6]. They are based on the calculation of $RMSD$ in terms of the singular values of a 3×3 matrix. The method of McLachlan, see [7,8] is based on quaternions and it finds the $RMSD$ in terms of the largest eigenvalue of a symmetric, traceless a four-dimensional matrix, see [4]. In these methods, the minimizing rotation is formed from the eigenvectors of the matrices and the minimum $RMSD$ is found from the eigenvalues of the key matrices, see [9–12]. The objective of the current research is to present a simple procedure of calculation of the $RMSD$ using both of the above-mentioned algorithms between pairs of aligned $3D$ structures. The result should be the minimum $RMSD$ value. The object-oriented programming language $C\#$ was used for the implementation of the tool.

2 Root Mean Squared Distance - RMSD

For comparing two sets of points (atoms of two proteins X and Y), where $X = \{x_1, x_2, \ldots, x_n\}$ and $Y = \{y_1, y_2, \ldots, y_n\}$ let us define one to one correspondence between the protein X and protein Y, where x_i corresponds to y_i for all $i \in [1, n]$, see [4,13]. Then we can define the $RMSD$ function by using the following formula, [14–16]

$$RMSD(X,Y) = \sqrt{\frac{1}{n} \sum_{i=1}^{n} \| x_i - y_i \|^2}. \tag{1}$$

When we find the correspondence between the set X and the set Y we can calculate the rigid transformation T by using the following equation:

$$RMSD(T(X),Y) = \sqrt{\frac{1}{n} \sum_{i=1}^{n} \| T(x_i) - y_i \|^2}. \tag{2}$$

After that we can find the minimal value of $RMSD(T(x), Y)$.

We want to compare two 3D structures represented as $X = (x_i)_{i=1}^{n}$ and $Y = (y_i)_{i=1}^{n}$ and we want to find a transformation $T : R^3 \to R^3$ and the minimum value of $RMSD(T(x), Y)$. If we consider $RMSD^*(x, y) = \sum_{i=1}^{n} \| x_i - y_i \|^2$ then the value of the $RMSD(x, y)$ is minimal when the $RMSD^*(x, y)$ is minimal. We can separate the problem of finding the transformation T into finding an optimal translation and an optimal rotation. Then T can be expressed as following:

$$T(x) = R(x) - t, \tag{3}$$

where R is a rotation and t is a translation.

2.1   Some Useful Softwares

In this section we consider three software types - Blazor, Chart.js and Math.Net Numerics which are useful for the development of the present new tool for calculating $RMSD$ between the pairs of aligned $3D$ structures. A short information about them is as follows:

Blazor is a free and open-source web framework. This program enables developers to create interactive web UIs using C #, HyperText Markup language ($HTML$) and Cascading Style Sheets (CSS) [17–22]. Blazor is a feature of ASP.NET, the popular web development framework that extends the .NET developer platform with tools and libraries for building web apps.

CHart.js is a free open-source flexible JavaScript library for data visualization. It is used by designers and developers and supports eight chart types: bar, line, area, pie, bubble, radar, polar and dot [23].

Math.NET Numerics aims to provide methods and algorithms for numerical computations in science, engineering and it is for everyday use. Covered topics include special functions, linear algebra, probability models, random numbers, interpolation, integration, regression, optimization problems and more [24].

2.2   Kabsch Algorithm [5, 6]

Kabsch algorithm computes the 3×3 rotation matrix for the given sets of atomic positions and a translation vector. This method produces the residual in terms of the singular value decomposition (SVD). If the points are not centred it is important before the operation to translate them in a way that their average to coincide with the origin. The algorithm works in three steps: a translation, the computation of the covariance matrix C and the computation of the optimal rotation matrix - U. A detailed description of the algorithm is presented in the following publications [5,6,9]. The following pseudocode represents the description of the algorithm, finding the rotation matrix and computing the minimal RMSD according to the considered algorithm.

2.3   Quaternion Algorithm [7, 8]:

For the presentation of the quaternion problem we follow the notes of [2, 7–11] and suggest the reader to refer to these notes for more detailes. A quaternion $Q = [s, q]$ is a combination of a scalar s with a three- component vector $q = \{x, y, x\}^T$. The squared norm of Q can be represented as $\mid Q \mid^2 = s^2 + q.q$ and a unit quaternion - $\widehat{Q}$ is a quanternion with its norm equal to 1, i.e.
$Q^{-1} = [s, -q]/(s^2 + q.q)$ - inverse A quaternion. The quaternion Q corresponds to a vector $v = (v_0, v_x, v_y, v_z)^T \in R^4$. So the unit quaternion $\widehat{Q}$ can be used to

Algorithm: RMSD

Input: *P* and *Q* - N x D matrices where *N* is the number of points and *D* is the dimension

Output: Root-mean-square deviation

1. sum ← 0
2. **for** i=1 **to** N **do**
3.   **for** j=1 **to** D **do**
4.   sum ← sum + pow(P[i,j]-Q[i,j],2)
5.   rmsd ← sqrt(sum/N)
6.   **end for**
7. **end for**
8. **Return** rmsd

Algorithm: KabschRMSD

Input: *P* and *Q* - N x D matrices where *N* is the number of points and *D* is the dimension

Output: Root-mean-square deviation

1. P ← KabschRotate (P,Q)
2.   rmsd ← Rmsd (P,Q)
3. **Return** rmsd

Algorithm: KabschRotate

Input: *P* and *Q* - N x D matrices where *N* is the number of points and *D* is the dimension

Output: Rotated matrix P

1. U ← Kabsch (P,Q)
2.   P ← P*U
3. **Return** P

Algorithm: Kabsch

Input: *P* and *Q* - N x D matrices where *N* is the number of points and *D* is the dimension

Output: The optimal rotation matrix U

1. C ← (transpose of P)*Q
2. svd ← singular value decomposition of C
3. svd U ← left singular vectors of svd, unitary matrix
4. S ← singular values (Σ) of svd in ascending value
5. VT ← transpose right singular vectors of svd (transpose of V, unitary matrix)
6.   **if** determinant of svdU* and the determinant of VT < 0
7.   last element of S ← -1*last element of S
8.   multiply all elements in last column of svdU by -1
9.   U ← svdU*VT
10.**Return** U

rotate the vector v into vector v'. The same rotation can be represented with a rotation matrix $R : v' = Rv$. The matrix R can be expressed by the components of the quaternion $\widehat{Q}$ i.e. $\widehat{R^T} = -\widehat{R}$. The quaternion $\widehat{Q}$ corresponds to a rotation with angle θ over an axis c and it is represented by the unit vector $v = (v_x, v_y, v_z)$ in the following way: $\widehat{Q} = [cos\frac{\theta}{2}, c.sin\frac{\theta}{2}]$ and $\widehat{Q^{-1}} = [cos\frac{\theta}{2}, -c.sin\frac{\theta}{2}]$.

The quaternion which corresponds to the optimal rotation is found as a leading eigenvector of a certain 4×4 matrix whose elements are formed from the rotation 3×3 matrix R, see [7–11]. In the implementation of the Quaternion algorithm [10] the matrix involved in the quaternion rotation is initially created. We give a pseudo-code which represents the rotation of the matrix P into Q and also the calculation of the $RMSD$ value based on [7].

Algorithm: QuaternionRMSD
Input: P and Q - N x D matrices where N is the number of points and D is the dimension
Output: Root-mean-square deviation
1. rot $\leftarrow$ QuaternionRotate (P,Q)
2.   P $\leftarrow$ P*rot
3.   rmsd $\leftarrow$ Rmsd(P,Q)
4. **Return** rmsd

Algorithm: QuaternionRotate
Input: X and Y - N x D matrices where N is the number of points and D is the dimension
Output: Rotation
1. W $\leftarrow$ new array of matrices with size N
2.   for $i = 1$ to N do
3.   W[i] $\leftarrow$ make W (Y[i,1], Y[i,2], Y[i,3])
4.   Q $\leftarrow$ new array of matrices with size N
5.   for $i = 1$ to N do
6.   Q[i] $\leftarrow$ Make Q (X[i,1], X[i,2], X[i,3])
7. $OtDotW[i] \leftarrow$ new array of matrices with size N
8.   for $i = 1$ to N do
9. $OtDotW[i] \leftarrow$ transpose of $Q[i] * W[i]$
10.   A $\leftarrow MakeA(OtDotW)$
11.   eigen $\leftarrow EVDofA$
12.   r $\leftarrow$ last column of eigen vectors of *eigen*
13.   rot $\leftarrow$ Quaternion transform (r)
14.   **end for**
15.   **end for**
16.   **end for**
17.**Return** rot

The following pseudo-code represents the achievement of an optimal rotation and a translation that can be zero when the centroids of each molecules are the same. The centroid is considered to be the mean position of all points along all the coordinate directions from a given vector.

Algorithm: Quaternion Transform
Input: r
Output: Optimal rotation
1. $WtRot \leftarrow$ transpose of Make W $(r[1], r[2], r[3], r[4])$
2. $QRot \leftarrow$ Make Q $r[1], r[2], r[3], r[4]$
3.   rot $\leftarrow WtRot * QRot$
3.   rmsd $\leftarrow$ Rmsd(P,Q)
4. Remove last row and last column of rot
5. **Return** rot

The developed program has a simple interface. The tool reads the data about the structure of two molecules from two files: *.pdb* (Protein Data Bank fails) or *.xyz*. The program returns a result - the minimum $RMSD$ value between the two structures of the molecules. The $RMSD$ value is calculated empoying the both algorithms - $KabschRMSD$ [5,6] and $QuaternionRMSD$ [7,8].

For the implementation of the tool the C # programming language [9–21] has been chosen. It is a modern object-oriented, general-purpose and multi-platform programming language created by Microsoft together with the *.NET* platform. If the $RMSD$ values are successfully calculated then the graph is visualized using Chart.js. The user interface of the developed $RMSD$ calculator is presented in Fig. 1.

The RMSD value is expressed in length units and in BIOINFORMATICS it is the Ångström (Å) which is equal to $10^{-10}m$, see [13]. It is assumed that the smaller the $RMSD$ value between two structures the more similar are these two structures. According to [26] a good score for structure similarity is approached at $RMSD < 3Å$. In order to compare the functionality of the $RMSD$ calculator, tests with different sizes of molecules were performed. The results of the program implementation with the given example from [27] are shown in Fig. 2. The developed tool allows to compute the $RMSD$ between two $3D$ structures and show the graphical results for three $RMSD$ values - NormalRMSD, KabschRMSD and QuaternionRMSD. In our previous studies on this topic the comparison between real structures of delta-opioid receptors (DOR) and mu-opioid receptors (MOR) were presented in [25–29].

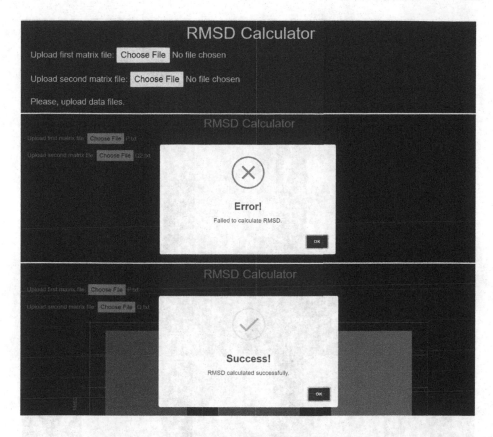

Fig. 1. RMSD calculator

The running time of the realized algorithms is linear $\theta(n)$ where n is the number of points in the structure. As a continuation of our previous work on $RMSD$ calculations, see [30] the present investigation is focused on the development of a program that calculates $RMSD$ between two $3D$ structures employing the Kabsch and the Quaternion algorithms. The developed tool has a simple graphical user interface that will be uploaded to a server and can be freely used by researchers in the field of bioinformatics.

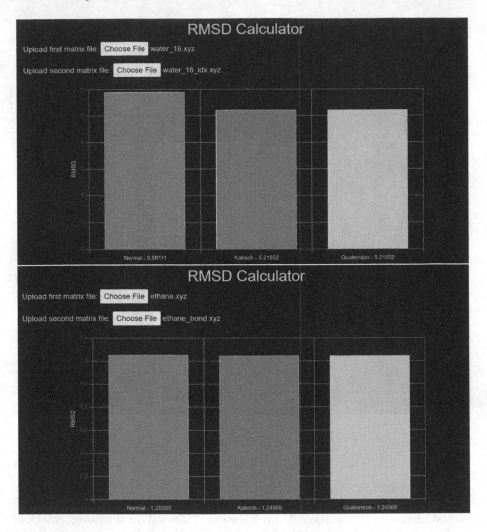

Fig. 2. Results of the RMSD calculations for different examples

3   Conclusion

Measuring the difference between $3D$ structures is particularly useful in computational chemistry and bioinformatics where it is necessary to compare different conformations of the molecules. We introduce a C# implementation for computing the $RMSD$ using Kabsch algorithm and quaternions.

Acknowledgments. The work of the second author (M.L.) is supported by the Bulgarian National Science Fund under Project KP-06-M62/1 "Numerical deterministic, stochastic, machine and deep learning methods with applications in computational, quantitative, algorithmic finance, biomathematics, ecology and algebra" from 2022.

References

1. Li, S.C.: The difficulty of protein structure alignment under the $RMSD$. Algorithms Mol. Biol. **8**, 1 (2013). https://doi.org/10.1186/1748-7188-8-1
2. Sehnal, D.: In: Algorithms for Comparing Molecule Conformations. Masaryk University, Faculty of Informatics (2007)
3. Jensen, F.: Introduction to Computational Chemistry. Wiley (2007)
4. Coutsias, E., Wester, M.: $RMSD$ and symmetry. J. Comput. Chem. **40**(15), 1496–150803 (2021). https://doi.org/10.1002/jcc.25802
5. Kabsch, W.: A solution for the best rotation to relate two sets of vectors. Acta Crystallographica Sec. A **32**(5), 922–923 (1976)
6. Kabsch, W.: A discussion of the solution for the best rotation to relate two sets of vectors. Acta Crystallographica Sec. A **34**(5), 827–828 (1978)
7. McLachlan, A.: A mathematical procedure for superimposing atomic coordinates of proteins, Acta Crystallographica, A **28**, 656–657
8. Walker, M., Shao, L., Volz, R.: Estimating $3 - D$ location parameters using dual number quaternions CVGIP: image understanding **54**(3), 358–367 (1991). https://doi.org/10.1016/1049-9660(91)90036-O
9. Theobald, D.: Rapid calculation of $RMSDs$ using a quaternion-based characteristic polynomial. Acta crystallographica, Sec. A Found. Crystallography **61**(Pt 4), 478–480 (2005). https://doi.org/10.1107/S0108767305015266
10. Coutsias, E., Seok, C., Dill, K.: Using quaternions to calculate $RMSD$. J. Comput. Chem. **25**, 1849–1857 (2004). https://doi.org/10.1002/jcc.20110
11. Popov, P., Grudinin, S.: Rapid determination of $RMSDs$ corresponding to macromolecular rigid body motions. J. Comput. Chem. **35**(12), 950–956 (2014). https://doi.org/10.1002/jcc.23569
12. Wallin, S., Farwer, J., Bastolla, U.: Testing similarity measures with continuous and discrete protein models. Proteins **50**, 144–157 (2003)
13. Root Mean Square Deviation (RMSD). https://en.wikipedia.org/wiki/Root-mean-square_deviation_of_atomic_positions
14. Eisstein, E.: Symmetric matrix. Wolfram MathWorld-A Wolfram Web Resource (2007). http://mathworld.wolfram.com/SymmetricMatrix.html
15. Kavraki, L.E.: Geometric Methods in Structural Computational Biology, Rice University, Houston, Texas (2007). http://cnx.org/content/col10344/1.6/
16. Schreiner, W., Karch, R., Knapp, B., Ilieva, N.: Relaxation estimation of RMSD in molecular dynamics immunosimulations. Comput. Math. Meth. Med. **173521** (2012). 9 pages
17. Blazor: https://dotnet.microsoft.com/en-us/apps/aspnet/web-apps/blazor
18. Miller, J.: C# in the Browser with Blazor. MSDN Mag. **33**(9)
19. Roth, D.: Get started building.NET web apps that run in the browser with Blazor. ASP.NET blog
20. Enström, J.: Web Development with Blazor: a hands-on guide for.NET developers to build interactive UIs with C#, Packt Publishing, ISBN 978-1800208728
21. Himschoot, P.: Microsoft Blazor: Building Web Applications in.NET 6 and Beyond. Apress (2021). ISBN 978-1484278444

22. Wright, T.: Blazor WebAssembly by example: a project-based quide to building web apps with.NET, Blazor Web Assembly and C#, Packt Pulishing (2021). ISBN 978-1800567511

23. CHartjs: https://www.chartjs.org/

24. Math.NET Numerics: https://numerics.mathdotnet.com/

25. Reva, B., Finkelstein, A., Skolnick, J.: What is the probability of a chance prediction of a protein structure with an RMSD of 6A? Fold Des **3**(2), 141–147 (1998)

26. Calculate Root-mean-square deviation (RMSD) of Two Molecules Using Rotation, Github. http://github.com/charnley/rmsd Version 1.5.1

27. Sapundzhi, F., Slavov, V.: RMSD calculations and computer modelling of protein structures. J. Chem. Technol. Metallurgy **55**(5), 935–938 (2020)

28. Sapundzhi, F., Prodanova, K., Lazarova, M.: Survey of the scoring functions for protein-ligand docking. AIP Conf. Proc. **2172**(100008), 1–6 (2019)

29. Topalska, R., Sapundzhi, F.: Chemical structure computer modelling. J. Chem. Technol. Metallurgy **55**(4), 714–718 (2020)

30. Sapundzhi, F., Popstoilov, M.: Proceedings of the 36th European Peptide Symposium, Barcelona, Spain, 28.08-02.09.2022, pp. 283–284 (2022)

On the Application of a Hierarchically Semi-separable Compression for Space-Fractional Parabolic Problems with Varying Time Steps

Dimitar Slavchev$^{(\boxtimes)}$ⓘ and Svetozar Margenovⓘ

Institute of Information and Communication Technologies at the Bulgarian Academy
of Sciences, Sofia, Bulgaria
{dimitargslavchev,margenov}@parallel.bas.bg

Abstract. Anomalous (fractional) diffusion is observed when the Brownian motion hypotheses are violated. It is modeled with the fractional Laplace operator, which can be defined in several ways. In this work we use the integral definition with the Riesz potential. For the discretization in space we apply the finite element method and for the discretization in time – a backward Euler scheme with varying time steps. The fractional Laplacian is a non-local operator and the arising stiffness matrix is dense. The time dependent problem is reduced to solving a sequence of linear systems whose matrices are constructed from the stiffness matrix, lumped mass matrix and the time step. When the time step changes we must refactorize the matrix before solving the current system. If the time step doesn't change we can solve with the matrix factorized on a previous time step change. When utilizing the generic method using a block LU factorization, the computational complexity of the forward elimination is $O(n^3)$ and $O(n^2)$ of the backward substitution. In this work we develop an alternative method based on the hierarchically semi-separable (HSS) compression. With this method we compress the matrix at the beginning only. The HSS compression has a computational complexity $O(n^2 r)$. Then, when the time step changes we need to apply ULV-like factorization with computational complexity of $O(nr^2)$. The solution step with the factorized matrix at each time step has computational complexity of $O(nr)$. Here, r is the maximum off-diagonal rank of the approximate matrix, which is computed during the compression process. For suitable problems r is much smaller than the number of unknowns n. The numerical experiments presented show the advantages of the developed HSS compression based solution method.

1 Introduction

There are many applications of the anomalous diffusion arises in natural sciences and engineering, including e.g.: flows in strongly heterogeneous porous media, superconductivity, diffusion of polymers in supercold media [1]; electrodiffusion

I. Georgiev et al. (Eds.): NMA 2022, LNCS 13858, pp. 289–301, 2023.
https://doi.org/10.1007/978-3-031-32412-3_26

of ions into nerve cells [2] and photon diffusion diagnostics [3]; image processing and machine learning [4]; spread of viral diseases, computer viruses and crime [5]; and others. More applications can be found in [6]. Anomalous diffusion can be modeled with the *fractional* Laplace operator $(-\Delta)^\alpha$, where α is the fractional power. There are several definitions of the fractional Laplacian, which are non-equivalent. In [7] Harizanov et al. discuss the differences between the integral and spectral definitions. In this paper we use the integral definition through the Riesz potential. The finite element method (FEM) is used for discretization of the parabolic problem in space. The resulting FEM stiffness matrix K is dense thus representing the non-locality of the fractional Laplacian. We follow the method outlined in [8] and employ the numerical implementation from [9] to obtain K. The calculation of the lumped mass matrix M_L as well as a backward Euler scheme with uniform time steps are discussed in [10].

Here we consider the case of variable time steps. After applying the FEM discretization in space we obtain a dense stiffness matrix $K \in \mathbb{R}^{n \times n}$. This is due to the *non-local* nature of the fractional Laplacian. A system of linear algebraic equations with the matrix $\widetilde{K}_j = M_L/\tau_j + K$ is solved at each backward Euler time step $j \in [1, m]$. Thus, the matrix $\widetilde{K}_j$ is factorized once every time τ_j changes.

The Gaussian elimination with block LU factorization is considered as a generic method for solving of linear algebraic equations with dense matrices. This method is well developed being implemented in several parallel high performance computing (HPC) software libraries, including e.g.: LAPACK [11], Intel®'s Math Kernel Library[1], AMD's ACML[2], Parallel Linear Algebra for Scalable Multi-core Architectures [12,13], Automatically Tuned Linear Algebra Subroutines [14] and others. However, block LU factorization is quite expensive because the computational complexity is $O(n^3)$ for the factorization step and $O(n^2)$ for the solution with the factorized matrix.

Another class of methods is based on approximation of the dense matrix using its *structure*. Here by *structure* we understand the low rank property of the off-diagonal blocks of the constructed approximation. One such method is based on a Hierarchically Semi-Separable compression. The HSS compression algorithm [15] is implemented in the STRUctured Matrices PACKage (STRUMPACK) [16]. The computational complexity of this method is as follows: HSS compression – $O(n^2 r)$; ULV-like factorization – $O(nr^2)$; and solution of the system with the factorized matrix – $O(nr)$. Here r is the maximum rank of the off-diagonal blocks. For matrices with suitable *structure* $r \ll n$.

In this paper, we analyze the parallel scalability and performance of the developed solution method for the space-fractional parabolic problem, comparing the HSS compression-based solver with the implementation where the block LU factorization solver from the MKL package is used.

---

[1] https://software.intel.com/en-us/intel-mkl.
[2] https://developer.amd.com/wordpress/media/2012/10/acml_userguide.pd.

2   Space-Fractional Parabolic Problem

The *fractional* Laplace operator $(-\Delta)^\alpha$ can be defined through the Riesz potential as:

$$(-\Delta)^\alpha u(x) = C(d,\alpha) \text{ P.V.} \int_{\mathbb{R}^d} \frac{u(x) - u(y)}{|x - y|^{d+2\alpha}} dy, \quad \alpha \in (0,1),$$

where P.V. denotes the principal value, d is the number of dimensions (the presented numerical experiments are for $d = 2$), $\alpha \in (0,1)$ and $C(d,\alpha)$ is the normalized constant

$$C(d,\alpha) = \frac{2^{2\alpha} \alpha \Gamma\left(\alpha + \frac{d}{2}\right)}{\pi^{d/2} \Gamma(1 - \alpha)}.$$

Here Γ is the gamma function. The thus defined fractional Laplacian can be regarded as an infinitesimal generator of an α-stable Lévi operator. We consider the following parabolic problem

$$\left| \begin{array}{ll} \dfrac{\partial u(x,t)}{\partial t} + (-\Delta)^\alpha u(x,t) = f(x,t), & (x,t) \in \Omega \times [0,T], \\[2mm] u(x,t) = 0, & (x,t) \in \Omega^c \times [0,T], \\[2mm] u(x,0) = u^0(x), & x \in \Omega, \end{array} \right. \tag{1}$$

where Ω is a bounded Lipschitz domain, Ω^c is its complement, $f(x,t)$ is a source function, and $u(x,t)$ is the unknown solution.

2.1   Discretizations in Space and Time

FEM discretization is applied in space. Consider $\mathcal{T}$ to be an admissible triangulation on Ω with $n_{\mathcal{T}}$ piecewise linear conforming elements. Let the nodal basis $\{\varphi_1, \ldots, \varphi_n\} \subset \mathbb{V}_h$ corresponds to the internal nodes $\{x_1, \ldots, x_n\}$ such that $\varphi_i(x_j) = \delta_i^j$. Thus, the parabolic equation (1) is reduced to the Cauchy problem

$$M_L \frac{d\mathbf{u}}{dt} + K\mathbf{u} = M_L \mathbf{f}, \quad 0 < t \leq T, \quad \mathbf{u}(0) = \mathbf{u}^0,$$

where $K = (K_{ij}) \in \mathbb{R}^{n \times n}$ is the stiffness matrix, $M_L = \text{diag}(m_L^i) \in \mathbb{R}^n$ is the lumped mass matrix (where m_L^i is the mass at node x_i), $\mathbf{f}(t) \in \mathbb{R}^n$ is the corresponding right hand vector and $\mathbf{u}(t) \in \mathbb{R}^n$ is the unknown solution vector.

The *fractional* diffusion stiffness matrix K is symmetric positive definite. Its coefficients can be written in the following form

$$K_{ij} = \frac{C(d,\alpha)}{2} \langle \varphi_i, \varphi_j \rangle_{H^\alpha(\mathbb{R})}.$$

It should be noted that the integrals involved in the computation of K_{ij} must be carried over the $\mathbb{R}^d$. In order to approximate them, a ball domain B is constructed around Ω, such that the distance from the boundary of $\overline{\Omega}$ to the compliment B^c is sufficiently large (for more details see [8]). Then an auxiliary triangulation $\mathcal{T}_A$ is introduced on $B \setminus \Omega$, such that the combined triangulation $\widetilde{\mathcal{T}} = \mathcal{T} \cup \mathcal{T}_A$ is admissible with $n_{\widetilde{\mathcal{T}}}$ elements. Finally, the coefficients of the stiffness matrix are written as

$$K_{ij} = \frac{C(d,\alpha)}{2} \sum_{l=1}^{n_{\widetilde{\mathcal{T}}}} \left(\sum_{m=1}^{n_{\widetilde{\mathcal{T}}}} I_{l,m}^{i,j} + 2J_l^{i,j} \right), \quad l,m \in [1, n_{\widetilde{\mathcal{T}}}],$$

where the integrals I and J have the form

$$I_{l,m}^{i,j} = \int_{T_l} \int_{T_m} \frac{\left(\varphi_i(x) - \varphi_i(y)\right)\left(\varphi_j(x) - \varphi_j(y)\right)}{|x-y|^{d+2\alpha}} dx\, dy$$

$$J_l^{i,j} = \int_{T_l} \int_{B^c} \frac{\varphi_i(x)\varphi_j(x)}{|x-y|^{d+2\alpha}} dy\, dx.$$

The above described method is introduced along with a complete d-dimensional FEM analysis by Acosta and Borthagaray in [8]. In our study, we used the open source MatLab code published in [9] to compute the stiffness matrix K.

In what follows we develop a method and algorithm for implementation of the backward Euler scheme

$$M_L \frac{\mathbf{u}^{j+1} - \mathbf{u}^j}{\tau_j} + K\mathbf{u}^{j+1} = M_L \frac{\mathbf{f}^{j+1} + \mathbf{f}^j}{2}, \quad j = 0, \ldots, m-1, \qquad (2)$$

applied to (3), where $\sum_{j=1}^m \tau_j = T$, m is the number of time steps.

In [10] we considered discretization in time with uniform time steps. Here, the focus is on the case of varying time steps motivated by the local refinement in time around points of lower regularity of the solution due to jumps of the right-hand side $f(x,t)$.

2.2   Test Problem

The parabolic equation (1) in $\Omega \times [0,T] = [-1,1] \times [-1,1] \times [0,0.1]$ is used as a test problem in our numerical experiments, where $u^0(x) = 0$. The right hand side $f(x,t) \in L^2(\Omega \times [0,T])$ is defined through the checkerboard function $\bar{f}(x)$ as follows

$$f(x,t) = \begin{cases} 100\bar{f}(x), & t \in (0, \widetilde{\tau}_1) \\ 200\bar{f}(x), & t \in (\widetilde{\tau}_2, \widetilde{\tau}_3) \\ 0, & otherwise \end{cases}, \text{ where } \bar{f}(x) = \begin{cases} 1 & x_1 x_2 > 0, x \in \Omega \\ -1 & x_1 x_2 \leq 0, x \in \Omega \\ 0 & x \notin \Omega \end{cases} \quad (3)$$

Here $\widetilde{\tau}_1 = 0.01$, $\widetilde{\tau}_2 = 0.05$ and $\widetilde{\tau}_3 = 0.06$.

For this test problem we construct a hybrid refined mesh in time. Examples of adaptively refined meshes for fractional diffusion problems, but in a different context, can be found in [17,18].

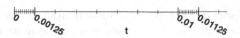

Fig. 1. Left end of the time interval showing smaller time steps immediately after 0 and $\tilde{\tau}_1$.

We start with a uniform mesh with $m_c = 80$ uniform time steps. Then the intervals with left ends $t = 0, \tilde{\tau}_1, \tilde{\tau}_2$ and $\tilde{\tau}_3$ are divided into $m_r = 10$ subintervals. Thus the total number of time steps is $m = m_c - M_R + 4 \times m_r = 116$. Figure 1 shows the time step changes in the left part of the time interval.

Although the described test problem uses a relatively small number of time step changes, it already allows to demonstrate the advantages of the developed method based on HSS compression.

3 Hierarchically Semi-separable Compression Based Solver

The first hierarchical compression method was proposed by Hackbusch, introducing the $\mathcal{H}$-matrices in [19]. The $\mathcal{H}$-matrices were originally applied to solve systems of linear algebraic equations arising from boundary element method discretizations. Nowadays, various types of hierarchical matrices exist, including $\mathcal{H}$, $\mathcal{H}^2$, and the hierarchically semi-separable compression matrices. In this work we use the HSS compression method developed by Martinsson [20] and its implementation within the STRUMPACK project [16,21]. The method solves a system of linear algebraic equations in three steps: (i) HSS compression, (ii) ULV-like factorization and (iii) solving the system with the factorized matrix.

3.1 Hierarchically Semi-separable Compression

A matrix A can be compressed into an HSS form H in the following way:

1. We divide A in a 2×2 block form. The off-diagonal Blocks $A_{ij}, i \neq j$ can be represented as a multiplication of U, B and V matrices called *generators*. It is assumed that these blocks are low rank and a rank finding decomposition like Singular Value Decomposition can be applied to them. Thus, A can be presented as:

$$A = \begin{bmatrix} A_{11} & A_{12} \\ A_{21} & A_{22} \end{bmatrix} = \begin{bmatrix} D_1 & U_1^{\text{big}} B_{12} V_2^{\text{big}} \\ U_1^{\text{big}} B_{12} V_2^{\text{big}} & D_2 \end{bmatrix}.$$

For matrices with good *structure* (meaning low rank off-diagonal blocks) U is "tall and slim", B is small and square and V is "short and wide", "big" is described below.

2. The diagonal blocks D_i are then divided in four blocks in a similar manner and this process continues recursively for as long as needed. The second hierarchical level of compression can be presented as

$$
A = \left[\begin{array}{cc} \begin{bmatrix} D_1 & U_1^{\text{big}} B_{1,2} V_2^{\text{big}*} \\ U_2^{\text{big}} B_{2,1} V_1^{\text{big}*} & D_2 \end{bmatrix} & U_3^{\text{big}} B_{3,6} V_6^{\text{big}*} \\ U_6^{\text{big}} B_{6,3} V_3^{\text{big}*} & \begin{bmatrix} D_4 & U_4^{\text{big}} B_{4,5} V_5^{\text{big}*} \\ U_5^{\text{big}} B_{5,4} V_4^{\text{big}*} & D_5 \end{bmatrix} \end{array} \right].
$$

3. A recursive property exist between the U and V generators at different levels of compression, such that

$$
U_3^{\text{big}} = \begin{bmatrix} U_1^{\text{big}} & 0 \\ 0 & U_2^{\text{big}} \end{bmatrix} U_3 \quad \text{and} \quad V_3^{\text{big}} = \begin{bmatrix} V_1^{\text{big}} & 0 \\ 0 & V_2^{\text{big}} \end{bmatrix} V_3.
$$

This property allows less memory to be used for the compressed matrix. The second level of compression can then be written as

$$
A = \left[\begin{array}{cc} \begin{bmatrix} D_1 & U_1^{\text{big}} B_{1,2} V_2^{\text{big}*} \\ U_2^{\text{big}} B_{2,1} V_1^{\text{big}*} & D_2 \end{bmatrix} & \begin{bmatrix} U_1^{\text{big}} & 0 \\ 0 & U_2^{\text{big}} \end{bmatrix} U_3 B_{3,6} V_6^* \begin{bmatrix} V_4^{\text{big}*} & 0 \\ 0 & V_5^{\text{big}*} \end{bmatrix} \\ \begin{bmatrix} U_4^{\text{big}} & 0 \\ 0 & U_5^{\text{big}} \end{bmatrix} U_6 B_{6,3} V_3^* \begin{bmatrix} V_1^{\text{big}*} & 0 \\ 0 & V_2^{\text{big}*} \end{bmatrix} & \begin{bmatrix} D_4 & U_4^{\text{big}} B_{4,5} V_5^{\text{big}*} \\ U_5^{\text{big}} B_{5,4} V_4^{\text{big}*} & D_5 \end{bmatrix} \end{array} \right].
$$

The efficiency of the compression is measured by the maximum off-diagonal rank r, calculated during the compression process. In the general case, HSS compression forms a matrix H that approximates A. STRUMPACK's algorithm requires the user to input two thresholds (absolute ε_{abs} and relative ε_{rel}), that are used during the compression process. Smaller thresholds lead to larger r (less efficient compression, more computing time and more memory required, but higher accuracy) while larger thresholds lead to smaller r (more efficient compression, less computation time and memory required, but less accuracy). The effectiveness of compression also depends on the correct *structure* of the matrix. In [22] we have examined several schemes for reordering the unknowns for the elliptic fractional diffusion problem, and here we employ the best one – the recursive bisection.

In the general case, the computational complexity of this step is $O(n^2 r)$. For some specific problems and under certain conditions, it can be reduced to $O(nr^2)$.

3.2   ULV-Like Factorization and Solution

After the original matrix A is compressed into HSS form H we can apply a special form of LU factorization, known as ULV-like factorization. The factorization in STRUMPACK uses the structure of the generators, unlike the original

ULV factorization that uses orthogonal transformations, hence the ULV-like name [16]. This factorization transforms the problem of eliminating n unknowns into the problem of eliminating r unknowns. Then, a standard LU decomposition is applied to the last $r \times r$ block. This process is visualized in Fig. 2. The computational complexity of the ULV-like factorization is $O(nr^2)$.

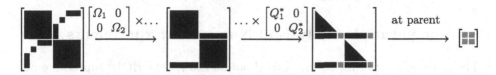

Fig. 2. ULV-like factorization.

After the ULV-like factorization the problem is reduced to solving two systems with sparse triangular matrices. The computational complexity of the solution (backward substitution) step is $O(rn)$.

Thus, the overall complexity of the method is dominated by the HSS compression step with a computational complexity of $O(rn^2)$.

4   Complexity of Parabolic Solvers and the Test Problem

In the case of parabolic problems, sequences of linear algebraic equations have to be solved. In [10] we considered a backward Euler discretization in time with uniform time step. As shown there, one can apply the compression and ULV-like factorization only once (computational complexity $O(n^2 r)$) and at each time step solve the already factorized system of linear algebraic equations (computational complexity $O(nr)$).

For the implementation of the scheme (2) we compute the HSS compression once with computational complexity $O(n^2 r)$ (HSS compression doesn't change the diagonal and $\widetilde{K}_j$ differ only on the diagonal). At each time step that changes from the previous we modify the diagonal of the compressed matrix and apply the ULV-like factorization with computational complexity of $O(nr^2)$. Finally, we solve the current system with the factorized matrix $\widetilde{K}_j$ at each time step j with computational complexity of $O(nr)$.

Now, let us take a closer look at the test problem (3). In this, case the total computational complexity $\mathcal{N}_{\mathrm{HSS}}$ can be written in the form:

$$\mathcal{N}_{\mathrm{HSS}} = \mathcal{N}_{\mathrm{HSS}}^{(1)} + 8 \times \mathcal{N}_{\mathrm{HSS}}^{(2)} + 116 \times \mathcal{N}_{\mathrm{HSS}}^{(3)},$$

where
$$\mathcal{N}_{\mathrm{HSS}}^{(1)} = O(n^2 r), \quad \mathcal{N}_{\mathrm{HSS}}^{(2)} = O(nr^2), \quad \mathcal{N}_{\mathrm{HSS}}^{(3)} = O(nr).$$

In comparison, when applying the block LU factorization, we need to perform the factorization each time the time step changes with computational complexity

$O(n^3)$ and solve a system with the factorized matrix at each time step with computational complexity $O(n^2)$. Then, the total computational complexity $\mathcal{N}_{\mathrm{BLU}}$ takes the form:

$$\mathcal{N}_{\mathrm{BLU}} = 8 \times \mathcal{N}_{\mathrm{BLU}}^{(1)} + 116 \times \mathcal{N}_{\mathrm{BLU}}^{(2)},$$

where

$$\mathcal{N}_{\mathrm{BLU}}^{(1)} = O(n^3), \quad \mathcal{N}_{\mathrm{BLU}}^{(2)} = O(n^2).$$

5   Comparative Analysis of Numerical Experiments

The numerical experiments are carried out on the AVITOHOL[3] supercomputer at IICT-BAS. AVITOHOL is made of 150 nodes, each with two Intel Xeon E5-2650v2 8C 2.6 GHz CPUs. A single node is used for the results presented. Each CPU has 8 cores, and therefore a maximum of 16 hardware threads can be used. The cores allow hyperthreading, but this doesn't lead to improvement for the considered problem and such results are not presented.

In Fig. 3 we present the sequential and parallel (with 16 threads) execution times. The HSS compression times are presented in Fig. 3a and Fig. 3d. This step is executed only once, when solving the sequence of systems of linear algebraic equations with STRUMPACK's solver only. The default value of the absolute threshold $\varepsilon_{\mathrm{abs}} = 10^{-8}$ is used, while the relative threshold $\varepsilon_{\mathrm{rel}} = 10^{-2}, 10^{-4}, 10^{-6}$ and 10^{-8} varies.

In Fig. 3b and Fig. 3e we present the sequential and parallel execution times for the factorization and in Fig. 3c and Fig. 3f – the backward substitution (solving) steps. Here we use factorization (ULV-like for the HSS based method and block LU for the Gaussian elimination from MKL) at eight time steps (when the interval's left end is at $t = 0, 0.1, 0.5$ and 0.6 (when the time step becomes smaller) or after ten small steps when the time step becomes bigger again). The solving steps with the factorized matrices are performed at each time step. Figure 3g and Fig. 3h present the total time required to run the entire algorithm. We remind that the block LU based solver does not have a separate compression step, and thus the times for the total and solving parts coincide. The HSS based solver has overall better computational times than the LU based solver from MKL for all values of the relative threshold $\varepsilon_{\mathrm{rel}}$ examined.

We should note that the parallel efficiency of STRUMPACK's HSS algorithm is lower than the block LU factorization. In Fig. 4 we present the parallel speed up with 8 and 16 threads of the HSS based algorithm. With 16 threads and for the larger problems STRUMPACK achieves parallel speed up of up to 10 for the compression step, up to 6 for the factorization, up to 5 for the solving and up to 8 for the total execution. The speed up in the case of the MKL implementation of block LU factorization is better, achieving values of up to 15. This can be explained by the more complex hierarchical structure of the HSS compression based algorithm.

---

[3] http://www.hpc.acad.bg/system-1/.

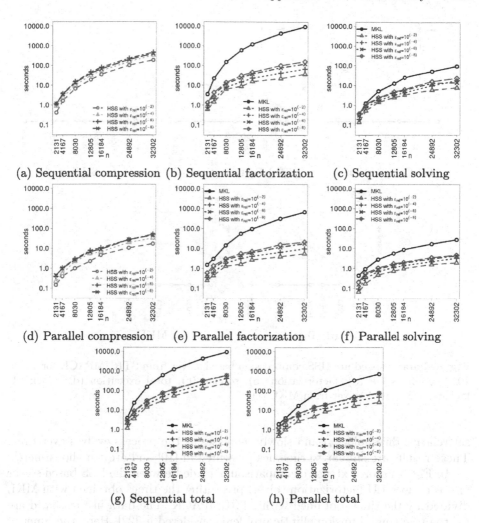

(a) Sequential compression (b) Sequential factorization   (c) Sequential solving

(d) Parallel compression   (e) Parallel factorization   (f) Parallel solving

(g) Sequential total   (h) Parallel total

Fig. 3. Sequential and parallel with 16 threads execution times.

The maximum off-diagonal rank r is calculated during the compression process. The rank is a measure of the overall efficiency of HSS compression, which is particularly evident from the asymptotic estimate of the $O(n^2 r)$ computational complexity of the algorithm. Here, we should also mention that the HSS compression is approximate. We can evaluate the accuracy by using as a reference solution the results obtained by the block LU factorization solver from MKL. The relative error obtained in this way is within the order of magnitude of the

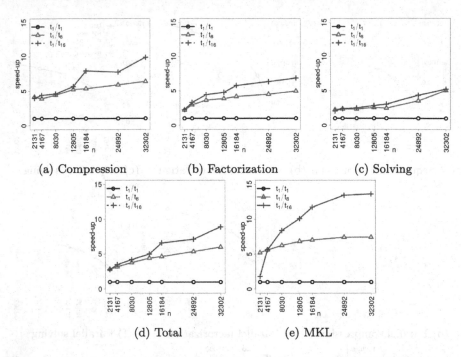

Fig. 4. Parallel speed up: HSS compression based solver from STRUMPACK for $\varepsilon_{rel} = 10^{-4}$; compression (a); factorization (b); solving (c); total execution (d); block LU factorization based solver from MKL (f).

set relative threshold ε_{rel} and slowly increases as the process evolves over time. These results are similar to the ones presented in [10] and are not shown here.

In Fig. 5 a more extensive comparison is made between the HSS-based solver and the block LU factorization solver, presenting the times obtained with MKL divided by the times obtained with STRUMPACK. The data in Figs. 5a, d are for the fractional diffusion elliptic problem considered in [22]. Here, the times of HSS compression based solver are better in two cases only: for the sequential runs when the bigger thresholds ($\varepsilon_{rel} = 10^{-2}$ and 10^{-4}) are set. Things change for the spatial-fractional parabolic problems, for which STRUMPACK shows much better execution times for both the sequential and parallel experiments. The experiments with uniform time step from [10] given in Figs. 5b, e clearly show the advantages of the HSS compression based solver. Now, in Figs. 5c, f we see that STRUMPACK outperforms MKL even more strongly when the time step is varying. The experimental results are in full agreement with the theoretical estimates of the computational complexity.

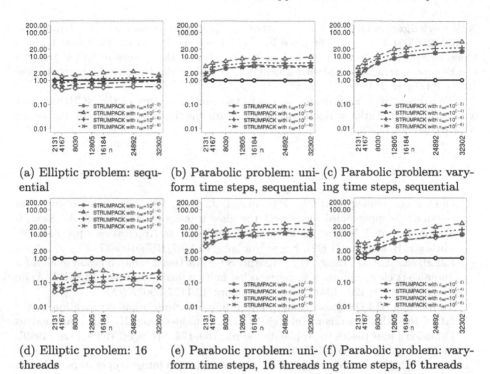

(a) Elliptic problem: sequential

(b) Parabolic problem: uniform time steps, sequential

(c) Parabolic problem: varying time steps, sequential

(d) Elliptic problem: 16 threads

(e) Parabolic problem: uniform time steps, 16 threads

(f) Parabolic problem: varying time steps, 16 threads

Fig. 5. Comparison of the performance of the HSS compression based solver with the block LU solver for: elliptic problem (a and d); parabolic problem with uniform time steps (b and e); and parabolic problem with varying time steps (c and f). The fractions $t_{\mathrm{MKL}}/t_{\mathrm{STRUMPACK}}$ are shown.

6   Concluding Remarks

We have presented and analyzed a method and algorithm for numerically solving fractional-in-space parabolic problems that use backward Euler discretization in time with varying time steps. The method is based on hierarchically semi-separable compression. It is shown that the HSS compression based solver has better computational complexity than the analogous solver based on block LU factorization. The presented numerical tests use the software implementations of HSS in STRUMPACK and, respectively, of the block LU factorization in MKL. Comparative analysis of numerical experiments confirms the advantages of the hierarchical method. For the sequential experiments, the HSS based solver has between ~16 and ~40 times faster execution times than the block LU solver, while the parallel times are between ~9 and ~24 times faster with 16 threads. These data are for largest discrete problem considered. The maximum off-diagonal rank is ~20 to ~80 times smaller than the number of unknowns which means a good *structure* of the stiffness matrix of the problem and explains the better performance of the HSS based solver over block LU factorization one.

Acknowledgements. The first author is partially supported by Grant No BG05M2OP001-1.001-0003, financed by the Science and Education for Smart Growth Operational Program (2014–2020) and co-financed by the European Union through the European structural and Investment funds, and by the Bulgarian NSF under the Grant KII-06-H52/4.

We acknowledge the provided access to the e-infrastructure of the NCHDC - part of the Bulgarian National Roadmap on RIs, with the financial support by the Grant No D01-387/18.12.2020.

References

1. Binder, K., Bennemann, Ch., Baschnagel, J., Paul, W.: Anomalous diffusion of polymers in supercooled melts near the glass transition. In: Pękalski, A., Sznajd-Weron, K. (eds.) Anomalous Diffusion From Basics to Applications, pp. 124–139. Springer, Heidelberg (1999). https://doi.org/10.1007/BFb0106837
2. Langlands, T.A.M., Henry, B.I., Wearne, S.L.: Fractional cable equation models for anomalous electrodiffusion in nerve cells: finite domain solutions. SIAM J. Appl. Math. **71**, 1168–1203 (2011). https://doi.org/10.1137/090775920
3. Taitelbaum, H.: Diagnosis using photon diffusion: from brain oxygenation to the fat of the atlantic salmon. In: Pękalski, A., Sznajd-Weron, K. (eds.) Anomalous Diffusion From Basics to Applications, pp. 160–174, Springer, Heidelberg (1999). https://doi.org/10.1007/BFb0106840
4. Rosasco, L., Belkin, M., De Vito, E.: On learning with integral operators. J. Mach. Learn. Res. 11, 905–934 (2010). http://jmlr.org/papers/v11/rosasco10a.html
5. Chaturapruek, S., Breslau, J., Yazdi, D., Kolokolnikov, T., McCalla, S.G.: Crime modeling with Lèvy flights. SIAM J. Appl. Math. **73**, 1703–1720 (2013). Society for Industrial and Applied Mathematics, 2021/12/06/ 2013. http://www.jstor.org/stable/24510700
6. Sun, H., Zhang, Y., Baleanu, D., Chen, W., Chen, Y.: A new collection of real world applications of fractional calculus in science and engineering. Commun. Nonlinear Sci. Numer. Simul. **64**, 213–231 (2018). https://doi.org/10.1016/j.cnsns.2018.04.019
7. Harizanov, S., Margenov, S., Popivanov, N.: Spectral fractional Laplacian with inhomogeneous Dirichlet data: questions, problems, solutions. In: Georgiev, I., Kostadinov, H., Lilkova, E. (eds.) BGSIAM 2018. SCI, vol. 961, pp. 123–138. Springer, Cham (2021). https://doi.org/10.1007/978-3-030-71616-5_13
8. Acosta, G., Borthagaray, J.P.: A fractional Laplace equation: regularity of solutions and finite element approximations. SIAM J. Num. Anal. **55**, 472–495 (2017). https://doi.org/10.1137/15M1033952
9. Acosta, G., Bersetche, F.M., Borthagaray, J.P.: A short FE implementation for a 2d homogeneous Dirichlet problem of a fractional Laplacian. Comput. Math. Appl. **74**, 784–816 (2017). https://doi.org/10.1016/j.camwa.2017.05.026
10. Slavchev, D., Margenov, S.: Performance study of hierarchical semi-separable compression solver for parabolic problems with space-fractional diffusion. In: Lirkov, I., Margenov, S. (eds.) LSSC 2021. LNCS, vol. 13127, pp. 71–80. Springer, Cham (2022). https://doi.org/10.1007/978-3-030-97549-4_8
11. Anderson, E., et al.: LAPACK Users' Guide. Society for Industrial and Applied Mathematics, 3rd. Philadelphia (1999)

12. Abalenkovs, M., et al.: Plasma 17 performance report. Technical Report 292, LAPACK Working Note, June 2017. http://www.netlib.org/lapack/lawnspdf/lawn292.pdf

13. Abalenkovs, M., et al.: Plasma 17.1 functionality report. Technical Report 293, LAPACK Working Note, June 2017. http://www.netlib.org/lapack/lawnspdf/lawn293.pdf

14. Clint Whaley, R., Petitet, A.: Minimizing development and maintenance costs in supporting persistently optimized BLAS. Software: Pract. Exp. **35**(2), 101–121 (2005). http://www.cs.utsa.edu/~whaley/papers/spercw04.ps

15. Xia, J.: Randomized sparse direct solvers. SIAM J. Matrix Anal. Appl. **34**, 197–227 (2013). https://doi.org/10.1137/12087116X

16. Rouet, F.-H., Li, X.S., Ghysels, P., Napov, A.: A distributed-memory package for dense hierarchically semi-separable matrix computations using randomization. ACM Trans. Math. Softw. 42, 27:1–27:35 (2016). ACM. https://doi.org/10.1145/2930660

17. Duan, B., Lazarov, R.D., Pasciak, J.E.: Numerical approximation of fractional powers of elliptic operators. IMA J. Num. Anal. **40**(3), 1746–1771 (2019). https://doi.org/10.1093/imanum/drz013

18. Harizanov, S., Lazarov, R., Margenov, S., Marinov, P., Pasciak, J.: Analysis of numerical methods for spectral fractional elliptic equations based on the best uniform rational approximation. J. Comput. Phys. **408**, 109285 (2020). https://doi.org/10.1016/j.jcp.2020.109285

19. Hackbusch, W.: A sparse matrix arithmetic based on $\mathcal{H}$-matrices. Part I: Introduction to $\mathcal{H}$-matrices. Computing **62**, 89–108 (1999). https://doi.org/10.1007/s006070050015

20. Martinsson, P.G.: A fast randomized algorithm for computing a hierarchically semiseparable representation of a matrix. SIAM J. Matrix Anal. Appl. **32**, 1251–1274 (2011). https://doi.org/10.1137/100786617

21. Xia, J., Chandrasekaran, S., Gu, M., Li, X.S.: Fast algorithms for hierarchically semiseparable matrices. Numerical Lin. Alg. Appl. **17**, 953–976 (2010). https://doi.org/10.1002/nla.691

22. Slavchev, D., Margenov, S., Georgiev, I.G.: On the application of recursive bisection and nested dissection reorderings for solving fractional diffusion problems using HSS compression. AIP Conference Proceedings **2302**, 120008 (2020). http://arxiv.org/abs/https://aip.scitation.org/doi/pdf/10.1063/5.0034506, https://doi.org/10.1063/5.0034506

Optimized Stochastic Approaches Based on Sobol Quasirandom Sequences for Fredholm Integral Equations of the Second Kind

Venelin Todorov[1,2]([✉]) [iD], Ivan Dimov[2], Rayna Georgieva[2] [iD],
and Tzvetan Ostromsky[2]

[1] Department of Information Modeling, Institute of Mathematics and Informatics,
Bulgarian Academy of Sciences, Acad. Georgi Bonchev Str., Block 8,
1113 Sofia, Bulgaria
vtodorov@math.bas.bg
[2] Department of Parallel Algorithms, Institute of Information and Communication
Technologies, Bulgarian Academy of Sciences, Acad. G. Bonchev Str., Block 25A,
1113 Sofia, Bulgaria
ivdimov@bas.bg, {venelin,rayna,ceco}@parallel.bas.bg

Abstract. In this paper three possible approaches to compute linear functionals of the solution of the Fredholm integral equation of the second kind are under consideration: a biased Monte Carlo method based on evaluation of truncated Liouville-Neumann series; transformation of this problem into the problem of computing a finite number of integrals, and an unbiased stochastic approach. In the second case several Monte Carlo algorithms for numerical integration have been applied including optimized stochastic approaches developed in our previous studies. The unbiased stochastic approach has been applied to a multidimensional numerical example. A comprehensive analysis about the reliability and the efficiency of the algorithms has been done.

Keywords: Fredholm integral equations of the second kind ·
Liouville-Neumann series · Optimized stochastic approaches based on
modified Sobol quasirandom sequences · Unbiased stochastic approach

1 Introduction

The existing Monte Carlo (MC) methods for integral equations (MCM-IE) are based on probabilistic representations of the Liouville-Neumann (LN) series

The presented work was supported by the Bulgarian National Science Fund under the Bilateral Project KP-06-Russia/17 "New Highly Efficient Stochastic Simulation Methods and Applications", Project KP-06-N52/5 "Efficient methods for modeling, optimization and decision making" and Project KP-06-N62/6 "Machine learning through physics-informed neural networks".

I. Georgiev et al. (Eds.): NMA 2022, LNCS 13858, pp. 302–313, 2023.
https://doi.org/10.1007/978-3-031-32412-3_27

for the second kind Fredholm integral equation [2,3,7]. The possible unbiased approaches deal with infinite series, while the biased MC approaches use probabilistic representations of truncated LN series. A well known and widely used biased method is the Markov chain MC (see, for example [10]). Usually, the Markov chain stops after a fixed number of steps.

A possible approach to deal with the problem of approximation of linear functionals of the solution of an integral equation is to transform it into approximate evaluation of finite number of integrals (FNI) (linear functionals of iterative functions) [3]. In this paper we extend the study of the properties of five different MC algorithms for multidimensional numerical integration: Crude MC algorithm [14], based on SIMD-oriented fast Mersenne Twister pseudo-random number generator [12,17], quasi-Monte Carlo (qMC) algorithm based on $\Lambda\Pi_\tau$ Sobol quasirandom sequences, MC algorithms (MCA-MSS-1 and MCA-MSS-2) based on modified Sobol quasirandom sequences [4–6], a stratified symmetrised MC algorithm (MCA-MSS-2-S) [4] and an unbiased stochastic approach [7].

Consider a Fredholm integral equation of the second kind:

$$u(\mathbf{x}) = \int_\Omega k(\mathbf{x},\mathbf{x}')u(\mathbf{x}')d\mathbf{x}' + f(\mathbf{x}) \text{ or } u = \mathcal{K}u + f \ (\mathcal{K} \text{ is an integral operator}),$$

where $k(\mathbf{x},\mathbf{x}') \in L_2(\Omega \times \Omega), f(\mathbf{x}) \in L_2(\Omega)$ are given functions and $u(\mathbf{x}) \in L_2(\Omega)$ is an unknown function, $\mathbf{x},\mathbf{x}' \in \Omega \subset R^n$ (Ω is a bounded domain).

Here the main problem from a mathematical point of view is to evaluate linear functionals of the solution $u(\mathbf{x})$ of the following type: $J(u) = \int \varphi(\mathbf{x})u(\mathbf{x})d\mathbf{x} = (\varphi,u)$, where $\varphi(\mathbf{x}) \in L_2(\Omega)$ is a given function. We can apply successive approximation method for solving integral equations:

$$u^{(i)} = \sum_{j=0}^{i} \mathcal{K}^{(j)}f = f + \mathcal{K}f + \ldots + \mathcal{K}^{(i-1)}f + \mathcal{K}^{(i)}f, \quad i = 1,2,\ldots \quad (1)$$

where $u^{(0)}(\mathbf{x}) \equiv f(\mathbf{x})$. It is known that the condition $\|\mathcal{K}\|_{L_2} < 1$ is a sufficient condition for convergence of the LN series [2,3]. Thus, when this condition is satisfied, the following statement holds: $u^{(i)} \longrightarrow u$ as $i \to \infty$ in sense of L_2 norm.

An approximation of the unknown value (φ,u) can be obtained using a truncated LN series (1) for a sufficiently large i: $(\varphi,u^{(i)}) = (\varphi,f) + (\varphi,\mathcal{K}f) + \ldots + (\varphi,\mathcal{K}^{(i-1)}f)+(\varphi,\mathcal{K}^{(i)}f)$. So, we transform the problem for solving integral equations into a problem for approximate evaluation of a finite number of multidimensional integrals. We will use the following denotation $(\varphi,\mathcal{K}^{(j)}f) = I(j)$, where $I(j)$ is a value, obtained after integration over $\Omega^{j+1} = \Omega \times \ldots \times \Omega, j = 0,\ldots,i$. It is obvious that the calculation of the estimate $(\varphi,u^{(i)})$ can be replaced by an evaluation of a sum of linear functionals of iterative functions of the type $(\varphi,\mathcal{K}^{(j)}f), j = 0,\ldots,i$, which can be presented as:

$$\begin{aligned}(\varphi,\mathcal{K}^{(j)}f) &= \int_\Omega \varphi(t_0)\mathcal{K}^{(j)}f(t_0)dt_0 = \\ &= \int_G \varphi(t_0)k(t_0,t_1)\ldots k(t_{j-1},t_j)f(t_j)dt_0\ldots dt_j,\end{aligned} \quad (2)$$

where $t = (t_0, \ldots, t_j) \in G \equiv \Omega^{j+1} \subset R^{n(j+1)}$. If we denote by $F_j(t)$ the integrand function $F(t) = \varphi(t_0)k(t_0, t_1) \ldots k(t_{j-1}, t_j)f(t_j)$, $t \in \Omega^{j+1}$, then we obtain the following expression for (2):

$$I(j) = (\varphi, \mathcal{K}^{(j)}f) = \int_G F_j(t)dt, \quad t \in G \subset R^{n(j+1)}. \tag{3}$$

Thus, we consider the problem for an approximate calculation of multiple integrals of the type (3). The definition of domain G given above is important for the further presented method of solving the integral equation as a set of finite number of multiple integrals. Actually, both representations (2) and (3) allow to define a biased algorithm. In this case the approximation $(\varphi, \mathcal{K}^{(j)}f)$ to the true inner product (φ, u) is presented as a set of i integrals $I(j)$ (see, (3)) with integrands $F_j(t)$, where t is $n(j+1)$ dimensional point from the newly defined domain G.

The above consideration shows that there are two classes of possible stochastic approaches - the first is the so-called biased approach when one is looking for a random variable which mathematical expectation is equal to the approximation of the solution problem by a truncated LN series (1) for a sufficiently large i. In the biased approaches there are two errors: a systematic one (a truncation error) R_{sys} and a stochastic one, namely R_N, which depends on the number N of values of the random variable or on the number of chains used in the estimate. An unbiased approach assumes that the formulated random variable is such that its mean value approaches the true solution of the problem. In the case of unbiased stochastic methods one should only analyse the probabilistic error. In the case of biased stochastic methods more careful error analysis is needed: balancing of both systematic and stochastic error should be done in order to minimize the computational complexity of the methods (for more details, see [3]).

2   Biased Stochastic Approaches

2.1   Monte Carlo Algorithms Based on Modified Sobol $\Lambda\Pi_\tau$ Sequences

$\Lambda\Pi_\tau$ sequences are *uniformly distributed sequences* (u.d.s.). Subroutines to compute these points can be found in [1], and with more details in [13]. Two randomized quasi-Monte Carlo algorithms based on Sobol $\Lambda\Pi_\tau$ sequences, namely MCA-MSS-1 and MCA-MSS-2, have already been studied and analyzed in [4]. The first algorithm has an optimal rate of convergence for integrands with continuous and bounded first derivatives, and the second one has an optimal rate of convergence for functions with continuous and bounded second derivatives.

2.2   Stratified Symmetrised Monte Carlo Approach

The main idea of the stratified symmetrised Monte Carlo approach is to generate a random point $\xi^{(l)} \in K_l$ uniformly distributed inside K_l and after that take the symmetric point $\xi^{(l)'}$ according to the central point $s^{(l)}$. Such a completely randomized approach simulates the concept of MCA-MSS-2 algorithm, but the *shaking* is with different radiuses ρ in each sub-domain. This algorithm is called MCA-MSS-2-S, because this approach looks like the stratified symmetrised Monte Carlo. Obviously, MCA-MSS-2-S is less expensive than MCA-MSS-2, but the drawback is that there is not such a control parameter like the radius ρ, which can be considered as a parameter randomly chosen in each sub-domain K_j. It is important to notice that the algorithm MCA-MSS-2-S has also an optimal (unimprovable) rate of convergence for the corresponding functional class with continuous and bounded second derivatives.

2.3 Specific Modification of the Algorithms for Solving Integral Equations

The following modifications have been done: when the function from the desired linear functional is the δ-function, the dimension of the corresponding quasirandom and pseudorandom sequences coincides with the number of transitions in LN series. For any other function, the dimension is larger than the number of transitions by one. Minimal distance between quasirandom points is calculated to compute each of the integrals with a different dimension. This procedure during the pre-processing approximation of the integrals increases the total computational time for MCA-MSS-1 and MCA-MSS-2 algorithms. We do one additional check: it is verified not only if the generated pseudorandom point lies inside the corresponding integration domain, but also if it lies inside the corresponding elementary multidimensional interval. Usually some points are rejected and that is why the total number of samples corresponding to MCA-MSS-2 is smaller than the number of samples for MCA-MSS-1 for some prescribed fixed values of coefficient ratio. One important feature of MCA-MSS-2 is that it is efficient when the number of divisions is the same for all directions, i.e. the integration domain is divided into multidimensional subcubes.

3 Unbiased Stochastic Approach

In this section an Unbiased Stochastic Approach (USA) is a generalization to integral equations [7] of the approach developed in [8] for solving systems of linear algebraic equations. The discrete Markov chain used in [8] is replaced by a continuous one with transition probabilities depending on the kernel $k(x, x')$. Here the case $0 \leq k(x, x') \leq 1$ is described, where $1 - k(x, x')$ is an absorption rate. The USA algorithm for more general kernels when the above condition may not be fulfilled has been described in details in [7].

The algorithm in this simplified case may be described as follows.

Algorithm 31 :

Unbiased stochastic algorithm for $0 \leq k(\mathbf{x}, y) \leq 1, \quad \mathbf{x}, y \in [0, 1]^n$

1. Initialization **Input** initial data: the kernel $k(\mathbf{x}, \mathbf{y})$, the function $f(\mathbf{x})$, and the number of random trajectories N.
2. Calculations:
   2.1. **Set** score=0
   2.2. **Set** $\mathbf{x}_0 = \mathbf{x}, test = 1$
      Do $j = 1, N$
         Do While $(test \neq 0)$
            $score = score + f(\mathbf{x}_0)$
            $U = rand[0, 1]^n, \quad V = rand[0, 1]$
            If $k(\mathbf{x}_0, U) < V$ **then**
               $test = 0$ **else**
               $\mathbf{x}_0 = U$
            Endif
         Endwhile
      Enddo
3. Compute the solution:
   $u(\mathbf{x}) = \frac{score}{N}$

4   Numerical Results

The numerical algorithms are tested on the following example [7,9]:

$$u(x) = \int_{\Omega} k(x, x') u(x') \mathrm{d}x' + f(x), \qquad \Omega \equiv [0, 1] \tag{4}$$

where

$$k(x, x') = x^2 e^{x'(x-1)}, \tag{5}$$

$$f(x) = x + (1 - x)e^x, \tag{6}$$

$$\varphi(x) = \delta(x - x_0). \tag{7}$$

The solution of this test problem is $u(x) = e^x$. We are interested in an approximate calculation of (φ, u), where $\varphi(x) = \delta(\mathbf{x} - \mathbf{x}_0)$, $x_0 = 0.5$.

We have performed the following biased stochastic algorithms: MC algorithm and QMC based on Sobol sequences for integral equations and several algorithms for integrals into which the problem of solving integral equations is transformed. For the above set of algorithms we use Crude MCA, quasi-Monte Carlo algorithm based on $\Lambda \Pi_\tau$ Sobol quasirandom sequences, MCA-MSS-1, MCA-MSS-2, and MCA-MSS-2-S.

The algorithms have been studied after 10 runs to average the final approximation and for various sample sizes chosen according to proper sample size for Sobol quasirandom sequences. The number of iterations i/d is fixed, but it is chosen according to the L_2-norm of the kernel $k(\mathbf{x}, \mathbf{x}')$. For an approximate

Table 1. Relative errors and computational time for computing $u(x_0)$ as a finite number of integrals from the Liouville-Neumann series at $x_0 = 0.5$ using MCA-MSS-1 and MCA-MSS-2 algorithms for integrals.

i	N	ρ	MCA-MSS-1					MCA-MSS-2		
			IE			FNI		IE		FNI
			Sobol	CMCM	Time	Sobol	CMCM	CMCM		
			Rel. err.	Rel. err.	(s)	Rel. R_N	Rel. R_N	Rel. R_N	Time (s)	Rel. R_N
1	128	2e-03	0.0516	0.0515	<1e-04	0.0003	0.0004	0.0514	<1e-04	0.0005
2	1017	2e-04	0.0144	0.0144	0.02	2e-06	2e-06	0.0144	0.02	2e-05
2	4 081	6e-05	0.0144	0.0144	0.14	2e-06	3e-06	0.0144	0.16	7e-06
3	32 749	8e-06	0.0040	0.0040	10.4	8e-07	8e-07	0.0040	10.73	1e-06
4	65 521	4e-06	0.0011	0.0011	54.9	1e-07	1e-07	0.0011	55.4	2e-07

computation of any integral $I(j)$, $j = 0, \ldots, i$ different number of samples are chosen to satisfy the error balancing requirements.

Because of the bias of the first two classes of algorithms there is a reason to present the relative errors, as well as, the approximation of the corresponding Liouville-Neumann series with a fixed length in Table 1, and Table 2, as well. The numerical results for computing the solution of the integral equation under consideration at different initial points using USA are presented in Table 3. We have used symbolic computations to determine the values of the systematic error R_{sys} for number of iterations up to 3. In Table 4 we present the computed values for the systematic error R_{sys} at point $x_0 = 0.1$ and $x_0 = 0.5$ just to have an idea about the magnitude of this kind of error. One can clearly see that the systematic error decreases when the number of iterations increase. At the same time, for such small number of iteration i the systematic error dominates in almost all cases. The reason for that is that the analysis of balancing of stochastic and systematic errors are done following the assumption that Crude MC method is used. Actually, the applied algorithms, especially randomized quasi-Monte Carlo algorithms MCA-MSS-1, MCA-MSS-2 based on *shaking* of Sobol points are of much higher quality and their stochastic errors R_N are much smaller than their systematic errors R_{sys}. Some typical values of R_N are, say 10^{-6} for relatively small number of random trajectories. These algorithms are very efficient when the norm of the kernel is relatively small. Then one may increase (but not too much) the number of iterations i, which will allow for a small number of N (and low computational complexity) to have a relatively high accuracy.

One should take into account that the choice of the sample size to compute integrals defined on subdomains (IDSs) for the algorithm MCA-MSS-2-S is a crucial question. The IDSs have different dimensionality and this fact affects the corresponding sample size. The following strategy is applied here: first, the number N_i is chosen to be approximately equal to 2^k (to give a possibility for a fair comparison with results obtained using Sobol quasirandom sequences) and to be close to the i-th power of an integer. We need this because the integration domain is the i-th dimensional unit cube. Then, $N_1 := N_i$, $N_l < N_i$, $l = 2, 3, \ldots, i - 1$, and N_l is the l-th power of an integer that is the number of

Table 2. Relative errors and computational time for computing $u(x_0)$ as a finite number (i) of integrals from the Liouville-Neumann series at $x_0 = 0.5$ using MCA-MSS-2-S algorithm for integrals.

i	N	Integral equation		Finite number of int, FNI
		Crude MC		Crude MC
		Rel. err.	Time (s)	Rel. R_N
1	128	0.05182	< 0.0001	4e-09
2	1 024	0.01437	< 0.0001	6e-08
2	4 096	0.01437	0.01	2e-08
3	46 656	0.00399	0.31	2e-08
4	4 096	0.00111	0.04	1e-06
4	531 441	0.00111	5.97	7e-09

sunintervals on each direction. Finally, we have chosen to compute each IDS with the same number of samples. It defines serious restrictions about the sample size applied during the numerical experiments. Nevertheless, we succeeded to achieve the goal of studying properties of MCA-MSS-2-S for solving integral equations based on the computation of a finite number of integrals.

The results presented in Table 5 definitely show that the Unbiased algorithm performs (in general) with a higher accuracy. One may also observe that for relatively small N (up to 1024) the results for the relative error for the USA and biased MC for FNI are similar (results presented in the third and forth column of the table), but when N grows the USA significantly outperforms the two biased MC for FNI. For this particular case the results obtained via FNI Sobol algorithm and FNI Crude algorithm are very close. The reason is that the SFMT

Table 3. Relative errors and computational time for computing $u(x_0)$ at different initial points using USA for integral equations.

N	$x_0 = 0.5$			$x_0 = 0.99$		
	i	Rel. err.	Time (s)	i	Rel. err.	Time (s)
128	1.3	0.02618	0.001	2.38	0.00237	0.001
512	1.28	0.01160	0.002	2.39	0.00714	0.002
1 024	1.28	0.01020	0.001	2.38	0.00199	0.005
2 048	1.27	0.00422	0.006	2.38	0.00213	0.005
16 384	1.27	0.00142	0.024	2.38	9e −06	0.044
65 536	1.27	0.00075	0.096	2.37	0.00041	0.142
131 072	1.27	0.00037	0.154	2.37	0.00024	0.281
262 144	1.27	0.00014	0.305	2.37	0.00026	0.558
4 194 304	1.27	0.00003	4.888	2.37	0.00011	8.802

Table 4. Relative systematic error R_{sys} computed for three number of iterations $i = 1, 2, 3$ at two points x_0

x_0	$i = 1$	$i = 2$	$i = 3$
$x = 0.1$	2.28 e−03	0.635 e−03	0.176 e−03
$x = 0.5$	51.8 e−03	14.3 e−03	3.99 e−03

is one of the best pseudorandom generators and it leads to similar results with well-distributed points like Sobol sequences. The main conclusion from the comparison between the results in the corresponding two columns in Table 5 is that the results have similar order and SFMT is an efficient pseudorandom number generator. The standard biased Markov chain MC (column two) performs a bit better than the two biased MC for FNI for small N, but, again, for large N the USA gives more accurate results. For the standard biased Markov chain MC we present the value ε (in brackets) for which the stochastic error is balanced with the systematic error. The numerical example used is specially constructed to be favored to the biased and unfavored to the USA. Because of the relatively small L_2 norm of the integral operator one need just a small number of iterations to get a good approximation of the solution. It means that the biased algorithms are only *slightly biased*.

Table 5. Comparison of the Unbiased Monte Carlo with the biased MC for computing $u(x_0)$. Standard Crude Markov chain Monte Carlo for integral equations is used. Results for the biased MC that use finite number (i) of multidimensional integrals from the Liouville-Neumann series are also presented. For the Unbiased MC the average number of moves before absorption is shown in brackets

N	Biased MC for Integral equation			Unbiased MC
	Markov chain MC	FNI Sobol seq.	FNI Crude MC	USA
	Rel. $R_N(\varepsilon)$	Rel. R_N (i)	Rel. R_N (i)	Rel. R_N (i)
128	0.05960 (0.4)	0.05158 (1)	0.05197 (1)	0.02618 (1.3)
512		0.01442 (2)	0.01415 (2)	0.01160 (1.28)
1024	0.00566 (0.14)	0.01439 (2)	0.01405 (2)	0.01020 (1.28)
2 048		0.01438 (2)	0.01431 (2)	0.00422 (1.27)
4 096	0.00294 (0.080)	0.01438 (2)	0.01422 (2)	
16 384		0.00400 (3)	0.00394 (3	0.00142 (1.27)
32 768	0.00092 (0.02)	0.00399 (3)	0.00404 (3)	
65 536	0.00076 (0.018)	0.00111 (3)	0.00108 (3)	0.00075 (1.27)
131 072		0.00111 (4)	0.00108 (4)	0.00037 (1.27)
262 144		0.00111 (4)	0.00109 (4)	0.00014 (1.27)

The tendencies concerning the behavior of biased MC algorithms here are similar to the tendencies observed in [4]. The following observations can be drawn based on the results of our computations:

- MCA-MSS-2 algorithm gives a slightly larger relative stochastic error R_N in comparison with MCA-MSS-1 related to the linear functional under consideration, and smaller relative stochastic error (almost 10 times smaller for the most cases) related to the finite number of integrals.
- MCA-MSS-2-S algorithm gives the smallest relative stochastic error and the shortest computational time for a fixed number of steps in Liouville-Neumann series in comparison with MCA-MSS-1 and MCA-MSS-2, but serious restrictions about the choice of sample size exist for its implementation.
- The main disadvantage of MCA-MSS-1 and MCA-MSS-2 algorithms is the high computational complexity due to computing the minimal distance between the generated original Sobol sequences.
- USA algorithm is the fastest and the most reliable for larger sample sizes than all other algorithms under consideration. The reason for that could be the fact that the algorithm is unbiased and also the probability for absorbtion of the random trajectory is chosen to depend on the problem data, namely, $k(\mathrm{x}, \mathrm{x}')$ and $f(\mathrm{x})$.
- The average number of the steps of the USA algorithm is smaller than 2 due to the fast convergence because i) the initial point at each chain is 0.5, and ii) the most of the kernel values are ranged in $(0, 1; 0, 3)$ in this case. Just to make a comparison: for MC algorithms $2, 3, 4$ jumps are necessary to reach the prescribed accuracy (the L_2 norm of the kernel is approximately equal to 0.3, i.e. one can expect a fast convergent iterative process).
- USA algorithm behavior has been studied for other initial points, especially for $x_0 = 0.99$; the average number of iteration steps for USA algorithm increased and was approximately equal to 2 (more precisely, it is around 2.4) for all sample sizes considered.

4.1 Stochastic Approach for Multidimensional Problems

Solving the multidimensional Fredholm integral equation of second kind is a great challenge to numerical algorithms. We are interested in the following multidimensional problem [7]:

$$u_n(\mathrm{x}) = \int_0^1 \cdots \int_0^1 \prod_{j=1}^n \{k(x^{(j)}, x^{(j)'})\} u(\mathrm{x}') \prod_{j=1}^n dx^{(j)'} + f_n(\mathrm{x}), \qquad (8)$$

$$\mathrm{x} \equiv (x^{(1)}, \ldots, x^{(n)}), \ \mathrm{x}' \equiv (x^{(1)'}, \ldots, x^{(n)'}), \ \prod_{j=1}^n dx^{(j)'} = d\mathrm{x}',$$

where

$$k(x^{(j)}, x^{(j)'}) = \beta_j \left(2x^{(j)'} - 1\right) \{x^{(j)}\}^2 \exp\{x^{(j)'}(x^{(j)} - 1)\}, \ \beta = \prod_{j=1}^n \beta_j, \qquad (9)$$

Table 6. Solution, Relative errors, Relative standard deviation, Average number of jumps before absorption and Computational time for computing $u(x_0)$ at $(1.0, 1.0, \ldots, 1.0)$ for different values of norm of the kernel controlled by the parameter β.

β	Parameter	Problem dimension			
		$n = 1$	$n = 3$	$n = 5$	$n = 10$
	Exact sol.	2.71828	20.0855	148.413	22026.4
	Rel. err.	6.4e−5	6.5e-5	4.1e−4	4.1 e−3
$0.1/\|K\|_{L_2}$	$\sigma/u(x_0)$	0.32	0.49	1.37	18.1
	i	1,22	1.55	1.68	1.82
	Comp. time	3.9	12.1	21.1	42.7
	Exact sol.	2.71828	20.0855	148.413	22026.4
	Rel. err.	3.2e−5	2.2e−3	1.0e−2	5.9e−3
$0.5/\|K\|_{L_2}$	$\sigma/u(x_0)$	1.1	4.0	11.8	192
	i	2.08	1.88	1.87	1.91
	Comp. time	7.11	15.1	23.1	44.4
	Exact sol.	2.71828	20.0855	148.413	22026.4
	Rel. err.	7.5e−3	1.6e−2	7.5e−3	7.1e−2
$0.95/\|K\|_{L_2}$	$\sigma/u(x_0)$	12.5	25.7	50.9	228
	i	2.48	1.99	1.93	1.93
	Comp. time	8.33	15.78	24.5	45.3

$$f_n(x) = e^{\sum_{j=1}^n x^{(j)}} + (-1)^{(j-1)}\beta_n(2e^{x^{(j)}} - 2 - x^{(j)}e^{x^{(j)}} - x^{(j)}), \quad (10)$$

$$\varphi(x) = \delta(x - x_0), \quad (11)$$

where $x_0 = (1.0, 1.0, \ldots 1.0)$ in all the numerical experiments. The solution of the above equation is a product of exponents $u(x) = \prod_{j=1}^n \exp\{x^{(j)}\} = \exp\{\sum_{j=1}^n x^{(j)}\}$. We use special parameter β to be able to deal with various difficulties, namely, with negative kernels and values of the kernel greater than one in absolute value. If the values of the parameter β increases, then we approach instability barrier when the variance is getting infinite. If we keep the same values of β, then the L_2 norm of the new kernel is getting smaller and smaller with the increasing of the dimensionality and thats why we actually increase the values of β to keep the same value of the norms. In such a way have a fair comparison of the results for different dimensions.

In this case only USA gives reliable results. The numerical results are presented in Table 6. We have chosen dimensions $n = 1, 3, 5, 10$. To study how the speed of convergence depends on the L_2 norm of the integral operator three different values of the parameter β were chosen. For all experiments the same number of Markov chains was used. The average length of the chains is between 1 and 3. The numerical approximations are less accurate when β increases. Nevertheless, one may observe that the speed of convergence is good not only for small values of β, but also for large values of β. The relative errors seems to be satisfactory even for large dimensions and for high values of β. For the smallest value of β, we observe an almost linear behaviour of the computational times with respect to the dimension. This shows the strength of USA to high dimensional problems.

5   Conclusion

Three possible approaches to compute linear functionals of the solution of integral equation have been analyzed: a biased Monte Carlo technique based on evaluation of truncated Liouville-Neumann series, a transformation of this problem into the problem of computing a finite number of integrals, and an unbiased stochastic approach. Five Monte Carlo algorithms for numerical integration have been applied in the second case, namely Crude Monte Carlo method based on a high quality SIMD-oriented Mersenne Twister pseudorandom number generator, Quasi-Monte Carlo based on $\Lambda\Pi_\tau$ Sobol quasirandom sequences, a stratified symmetrised MC algorithm MCA-MSS-2-S, a randomized Quasi-Monte Carlo based on a special procedure of *shaking* $\Lambda\Pi_\tau$ Sobol quasirandom points with a convergence rate $O\left(N^{-\frac{1}{2}-\frac{1}{d}}\right)$, and a randomized Quasi-Monte Carlo with a convergence rate $O\left(N^{-\frac{1}{2}-\frac{2}{d}}\right)$. We have shown that the procedure of balancing of both stochastic R_N and systematic R_{sys} errors is very important for the quality of the biased algorithms. In almost all numerical experiments performed, the unbiased stochastic algorithm performs with relatively high accuracy and low computational complexity in comparison with the best available biased algorithms.

Acknowledgements. The authors thanks to Prof. Sylvain Maire for the useful discussion regarding the USA method. The presented work was supported by the Bulgarian National Science Fund under the Bilateral Project KP-06-Russia/17 "New Highly Efficient Stochastic Simulation Methods and Applications". Venelin Todorov is supported by the BNSF under Projects KP-06-N52/5 "Efficient methods for modeling, optimization and decision making" and KP-06-N62/6 "Machine learning through physics-informed neural networks".

References

1. Bradley, P., Fox, B.: Algorithm 659: implementing Sobol's quasi random sequence generator. ACM Trans. Math. Software **14**(1), 88–100 (1988)
2. Curtiss, J.H.: Monte Carlo methods for the iteration of linear operators. J. Math Phys. **32**, 209–232 (1954)
3. Dimov, I.: Monte Carlo methods for applied scientists., World Scientific, New Jersey, London, Singapore, World Scientific (2008). 291p., ISBN-10 981–02-2329-3
4. Dimov, I.T., Georgieva, R.: Multidimensional sensitivity analysis of large-scale mathematical models. In: Iliev, O.P., et al. (eds.) Numerical Solution of Partial Differential Equations: Theory, Algorithms, and Their Applications, Mathematics

and Statistics, vol. 45, pp. 137–156. Springer, New York (2013). https://doi.org/
10.1007/978-1-4614-7172-1_8, ISBN: 978-1-4614-7171-4 (book chapter)

5. Dimov, I., Georgieva, R.: Monte Carlo method for numerical integration based on Sobol's sequences. In: Dimov, I., Dimova, S., Kolkovska, N. (eds.) NMA 2010. LNCS, vol. 6046, pp. 50–59. Springer, Heidelberg (2011). https://doi.org/10.1007/978-3-642-18466-6_5

6. Dimov, I.T., Georgieva, R., Ostromsky, Tz., Zlatev, Z.: Advanced algorithms for multidimensional sensitivity studies of large-scale air pollution models based on Sobol sequences, Computers and Math. Appl. **65**(3), 338–351 (2013). "Efficient Numerical Methods for Scientific Applications", Elsevier

7. Dimov, I.T., Maire, S.: A new unbiased stochastic algorithm for solving linear Fredholm equations of the second kind. Adv. Comput. Math. **45**(3), 1499–1519 (2019). https://doi.org/10.1007/s10444-019-09676-y

8. Dimov, I.T., Maire, S., Sellier, J.M.: A new walk on equations monte carlo method for solving systems of linear algebraic equations. Appl. Math. Modelling (2014). https://doi.org/10.1016/j.apm.2014.12.018

9. Farnoosh, R., Ebrahimi, M.: Monte Carlo method for solving Fredholm integral equations of the second kind. Appl. Math. Comput. **195**, 309–315 (2008)

10. Kalos, M.H., Whitlock, P.A.: Monte Carlo Methods. Wiley-VCH (2008). ISBN 978-3-527-40760-6

11. Niederreiter, H.: Low-discrepancy and low-dispersion sequences. J. Number Theory **30**, 51–70 (1988)

12. Saito, M., Matsumoto, M.: SIMD-oriented fast Mersenne Twister: a 128-bit pseudorandom number generator. In: Keller, A., Heinrich, S., Niederreiter, H. (eds.) Monte Carlo and Quasi-Monte Carlo Methods 2006, pp. 607–622. Springer, Heidelberg (2008). https://doi.org/10.1007/978-3-540-74496-2_36

13. Sobol, I., Asotsky, D., Kreinin, A., Kucherenko, S.: Construction and comparison of high-dimensional Sobol' generators. Wilmott J. 67–79 (2011)

14. Sobol, I.M.: Monte Carlo Numerical Methods. Nauka, Moscow (1973). (in Russian)

15. Sobol, I.M.: On quadratic formulas for functions of several variables satisfying a general Lipschitz condition. USSR Comput. Math. and Math. Phys. **29**(6), 936–941 (1989)

16. Weyl, H.: Ueber die Gleichverteilung von Zahlen mod Eins. Math. Ann. **77**(3), 313–352 (1916)

17. www.math.sci.hiroshima-u.ac.jp/~m-mat/MT/SFMT/index.html

Intuitionistic Fuzzy Knapsack Problem Trough the Index Matrices Prism

Velichka Traneva[✉][ID], Petar Petrov[ID], and Stoyan Tranev[ID]

"Prof. Asen Zlatarov" University,"Prof. Yakimov" Blvd, Bourgas 8000, Bulgaria
veleka13@gmail.com , tranev@abv.bg

Abstract. The Knapsack problem is a NP-hard combinatorial optimization problem that is used in business modelling. The objective of the problem is to select a subset of items so that the total profit of selling them is maximized without exceeding the capacity of the knapsack. Contemporary pandemic environment with galloping inflation predetermines uncertainty in the parameters of this problem. Intuitionistic fuzzy logic, which is an extension of fuzzy logic, is a flexible tool for dealing with ambiguity. In this work, an intuitionistic fuzzy knapsack problem is defined with intuitionistic fuzzy values of weights and profits of items as well as the knapsack capacity. Here, an index-matrix approach is proposed for its optimal solution by extending the classical dynamic optimization algorithm and intuitionistic fuzzy propositional logic. In the algorithm, three scenarios are proposed to the decision maker for the final choice - pessimistic, optimistic and average. A software implementation to represent the proposed intuitionistic fuzzy algorithm is also developed and its effectiveness is demonstrated on an example.

Keywords: Index matrices · Intuitionistic fuzzy logic · Knapsack problem

The 0-1 Knapsack problem (KP) has been studied trough different approaches [21] and is $NP - hard$. The objective is to optimize the total utility value of all chosen items by the decision-maker to the capacity of knapsack [21]. These problems can model many industrial applications in transportation and maritime shipping, in resourse allocation [1], in investment management [18], capital budgeting, cargo loading, etc. The early works on KP date as far back as 1897 [22]. The name "Knapsack problem" dates back to the early works of George Dantzig [13] in the fifties of the last century. In 1957, Bellman's dynamic programming theory introduced the first algorithms to exactly solve the 0-1 KP [10]. The dynamic programming approach to the KP was investigated by Gilmore and Gomory in the 1966 [16]. In 1967, Kolesar introduced the first branch-and-bound algorithm for the KP [21]. In 1979 Martello and Toth described exact algorithm for the 0-1 KP [21].

The book [21] has developed an algorithmic approach for approximate algorithms, dynamic programming techniques and branch-and-bound algorithms for KP. Different approaches based on greedy method, linear programming relaxation method, dynamic programming method, branch-and-bound method, approximation methods have been

Work is supported by the Asen Zlatarov University through project Ref. No. NIX-449/2021 "Index matrices as a tool for knowledge extraction".

I. Georgiev et al. (Eds.): NMA 2022, LNCS 13858, pp. 314–326, 2023.
https://doi.org/10.1007/978-3-031-32412-3_28

defined in the literature to solve a KP, where the data under consideration is given precisely [18]. But, in real life situation, precision of data is not always guaranteed and the value of weights or prices of items are given imprecisely. In such situations, the KP gets extended to the fuzzy (FKP) or intuitionistic fuzzy KP (IFKP) and in such cases, the fuzzy sets (FSs) [38] and intuitionistic fuzzy sets (IFSs) theories [2] can be used to solve the problem. When the term "fuzzy knapsack" is searched in the title, abstracts, and keywords of articles, the number of the scientific documents that the Scopus found became 140. The term "fuzzy knapsack" participates in the titles of 35 articles, "intuitionistic fuzzy knapsack" participates only in the title of 1 article in the Scopus [15]. The book publications on fuzzy or intuitionistic fuzzy KP are very limited when compared with articles and conference papers - only one chapter of Singh in [27]. A multiple choice KP and its generalisation with fuzzy coefficients are presented in [19,24] by approximate algorithm for solving the problem. In [20], a FKP is presented, where the weights of items are represented by triangular fuzzy numbers. The 0-1 FKP is also discussed by Kasperski and Kulej in [17]. Apart from various exact algorithms, some algorithms based on meta-heuristics, hyper-heuristics and evolutionary optimization have been proposed to solve FKP [12,21,25]. In [11,27], a dynamic programming approach has been given for solving FKP. Bi-objective FKP has been solved by using Dynamic Programming algorithm [28]. Ant Colony Optimization algorithm on Multiple-Constraint Knapsack Problem (MKP) using intuitionistic fuzzy (IF) pheromone updating is presented in [15].

The concept of an index matrix (IM) was introduced in 1987 in [3] to enable two matrices with different dimensions to be summed. This apparatus was used to determine an optimal solution of the intuitionistic fuzzy (IF) transportation problem [31,33,34], IF salesman problem [35] and IF assignment problem [36]. In this study, we present an index-matrix interpretation to the IFKP. Here, an the algorithm for its optimal solution, based on IMs and intuitionistic fuzzy propositional logic concepts, is proposed by extending the classical dynamic optimization algorithm [18,39]. In the algorithm, three scenarios are proposed to the decision maker for the final choice - pessimistic, optimistic and average. A software implementation to represent the proposed intuitionistic fuzzy algorithm is also developed and its effectiveness is demonstrated on an example. The rest of this work contains the following sections: Sect. 1 describes the definitions of IMs and intuitionistic fuzzy propositional logic. In Sect. 2, we describe a type of 0-1 IFKP and software for its implementation, and then apply it to an example. Section 3 marks some conclusions and some aspects for future research.

1 Basic Definitions of the Concepts of Index Matrices and Intuitionist Fuzzy Logic

1.1 Short Remarks on Intuitionistic Fuzzy Pairs (IFPs)

The intuitionistic fuzzy pair has the form as $\langle a, b \rangle = \langle \mu(p), v(p) \rangle$, where $a, b \in [0, 1]$ and $a + b \leq 1$, that is used as an evaluation of a proposition p (see [6,8]). $\mu(p)$ and $v(p)$ respectively determine the "truth degree" (degree of membership) and "falsity degree" (degree of non-membership).

Let us have two IFPs $x = \langle a,b \rangle$ and $y = \langle c,d \rangle$. The following operations and relations are defined in [4,8,9,14,26,29]:

$$\neg x = \langle b,a \rangle; x \wedge_1 y = \langle \min(a,c), \max(b,d) \rangle;$$
$$x \vee_1 y = \langle \max(a,c), \min(b,d) \rangle;$$
$$x \wedge_2 y = x + y = \langle a+c-a.c, b.d \rangle; \qquad (1)$$
$$x \vee_2 y = x.y = \langle a.c, b+d-b.d \rangle;$$
$$x - y = \langle \max(0, a-c), \min(1, b+d, 1-a+c) \rangle.$$

$$x : y = \begin{cases} \langle \min(1, a/c), \min(\max(0, 1-a/c), \\ \max(0, (b-d)/(1-d))) \rangle \text{ if } c \neq 0 \,\&\, d \neq 1 \\ \langle 0,1 \rangle \text{ otherwise} \end{cases} \qquad (2)$$

$$x \geq y \text{ iff } a \geq c \text{ and } b \leq d; \qquad x \geq_\square y \text{ iff } a \geq c;$$
$$x \geq_\diamond y \text{ iff } b \leq d; \qquad x = y \text{ iff } a = c \text{ and } b = d \qquad (3)$$
$$x \geq_R y \qquad \text{ iff } R_{\langle a,b \rangle} \leq R_{\langle c,d \rangle},$$

where $R_{\langle a,b \rangle} = 0.5(2-a-b)(1-a)$ [29].

We say that the IFP x is in α-proximity with IFP y according to the intuitionistic fuzzy (IF) distance measure, proposed in [29]: x is in α-proximity to y, if the IF distance between x and y $d(x,y) = 0.5(|a-b| + |c-d| + |c+d-a-b|) \leq \alpha$, where $\alpha \in [0;1]$.

1.2 Definition, Operations and Relations with Intuitionistic Fuzzy IMs

The concept of index matrices (IMs) was introduced in 1987 in [3]. This theory was extended with different operations, relations and operators over IMs [5,32]. Let $\mathscr{I}$ be a fixed set of indices. By two-dimensional IF index matrix (2-D IFIM) $A = [K, L, \{\langle \mu_{k_i,l_j}, \nu_{k_i,l_j} \rangle\}]$ with index sets K and L $(K, L \subset \mathscr{I})$, we denote the object [5]:

$$A = \begin{array}{c|cccccc}
 & l_1 & \cdots & l_j & \cdots & l_n \\
\hline
k_1 & \langle \mu_{k_1,l_1}, \nu_{k_1,l_1} \rangle & \cdots & \langle \mu_{k_1,l_j}, \nu_{k_1,l_j} \rangle & \cdots & \langle \mu_{k_1,l_n}, \nu_{k_1,l_n} \rangle \\
\vdots & \vdots & \ddots & \vdots & \ddots & \vdots \\
k_m & \langle \mu_{k_m,l_1}, \nu_{k_m,l_1} \rangle & \cdots & \langle \mu_{k_m,l_j}, \nu_{k_m,l_j} \rangle & \cdots & \langle \mu_{k_m,l_n}, \nu_{k_m,l_n} \rangle
\end{array}$$

In [3,5] a lot of operations are defined over the different types of IMs. When the elements of a given IM are real numbers and the operations are the standard ones, we obtain the partial cases of the standard matrices, but there are a lot of operations, relations and operators, defined over the different types of IMs, that do not have analogues in the classical matrix theory. The symbol "$\perp$" is used for lack of component in the definitions. Let us recall some operations over IMs $A = [K, L, \{\langle \mu_{k_i,l_j}, \nu_{k_i,l_j} \rangle\}]$ and $B = [P, Q, \{\langle \rho_{p_r,q_s}, \sigma_{p_r,q_s} \rangle\}]$

Addition-$(\circ, *)$ [5]: $A \oplus_{(\circ, *)} B = [K \cup P, L \cup Q, \{\langle \phi_{t_u,v_w}, \psi_{t_u,v_w} \rangle\}]$, where $\langle \circ, * \rangle \in \{\langle \max, \min \rangle, \langle \min, \max \rangle, \langle \text{ average, average} \rangle\}$.

Termwise Subtraction-(max,min) [5]: $A -_{(\max,\min)} B = A \oplus_{(\max,\min)} \neg B$.

Termwise Multiplication [5]:

$$A \otimes_{(\circ,*)} B = [K \cap P, L \cap Q, \{\langle \phi_{t_u,v_w}, \psi_{t_u,v_w}\rangle\}],$$

where $\langle \phi_{t_u,v_w}, \psi_{t_u,v_w}\rangle = \langle \circ(\mu_{k_i,l_j}, \rho_{p_r,q_s}), *(\nu_{k_i,l_j}, \sigma_{p_r,q_s})\rangle.$

Reduction [5]: The operations $(k,\perp)$-reduction of an IM A is defined by:

$$A_{(k,\perp)} = [K - \{k\}, L, \{c_{t_u,v_w}\}], \text{where}$$

$c_{t_u,v_w} = a_{k_i,l_j} (t_u = k_i \in K - \{k\}, v_w = l_j \in L).$

Projection [5]: Let $M \subseteq K$ and $N \subseteq L$. Then, $pr_{M,N}A = [M, N, \{b_{k_i,l_j}\}]$, where for each $k_i \in M$ and each $l_j \in N$, $b_{k_i,l_j} = a_{k_i,l_j}.$

Substitution [5]:

$$\left[\frac{p}{k}; \perp\right] A = \left[(K - \{k\}) \cup \{p\}, L, \{a_{k,l}\}\right],$$

Internal subtraction of IMs' Components [30]: $IO_{-_{(\max,\min)}}(\langle k_i, l_j, A\rangle, \langle p_r, q_s, B\rangle) = [K, L, \{\langle \gamma_{t_u,v_w}, \delta_{t_u,v_w}\rangle\}].$

Index Type Operations [30,32]:

$$AGIndex_{(\max_R),(\measuredangle)}(A) = \langle k_i, l_j\rangle, \tag{4}$$

where $\langle k_i, l_j\rangle$ (for $1 \leq i \leq m, 1 \leq j \leq n$) is the index of the maximum IFP of A in the sense of the relation (3) that has no empty value.

$$Index_{(\max_R),k_i}(A) = \{\langle k_i, l_{v_1}\rangle, \ldots, \langle k_i, l_{v_x}\rangle, \ldots, \langle k_i, l_{v_V}\rangle\}, \tag{5}$$

where $\langle k_i, l_{v_x}\rangle$ (for $1 \leq i \leq m, 1 \leq x \leq V$) are the indices of the maximum element of k_i-th row of A.

$$Index_{(\max_R),l_j}(A) = \{\langle k_{w_1}, l_j\rangle, \ldots, \langle k_{w_y}, l_j\rangle, \ldots, \langle k_{w_W}, l_j\rangle\}, \tag{6}$$

where $\langle k_{w_y}, l_j\rangle$ (for $1 \leq y \leq W, 1 \leq j \leq n$) are the indices of the maximum IFFP of l_j-th column of A.

Transposition [5]: A' is the transposed IM of A.

Aggregation Operations Let us use the operations $\#_q, (q \leq i \leq 3)$ from [37] for scaling aggregation operations over two IFPs $x = \langle a, b\rangle$ and $y = \langle c, d\rangle$:
$x\#_1 y = \langle \min(a,c), \max(b,d)\rangle$; $x\#_2 y = \langle average(a,c), average(b,d)\rangle$;
$x\#_3 y = \langle \max(a,c), \min(b,d)\rangle.$

Let $k_0 \notin K$ will be fixed index. The definition of the aggregation operation by the dimension K is [5,37]:

$$\alpha_{K,\#_q}(A, k_0) = \quad \begin{array}{c|ccc} h_g \in H & l_1 & \cdots & l_n \\ \hline k_0 & \displaystyle\#_q_{i=1}^{m} \langle \mu_{k_i,l_1,h_g}, \nu_{k_i,l_1,h_g}\rangle & \cdots & \displaystyle\#_q_{i=1}^{m} \langle \mu_{k_i,l_n,h_g}, \nu_{k_i,l_n,h_g}\rangle \end{array},$$

where $1 \leq q \leq 3$.

Aggregate Global Internal Operation [30]: $AGIO_{\oplus_{(\#_q)}}(A)$.

If $q = 1$ then we obtain pessimistic scenario and this operation finds the pessimistic value of the addition of all elements of A, if $q = 2$ then we obtain averaged scenario and the operation finds the averaged value of the addition of all elements of A and if $q = 3$ - the operation finds the optimistic value of the addition of all elements of A. The decision maker chooses which forecast to prefer.

Operation "Purge" of IM A

Here, we define a new operation "Purge" by a dimension K as follows:

$$Purge_K(A) \tag{7}$$

reduces each k_x-th row of A, if $a_{k_x,l_j} \leq a_{k_y,l_j}$, but $a_{k_x,l_e} \geq a_{k_y,l_e}$ for $1 \leq x \leq m, 1 \leq y \leq m$, $1 \leq j \leq n$ and $1 \leq e \leq n$.

The definition of the operation "Purge" by dimension L is similar:
$Purge_L(A)$ reduces each l_j-th column of A, if $a_{k_x,l_j} \leq a_{k_x,l_e}$, but $a_{k_y,l_j} \geq a_{k_y,l_e}$ for $1 \leq x \leq m, 1 \leq y \leq m, 1 \leq j \leq n$ and $1 \leq e \leq n$.

The Non-strict Relation "Inclusion About Value"

The form of this type of relations between two IMs A and B is as follows [5]:
$A \subseteq_v B$ iff $(K = P)\&(L = Q)\&(\forall k \in K)(\forall l \in L)(a_{k,l} \leq b_{k,l})$.

2   Index-Matrix Interpretation of a 0-1 Intuitionistic Fuzzy KP

Here, we solve a type of two-dimensional 0-1 intuitionistic fuzzy KP by the IMs concept (IFKP). Its formulation is:

A transportation company supplies products to a costumer. The global pandemic situation and rising inflation caused by the pandemic are leading to unclear and rapidly changing parameters. m products $\{k_1, \ldots, k_i, \ldots, k_m\}$ are given, each having an associated weight $a_{k_i,w}$ (for $i = 1, \ldots, m$) and profit $p_{k_i,w}$ (for $i = 1, \ldots, m$) under the form of IFPs. The company's vehicle has IF capacity $C = \langle \rho, \sigma \rangle$. The purpose of the problem is to select a subset of items such that the IF profit of the selected items is optimal following three decision-making scenarios - optimistic (with a maximum degree of membership), averaged (with membership degrees averaged) and pessimistic (with a minimum degree of membership), and the subset need to be packed into the given vehicle with a capacity at least equal to the total weight of the items in the subset.

The mathematical model of the above problem is as follows:

$$\text{maximize } \sum_{i=1}^{m} p_{k_i,w} x_{k_i,w}$$
$$\text{Subject to: } \sum_{i=1}^{m} x_{k_i,w} \leq W, \qquad i = 1, 2, \ldots, m; \tag{8}$$

$$x_{k_i,w} \in \begin{cases} \langle 1, 0 \rangle, & \text{if the product } k_i \text{ is selected} \\ \langle 0, 1 \rangle & \text{otherwise} \end{cases}$$

Note: The operations "addition" and "multiplication", used in the problem (8) are those for IFPs, defined in Sect. 2.1.

In this section we extend the classical dynamic programming approach from [39] to the 0-1 IFKP using the IMs and IFPs concepts.

2.1 Format of Calculations of Index Matrix Approach to the 0-1 KP, Based on Dynamic Programming Approach

The dynamic programming to the KP is described in [10, 18, 21]. This approach can be performed in running time of $O(mW)$ [18] and the space requirement is the same.

Here, we extend the dynamic programming approach from [39], such that it can be applied over IFKP (9) from the point of view of the theory of IMs.

Step 1. Let us construct the following IMs, in accordance with the problem (9):

$$A[K, L, \{a_{k_i, p(w)}\}] = \begin{array}{c|cc} & p & w \\ \hline k_1 & \langle \mu_{k_1, p}, v_{k_1, p} \rangle & \langle \mu_{k_1, w}, v_{k_1, w} \rangle \\ \vdots & \vdots & \vdots \\ k_m & \langle \mu_{k_m, p}, v_{k_m, p} \rangle & \langle \mu_{k_m, w}, v_{k_m, w} \rangle \end{array}, \tag{9}$$

where $K = \{k_1, \ldots, k_i, \ldots, k_m\}, i = 1, \ldots, m;$ $L = \{p, w\}$ with the IF elements. $\{a_{k_i, p}, a_{k_i, w}\}$ are respectively IF profit and weight of k_i-th object. The data values for profit and weight of k_i-th object are transformed into IFPs as demonstrated in [35]. Let us have the set of intervals $[i_1, i_I]$ for $1 \le i \le m$ and let

$$A_{min,i} = \min_{i_1 \le j \le i_I} x_{i,j} < \max_{i_1 \le j \le i_I} x_{i,j} = A_{max,i}.$$

For the interval $[i_1, i_I]$ we construct IFPs [4] as follows:

$$\mu_{i,j} = \frac{x_{i,j} - A_{min,i}}{A_{max,i} - A_{min,i}}, v_{i,j} = \frac{A_{max,i} - x_{i,j}}{A_{max,i} - A_{min,i}}.$$

The conditions $0 \le \mu_{i,j}, v_{i,j} \le 1, \le 0 \le \mu_{i,j} + v_{i,j} \le 1$ are satisfied. We can use the expert approach described in detail in [4] to convert the data to IFPs. The IFP $\{a_{k_i, p}\}$ (or $\{a_{k_i, w}\}$) presents the degrees of perception (the positive evaluation of an expert for the profit (or weight) of k_i-th object divided by the (maximum-minimum) evaluation of profit (or weight) and non-perception (the negative evaluation of the of the expert for the k_i-th object by the its profit (or weight) divided by the (maximum-minimum) evaluation) for the profit (or weight) of k_i-th object. The hesitation degree $\pi_{k_i, p} = 1 - \mu_{k_i, p} - v_{k_i, p}$ $(\pi_{k_i, w} = 1 - \mu_{k_i, w} - v_{k_i, w})$ corresponds to the uncertain evaluation of the expert for the profit (weight) of the k_i-th object. At this step, a check is made on the input data for the weight of the objects for not exceeding the capacity of the knapsack C.

for $i = 1$ to m
{ If $a_{k_i, w} > C$ then $A_{(k_i, \perp)}$ }

Let we denote by $|K| = m$ the number of elements of the set K, then $|L| = 2$. We also define

$$X[K, L] = \begin{array}{c|cc} & p & w \\ \hline k_1 & x_{k_1, p} & x_{k_1, w} \\ \vdots & \vdots & \vdots \\ k_m & x_{k_m, p} & x_{k_m, w} \end{array}, \tag{10}$$

and for $1 \le i \le m$: $\{x_{k_i, w}, x_{k_i, p}\} \in \begin{cases} \langle 1, 0 \rangle, & \text{if the product } k_i \text{ is selected} \\ \langle 0, 1 \rangle & \text{otherwise} \end{cases}$

Let in the beginning of the algorithm, all elements of IM X are equal to $\langle 0,1 \rangle$.

Construct IM $S^0[u_0, L] = \dfrac{\begin{array}{c|cc} & p & w \\ \hline u_0 & s^0_{u_0,p} & s^0_{u_0,w} \end{array}}{} = \dfrac{\begin{array}{c|cc} & p & w \\ \hline u_0 & \langle 0,0 \rangle & \langle 0,0 \rangle \end{array}}{}$

Step 2. for $i = 1$ to m do

{Create IMs $R_i[k_i, L] = pr_{k_i, L}A$;

$SH_1^{i-1} = \left[\dfrac{u_i}{u_{i-1}}; \perp \right] S^{i-1}$

for $h = 1$ to $i + 1$ do

$\{ SH_1^{i-1} = SH_1^{i-1} \oplus_{(\max,\min)} \left[\dfrac{u_h}{k_i}; \perp \right] R_i \}$

$S^i[U^i, L] = S^{i-1} \oplus_{(\max,\min)} SH_1^{i-1}$;

for $h = 1$ to $i + 1$ do

{ *Check the conditions for the capacity of the knapsack*

If $s^i_{h,w} > C$ then $S^i_{(h,\perp)}$ }

The "Purge" operation is running by $S^i = Purge_{U^i} S^i$

}

Go to *Step 3*.

Step 3. At this step, the index of the largest IF profit is found by

$\{ Index_{(\max_R),p}(A) = \langle u_g, p \rangle$

for $i = m$ to 1 do

{Find the α-nearest elements of $s^i_{u_g,p}$ (or $s^i_{u_g,w}$) ($\alpha = 0.5$) of S^i and choose the closest element from them - $s^i_{u_{g*},p}$ (or $s^i_{u_{g*},w}$).

If $\{ s^i_{u_{g*},p} $ (or $s^i_{u_{g*},w}) \} \in S^i$ and $\{ s^i_{u_{g*},p} $ (or $s^i_{u_{g*},w}) \} \notin S^{i-1}$ then

$\{ x_{k_i,p} = \langle 1,0 \rangle$ and $x_{k_i,w} = \langle 1,0 \rangle$

$s^i_{u_g,p} = s^i_{u_{g*},p} - a_{k_i,p}; \ s^i_{u_g,w} = s^i_{u_{g*},w} - a_{k_i,w} \}$

}

Go to *Step 4*.

Step 4. The optimal profit is: $AGIO_{\oplus_{(\#q)}} \left(pr_{K,p}A \otimes_{(\min,\max)} pr_{K,p}X \right)$.

If $q = 1$ then we obtain pessimistic scenario and the operation finds the pessimistic value of the addition of all elements of $pr_{K,p}A \otimes_{(\min,\max)} pr_{K,p}X$, if $q = 2$ - the operation finds the averaged value of the addition of all elements of $pr_{K,p}A \otimes_{(\min,\max)} pr_{K,p}X$ and if $q = 3$ then we obtain the optimistic value of the addition of all elements of $pr_{K,p}A \otimes_{(\min,\max)} pr_{K,p}X$. The decision maker chooses which forecast to prefer.

The optimal weight is:

$$AGIO_{\oplus_{(\#q)})} \left(pr_{K,w}A \otimes_{(\min,\max)} pr_{K,w}X \right).$$

The time complexity of the proposed algorithm is consistent with the standard dynamic programming algorithm [21, 39] - $O(m.C)$.

2.2 A Software Program for Optimal IFKP

In order to apply IFKP algorithm on actual data more easily, we have developed a command line utility, written in C++. For this aim, we have used an IM structure containing std::pair types from STL and using that structure the basic IM operations [23] are implemented. The program takes as input the IFPs for the items and the knapsack capacity For example, this is how it would be used for a matrix with IF pairs (see Fig. 1):

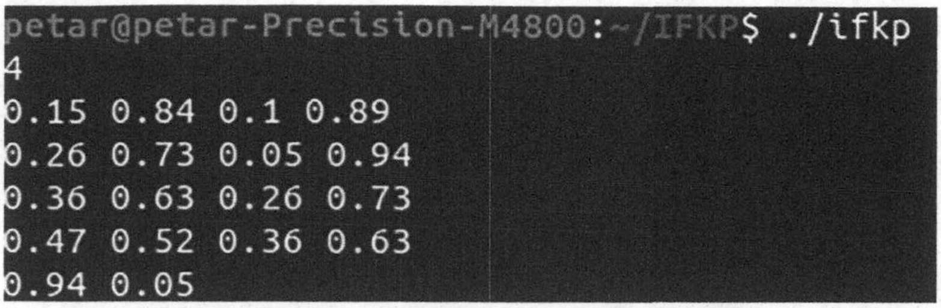

```
petar@petar-Precision-M4800:~/IFKP$ ./ifkp
4
0.15 0.84 0.1 0.89
0.26 0.73 0.05 0.94
0.36 0.63 0.26 0.73
0.47 0.52 0.36 0.63
0.94 0.05
```

Fig. 1. An input IFIM for the IF knapsack problem.

Once the program has completed, a detailed output of each iteration of the algorithm is shown and a solution is suggested on the Fig. 2:

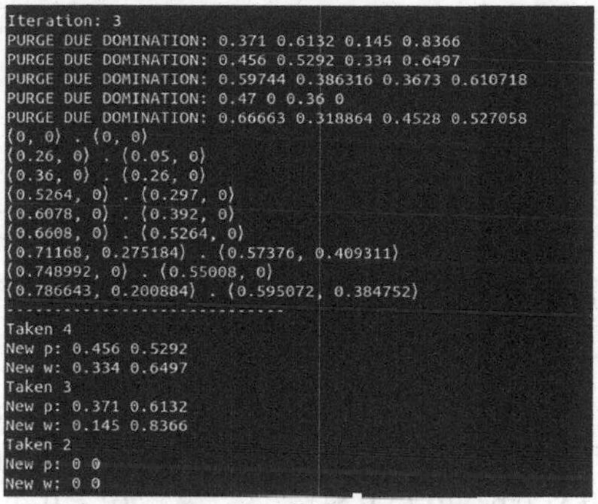

```
Iteration: 3
PURGE DUE DOMINATION: 0.371 0.6132 0.145 0.8366
PURGE DUE DOMINATION: 0.456 0.5292 0.334 0.6497
PURGE DUE DOMINATION: 0.59744 0.386316 0.3673 0.610718
PURGE DUE DOMINATION: 0.47 0 0.36 0
PURGE DUE DOMINATION: 0.66663 0.318864 0.4528 0.527058
(0, 0) . (0, 0)
(0.26, 0) . (0.05, 0)
(0.36, 0) . (0.26, 0)
(0.5264, 0) . (0.297, 0)
(0.6078, 0) . (0.392, 0)
(0.6608, 0) . (0.5264, 0)
(0.71168, 0.275184) . (0.57376, 0.409311)
(0.748992, 0) . (0.55008, 0)
(0.786643, 0.200884) . (0.595072, 0.384752)
--------------------------------
Taken 4
New p: 0.456 0.5292
New w: 0.334 0.6497
Taken 3
New p: 0.371 0.6132
New w: 0.145 0.8366
Taken 2
New p: 0 0
New w: 0 0
```

Fig. 2. The output for the IF knapsack problem.

2.3   A Case Study

In this section, the proposed IFKP approach is applied, using the "IFKP" software utility, to an example to prove its effectiveness .

Let us consider the following problem as an application of the algorithm, presented in the Sect. 2: A transportation company supplies products to a costumer. The global pandemic situation and rising inflation caused by the pandemic are leading to unclear and rapidly changing parameters. 4 products $\{k_1, k_2, k_3, k_4\}$ are given, each having an associated weight $a_{k_i,w}$ (for $i = 1, ..., 4$) and profit $p_{k_i,w}$ (for $i = 1, ..., 4$) under the form of IFPs. The company's vehicle has capacity $C = \langle 0.94, 0.05 \rangle$. The purpose of the problem is to select a subset of items such that the IF profit of the selected items is optimal following three decision-making scenarios - optimistic (with a maximum degree of membership), averaged (with membership degrees averaged) and pessimistic (with a minimum degree of membership), and the subset need to be packed into the given vehicle with a capacity at least equal to the total weight of the items in the subset.

Solution of the problem:

Step 1. Let us construct the following IMs, in accordance with the problem:

$$A[K,L] = \begin{array}{c|cc} & p & w \\ \hline k_1 & \langle 0.15,0.84\rangle & \langle 0.1,0.89\rangle \\ k_2 & \langle 0.26,0.73\rangle & \langle 0.05,0.94\rangle \\ k_3 & \langle 0.36,0.63\rangle & \langle 0.26,0.73\rangle \\ k_4 & \langle 0.47,0.52\rangle & \langle 0.36,0.63\rangle \end{array}, \tag{11}$$

where $K = \{k_1,k_2,k_3,k_4\}$, $L = \{p,w\}$ and $\{a_{k_i,p}, a_{k_i,w}\}$ are respectively IF profit and weight of k_i-th object.

We also define

$$X[K,L] = \begin{array}{c|cc} & p & w \\ \hline k_1 & \langle 0,1\rangle & \langle 0,1\rangle \\ \vdots & \vdots & \vdots \\ k_4 & \langle 0,1\rangle & \langle 0,1\rangle \end{array}, \tag{12}$$

Construct IM $S^0[u_0,L] = \begin{array}{c|cc} & p & w \\ \hline u_0 & s^0_{u_0,p} & s^0_{u_0,w} \end{array} = \begin{array}{c|cc} & p & w \\ \hline u_0 & \langle 0,0\rangle & \langle 0,0\rangle \end{array}$

Step 2. The algorithm calculates sequentially the following IMs:

$$S^0_1[u_1,L] = \begin{array}{c|cc} & p & w \\ \hline u_1 & \langle 0.15,0.84\rangle & \langle 0.1,0.89\rangle \end{array}; S^1[U_1,L] = \begin{array}{c|cc} & p & w \\ \hline u_0 & \langle 0,0\rangle & \langle 0,0\rangle \\ u_1 & \langle 0.15,0.84\rangle & \langle 0.1,0.89\rangle \end{array}$$

$$S^1_1[U*_1,L] = \begin{array}{c|cc} & p & w \\ \hline u_2 & \langle 0.26,0\rangle & \langle 0.05,0\rangle \\ u_3 & \langle 0.37,0.62\rangle & \langle 0.15,0.84\rangle \end{array}; S^2[U_2,L] = \begin{array}{c|cc} & p & w \\ \hline u_0 & \langle 0,0\rangle & \langle 0,0\rangle \\ u_1 & \langle 0.15,0.84\rangle & \langle 0.1,0.89\rangle \\ u_2 & \langle 0.26,0\rangle & \langle 0.05,0\rangle \\ u_3 & \langle 0.37,0.613\rangle & \langle 0.15,0.84\rangle \end{array}$$

The operation "Purge" reduces the u_1-th row of $S^2[U_2,L]$ (PURGE DUE DOMINATION: 0.15 0.84 0.1 0.89) as follows:

$$S^2[U_2,L] = \begin{array}{c|cc} & p & w \\ \hline u_0 & \langle 0,0\rangle & \langle 0,0\rangle \\ u_2 & \langle 0.26,0\rangle & \langle 0.05,0\rangle \\ u_3 & \langle 0.37,0.613\rangle & \langle 0.15,0.84\rangle \end{array}$$

Then IMs is created:

$$S^2_1[U*_2,L] = \begin{array}{c|cc} & p & w \\ \hline u_4 & \langle 0.36,0\rangle & \langle 0.26,0\rangle \\ u_5 & \langle 0.53,0\rangle & \langle 0.30,0\rangle \\ u_6 & \langle 0.6,0.39\rangle & \langle 0.37,0.62\rangle \end{array}; S^3[U_3,L] = \begin{array}{c|cc} & p & w \\ \hline u_0 & \langle 0,0\rangle & \langle 0,0\rangle \\ u_2 & \langle 0.26,0\rangle & \langle 0.05,0\rangle \\ u_3 & \langle 0.37,0.613\rangle & \langle 0.15,0.84\rangle \\ u_4 & \langle 0.36,0\rangle & \langle 0.26,0\rangle \\ u_5 & \langle 0.46,0.53\rangle & \langle 0.33,0.65\rangle \\ u_6 & \langle 0.53,0\rangle & \langle 0.3,0.61\rangle \end{array}$$

$$S_1^3[U*_3, L] = \begin{array}{c|cc} & p & w \\ \hline u_7 & \langle 0.47, 0 \rangle & \langle 0.36, 0 \rangle \\ u_8 & \langle 0.61, 0 \rangle & \langle 0.4, 0 \rangle \\ u_9 & \langle 0.67, 0.32 \rangle & \langle 0.46, 0.53 \rangle \\ u_{10} & \langle 0.66, 0 \rangle & \langle 0.53, 0 \rangle \\ u_{11} & \langle 0.75, 0 \rangle & \langle 0.55, 0 \rangle \\ u_{12} & \langle 0.79, 0.2 \rangle & \langle 0.6, 0.39 \rangle \end{array};$$

The final IM after the operation "Purge" is the following":

$$S^4[U_4, L] = \begin{array}{c|cc} & p & w \\ \hline u_0 & \langle 0, 0 \rangle & \langle 0, 0 \rangle \\ u_4 & \langle 0.36, 0 \rangle & \langle 0.26, 0 \rangle \\ u_5 & \langle 0.53, 0 \rangle & \langle 0.30, 0 \rangle \\ u_6 & \langle 0.6, 0.39 \rangle & \langle 0.37, 0.62 \rangle \\ u_8 & \langle 0.61, 0 \rangle & \langle 0.4, 0 \rangle \\ u_{10} & \langle 0.71, 0, 28 \rangle & \langle 0.57, 0.41 \rangle \\ u_{11} & \langle 0.75, 0 \rangle & \langle 0.55, 0 \rangle \\ u_{12} & \langle 0.79, 0.2 \rangle & \langle 0.6, 0.39 \rangle \end{array}$$

Step 3. At this step index of the largest IF profit is found by

$$Index_{(\max_R), p}(A) = \langle u_{12}, p \rangle$$

The IFPs $\langle 0.79, 0.2 \rangle$ (or $\langle 0.6, 0.39 \rangle$) $\in S^4$, but they $\notin S^4$, therefore $x_{k_4, p} = \langle 1, 0 \rangle$ and $x_{k_4, w} = \langle 1, 0 \rangle$. Then $p* = \langle 0.79, 0.2 \rangle - \langle 0.47, 0.52 \rangle = \langle 0.33, 0.65 \rangle$ and $w* = \langle 0.6, 0.39 \rangle - \langle 0.36, 0.63 \rangle = \langle 0.33, 0.65 \rangle$.

Find the the 0.5-nearest elements of $p*$ and $w*$ in S^3 - $\langle 0.36, 0 \rangle$ and $\langle 0.26, 0 \rangle$. These elements do not belong to S^2, therefore $x_{k_3, p} = \langle 1, 0 \rangle$ and $x_{k_3, w} = \langle 1, 0 \rangle$. Then $p* = \langle 0.36, 0 \rangle - \langle 0.36, 0.63 \rangle = \langle 0.37, 0.61$ and $w* = \langle 0.26, 0 \rangle - \langle 0.26, 0.73 \rangle = \langle 0.15, 0.84 \rangle$.

Then the algorithm finds the 0.5-nearest elements of $p*$ and $w*$ in S^2 - $\langle 0, 0 \rangle$ and $\langle 0, 0 \rangle$. These elements do not belong to S^1, therefore $x_{k_2, p} = \langle 0, 1 \rangle$ and $x_{k_2, w} = \langle 0, 1 \rangle$. Analogously $x_{k_1, p} = \langle 0, 1 \rangle$ and $x_{k_1, w} = \langle 0, 1 \rangle$.

Step 4. The optimal IF profit is:

$$AGIO_{\oplus_{(\#_q)}} \left(pr_{K, p} A \otimes_{(\min, \max)} pr_{K, p} X \right). \tag{13}$$

In an optimistic scenario, the optimal IF profit has degree of membership 0.86 and degree of non-membership 0.13, forming the IFP $\langle 0.86, 0.13 \rangle$. In a pessimistic scenario, the optimal IF profit is $\langle 0.36, 0.63 \rangle$. The degree of acceptance of the solution is equal to 0.36 and the its degree of non-acceptance is equal to 0.63. The degree of acceptance of profit in an averaged scenario is equal to 0.57 and the degree of non-acceptance is equal to 0.41. For the obtained optimistic optimal solution of IFKP, the distance between the optimal solution to the pair $\langle 1, 0 \rangle$ is equal to $R_{\langle 0.5; 0.2 \rangle} = 0.07$. The example illustrates the reliability of the proposed software in Sect. 2.3 to the studied IFKP.

3 Conclusion

In this work it is proposed for the first time to extend the classical dynamic approach to the KP [21, 39] such that it can be applied for finding an optimal solution of a type of 0-1 IFKP using the concepts of the IMs and IFSs. The index matrix prism for solution of IFKP can be applied to problems with imprecise parameters and can be extended in order to obtain the optimal solution for other types of n-dimensional KPs by using n-dimensional index matrices for storage and analysis of data [7]. The developed software, which implements this IFKP approach, is applied to a numerical example. In the future, we will extend IFKP approach to the n-dimensional intuitionistic fuzzy KPs [7] and will apply the proposed approach for various types n-dimensional IFKPs.

References

1. Aisopos, F., Tserpes, K., Varvarigou, T.: Resource management in software as a service using the knapsack problem model. Int. J. Product. Econ. **141**(2), 465–477 (2013)
2. Atanassov K. T.: Intuitionistic Fuzzy Sets, VII ITKR Session, Sofia, 20–23 June 1983 (Deposed in Centr. Sci.-Techn. Library of the Bulg. Acad. of Sci., 1697/84) (in Bulgarian). Reprinted: Int. J. Bioautomation **20**(S1), S1–S6 (2016)
3. Atanassov, K.: Generalized index matrices. Comptes rendus de l'Academie Bulgare des Sci. **40**(11), 15–18 (1987)
4. Atanassov, K.: On Intuitionistic Fuzzy Sets Theory. STUDFUZZ, vol. 283. Springer, Heidelberg (2012). https://doi.org/10.1007/978-3-642-29127-2
5. Atanassov, K.: Index Matrices: towards an Augmented Matrix Calculus. Studies in Computational Intelligence, vol. 573. Springer, Cham (2014). https://doi.org/10.1007/978-3-319-10945-9
6. Atanassov, K.: Intuitionistic Fuzzy Logics. Studies in Fuzziness and Soft Computing, vol. 351. Springer, Cham (2017). https://doi.org/10.1007/978-3-319-48953-7
7. Atanassov, K.: n-Dimensional extended index matrices Part 1. Adv. Stud. Contemp. Math. **28**(2), 245–259 (2018)
8. Atanassov, K., Szmidt, E., Kacprzyk, J.: On intuitionistic fuzzy pairs. Notes Intuition. Fuzzy Sets **19**(3), 1–13 (2013)
9. Atanassov, K.: Remark on an intuitionistic fuzzy operation "division". Issues in IFSs and GNs **14**, 113–116 (2018–2019)
10. Bellman, R.: Dynamic Programming. Princeton University Press (1957)
11. Chakraborty, D., Singh, V.: On solving fuzzy knapsack problem by multistage decision making using dynamic programming (2014)
12. Changdar, C., Mahapatra, G., Pal, R.K.: An improved genetic algorithm based approach to solve constrained knapsack problem in fuzzy environment. Expert Syst. Appl. **42**(4), 2276–2286 (2015)
13. Dantzig, G.: Linear programming and extensions. Princeton University Press (1963)
14. De, S.K., Bisvas, R., Roy, R.: Some operations on IFSs. Fuzzy Sets Syst. **114**(4), 477–484 (2000)
15. Fidanova, S., Atanassov, K.: ACO with intuitionistic fuzzy pheromone updating applied on multiple-constraint knapsack problem. Mathematics **9**, 1456 (2021). https://doi.org/10.3390/math9131456
16. Gilmore, P., Gomory, R.: The theory and computation of knapsack functions. Oper. Res. **14**, 1045–1074 (1966)

17. Kasperski, A., Kulej, M.: The 0–1 knapsack problem with fuzzy data. Fuzzy Optim. Decis. Making **6**(2), 163–172 (2007)
18. Kellerer, H., Pferschy, U., Pisinger, D.: Knapsack problems. Springer, Berlin (2004). https://doi.org/10.1007/978-3-540-24777-7
19. Kuchta, D.: A generalisation of an algorithm solving the fuzzy multiple choice knapsack problem. Fuzzy Sets Syst. **127**(2), 131–140 (2002)
20. Lin, F., Yao, J.-S.: Using fuzzy numbers in knapsack problems. Eur. J. Oper. Res. **135**(1), 158–176 (2001)
21. Martello, S., Toth, P.: Knapsack problems. Algorithms and computer implementations. John Wiley & sons (1990)
22. Mathews, G.B.: On the partition of numbers. Proc. Lond. Math. Soc. **28**, 486–490 (1987). https://doi.org/10.1112/plms/s1-28.1.486
23. Mavrov, D.: An application for performing operations on two-dimensional index matrices, annual of "informatics" section. Union Scient. Bulgaria **10**, 66–80 (2019)
24. Okada, S., Gen, M.: Fuzzy multiple choice knapsack problem. Fuzzy Sets Syt. **67**, 71–80 (1994)
25. Olivas, F., Amaya, I., Ortiz-Bayliss, J.C., Conant-Pablos, S.E., Terashima-Marın, H.: Enhancing hyperheuristics for the knapsack problem through fuzzy logic. Computational Intelligence and Neuroscience (2021)
26. Riecan, B., Atanassov, A.: Operation division by n over intuitionistic fuzzy sets. NIFS **16**(4), 1–4 (2010)
27. Singh, V.: An approach to solve fuzzy knapsack problem in investment and business model. In: Nogalski, B., Szpitter, A., Jaboski, A., Jaboski, M. (eds.). Networked Business Models in the Circular Economy (2020). https://doi.org/10.4018/978-1-5225-7850-5.ch007
28. Singh, V.P., Chakraborty, D.: A dynamic programming algorithm for solving bi-objective fuzzy knapsack problem. In: Mohapatra, R.N., Chowdhury, D.R., Giri, D. (eds.) Mathematics and Computing. SPMS, vol. 139, pp. 289–306. Springer, New Delhi (2015). https://doi.org/10.1007/978-81-322-2452-5_20
29. Szmidt, E., Kacprzyk, J.: Amount of information and its reliability in the ranking of Atanassov's intuitionistic fuzzy alternatives. In: Rakus-Andersson, E., Yager, R., Ichalkaranje, N., Jain, L.C. (eds.). Recent Advances in Decision Making, SCI, vol. 222, pp. 7–19. Springer, Heidelberg (2009). https://doi.org/10.1007/978-3-642-02187-9_2
30. Traneva, V.: Internal operations over 3-dimensional extended index matrices. Proceed. Jangjeon Math. Soc. **18**(4), 547–569 (2015)
31. Traneva, V., Marinov, P., Atanassov, K.: Index matrix interpretations of a new transportation-type problem. Comptes rendus de l'Academie Bulgare des Sci. **69**(10), 1275–1283 (2016)
32. Traneva, V., Tranev, S.: Index Matrices as a Tool for Managerial Decision Making. Publ, House of the Union of Scientists, Bulgaria (2017)
33. Traneva, V., Tranev, S.: Intuitionistic fuzzy transportation problem by zero point method. In: Proceedings of the 15th Conference on Computer Science and Information Systems (FedCSIS), pp. 345–348 (2020). https://doi.org/10.15439/2020F6
34. Traneva, V., Tranev, S.: An intuitionistic fuzzy zero suffix method for solving the transportation problem. In: Dimov, I., Fidanova, S. (eds.) HPC 2019. SCI, vol. 902, pp. 73–87. Springer, Cham (2021). https://doi.org/10.1007/978-3-030-55347-0_7
35. Traneva, V., Tranev, S.: An intuitionistic fuzzy approach to the travelling salesman problem. In: Lirkov, I., Margenov, S. (eds.) LSSC 2019. LNCS, vol. 11958, pp. 530–539. Springer, Cham (2020). https://doi.org/10.1007/978-3-030-41032-2_61
36. Traneva, V., Tranev, S., Atanassova, V.: An intuitionistic fuzzy approach to the Hungarian algorithm. In: Nikolov, G., Kolkovska, N., Georgiev, K. (eds.) NMA 2018. LNCS, vol. 11189, pp. 167–175. Springer, Cham (2019). https://doi.org/10.1007/978-3-030-10692-8_19

37. Traneva, V., Tranev, S., Stoenchev, M., Atanassov, K.: Scaled aggregation operations over two- and three-dimensional index matrices. Soft. Comput. **22**, 5115–5120 (2019)
38. Zadeh, L.: Fuzzy Sets. Inf. Control **8**(3), 338–353 (1965)
39. Knapsack problem using dynamic programming. https://codecrucks.com/knapsack-problem-using-dynamic-programming/. Accessed 18 Sept 2022

Numerical Solutions of the Boussinesq Equation with Nonlinear Restoring Force

Veselina Vucheva[1]([✉]), Vassil M. Vassilev[2], and Natalia Kolkovska[1]

[1] Institute of Mathematics and Informatics, Bulgarian Academy of Sciences,
Sofia, Bulgaria
{vucheva,natali}@math.bas.bg
[2] Institute of Mechanics, Bulgarian Academy of Sciences, Sofia, Bulgaria
vasilvas@imbm.bas.bg

Abstract. In this work we consider the Boussinesq equation with non-linear restoring force. Equations of this type model the transverse or longitudinal vibration of an elastic rod subject to a constant tangential follower force and laying on a nonlinear elastic foundation due to which a cubic term appears in addition to the linear terms corresponding to a purely Winkler-Pasternak foundation. The dynamical behavior of such mechanical systems is not well-studied in the current literature. Here, we give exact solitary wave solutions (solitons) of the regarded nonlinear equation in explicit analytic form. We propose and study finite difference schemes to solve the considered problem. The nonlinear terms are approximated in two different ways. Both schemes have second order of approximation in space and time steps. The extensive numerical experiments show second order of convergence for single solitary wave and the interaction between two solitary waves.

Keywords: Boussinesq equation · Numerical and analytical
solutions · Solitons

1 Introduction

We consider the following family of equations

$$u_{tt} - \beta u_{xx} - u_{ttxx} + u_{xxxx} - f(u)_{xx} + k_1 u + k_3 u^3 = 0, \qquad (1)$$

in two independent variables x and t, and one sought function $u(x,t)$. Here and in what follows, the subscripts denote derivatives with respect to the indicated variables, $f(u) = \alpha u^2$, where α, β, k_1 and k_3 are real constants. Equations of this type describe the dynamical behavior of various mechanical systems. Of this form is, for instance, the so-called "Boussinesq paradigm equation" (cf. [3]) governing the one-dimensional quasi-stationary flow of inviscid liquid in a shallow layer with free surface. Equations of this family also appears in the study of transverse and longitudinal vibrations of elastic rods laying on a foundation of Winkler-Pasternak type (see, e.g., [2,4,6–8]) due to which they are subject to a linear restoring force. The novelty in the present work is that we, following Birman [1] (see also [9]), take into account the influence of a nonlinear elastic foundation characterized by an additional cubic relation between the displacement and the foundation reaction. Let us remark that in this context $u(t, x)$ is interpreted as the transverse displacement function, x and t being the spatial coordinate and the time, respectively.

The Boussinesq paradigm equation, i.e. Eq. (1) with $k_1 = 0$ and $k_3 = 0$, subject to various boundary and initial conditions, is solved numerically by different methods, such as finite element method, finite difference method, spectral and pseudo-spectral methods. There is a vast available literature on the topic (see, e.g., [3,5] and the references therein). According to our best knowledge, the solution of (1) with $k_1 \neq 0$ and $k_3 \neq 0$ is not theoretically or numerically studied. The main difficulty here is the presence of two non-linear terms of different kind – the first one is a single polynomial, while the second polynomial term is differentiated twice.

In this paper, for particular values of the input parameters k_1, k_3, α and β we derive exact solitary wave solutions to (1) obeying the asymptotic boundary conditions

$$u(x,t) \to 0, \quad u_x(x,t) \to 0, \quad u_{xx}(x,t) \to 0, \quad \ldots \quad \text{as} \quad |x| \to \infty. \quad (2)$$

We obtain a functional, which is conserved in time on the smooth solutions of (1) with $f(u) = 0$ that satisfy the boundary conditions (2). We propose and study an explicit and an implicit finite difference scheme (FDS) for the numerical solution of (1) obeying given initial data and boundary conditions (2). Both schemes have second order of approximation with respect to space and time steps. In the particular case $f(u) = 0$ in (1) we develop a discrete functional, which is preserved exactly in time for the solution of the implicit scheme. The performed numerical experiments show second order of convergence of the numerical solution to the exact one for the obtained in this paper exact solutions.

2   Exact Solitary Wave Solutions

Evidently, the traveling wave solutions $u(x,t) = \varphi(\xi)$, $\xi = x - ct$ with velocity c to (1) satisfy the equation

$$(1 - c^2)\varphi_{\xi\xi\xi\xi} - (\beta - c^2)\varphi_{\xi\xi} - \alpha \left(\varphi^2\right)_{\xi\xi} + k_1\varphi + k_3\varphi^3 = 0. \quad (3)$$

In the following two propositions, which can be easily verified by straightforward computations, we give, solving the above Eq. (3), sufficient conditions for the existence of solitary wave solutions to equations of form (1) along with the explicit analytic expressions of the latter.

Proposition 1. Let α, β, k_1, k_3 and c be such that:
(i) $c \neq \pm 1$, $\alpha \neq 0$, $4\left(c^2 - \beta\right)^2 + 25k_1\left(c^2 - 1\right) \neq 0$,

$$\Delta_1 = \beta^2 + c^4 + c^2(4k_1 - 2\beta) - 4k_1 \geq 0,$$

$$\frac{c^2 - \beta \pm \sqrt{\Delta_1}}{c^2 - 1} > 0,$$

and

$$k_3 = -\frac{20k_1\alpha^2 \left[50k_1\left(c^2 - 1\right) + 17\left(c^2 - \beta\right)^2 \mp 15\left(c^2 - \beta\right)\sqrt{\Delta_1}\right]}{3\left[4\left(c^2 - \beta\right)^2 + 25k_1\left(c^2 - 1\right)\right]^2}.$$

Then

$$u(x,t) = \frac{9\left(c^2 - \beta\right) \pm 15\sqrt{\Delta_1}}{16\alpha}\,\mathrm{sech}^2\left(\frac{x - ct}{2\sqrt{2}}\sqrt{\frac{c^2 - \beta \pm \sqrt{\Delta_1}}{c^2 - 1}}\right) \qquad (4)$$

is a real-valued traveling wave solution of the respective equation of the form (1), satisfying the boundary conditions (2);
(ii) $c \neq \pm 1$, $\pm\sqrt{\Delta_2} - 3\alpha \neq 0$, $75\left(c^2 - 1\right)k_3 + 5\alpha\left(17\alpha \mp 3\sqrt{\Delta_2}\right) \neq 0$

$$\frac{\left(5\alpha \pm \sqrt{\Delta_2}\right)\left(c^2 - \beta\right)}{\left(\pm\sqrt{\Delta_2} - 3\alpha\right)\left(c^2 - 1\right)} > 0$$

and

$$k_1 = -\frac{12k_3\left(c^2 - \beta\right)^2}{75\left(c^2 - 1\right)k_3 + 5\alpha\left(17\alpha \mp 3\sqrt{\Delta_2}\right)},$$

where

$$\Delta_2 = 30\left(c^2 - 1\right)k_3 + 25\alpha^2 \geq 0.$$

Then

$$u(x,t) = \frac{3\left(c^2 - \beta\right)}{\pm\sqrt{\Delta_2} - 3\alpha}\,\mathrm{sech}^2\left(\frac{x - ct}{2\sqrt{5}}\sqrt{\frac{\left(5\alpha \pm \sqrt{\Delta_2}\right)\left(c^2 - \beta\right)}{\left(\pm\sqrt{\Delta_2} - 3\alpha\right)\left(c^2 - 1\right)}}\right) \qquad (5)$$

is a real-valued traveling wave solution of the respective equation of the form (1), satisfying the boundary conditions (2).

Solitary wave-like solutions (solitons) of the form (4) and (5) are depicted in Figs. 1 and 2, respectively.

Proposition 2. *Let α, β, k_1 and k_3 be such that $\alpha \neq 0$, $\beta \neq 1$ and*

$$\frac{k_1}{\beta - 1} > 0.$$

Then

$$u(t,x) = \frac{3\left(\beta - 1\right)}{8\alpha}\,\mathrm{sech}^2\left(\frac{x + t}{2}\sqrt{\frac{k_1}{\beta - 1}}\right) \qquad (6)$$

is a unit speed real-valued traveling wave solution of the respective equation of form (1), satisfying the boundary conditions (2) if the parameter k_3 meets the condition

$$k_3 = -\frac{40k_1\alpha^2}{3\left(\beta - 1\right)^2}.$$

Solitary wave-like solutions (solitons) of the form (6) are depicted in Fig. 3.

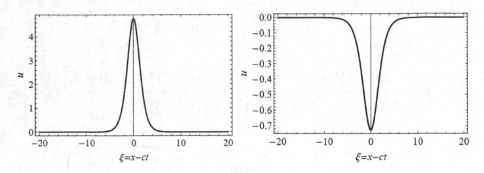

Fig. 1. Solitons of the form (4) for $\alpha = 1$, $\beta = 1$ and: $c = 1.8$, $k_1 = 1$, $k_3 = -0.082$ (left); $c = 0.6$, $k_1 = 0.1$, $k_3 = -0.354$ (right).

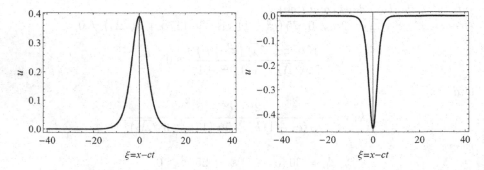

Fig. 2. Solitons of the form (5) for: $\alpha = 0$, $\beta = 1$, $c = 0.7$, $k_1 = 0.0816$, $k_3 = -1$ (left); $\alpha = 1$, $\beta = 1$ $c = 0.7$, $k_1 = 0.1114$, $k_3 = -1$ (right).

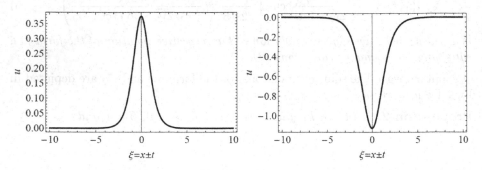

Fig. 3. Solitons of the form (6) for: $\alpha = 1$, $\beta = 2$, $k_1 = 3$ and $k_3 = -40$ (left); $\alpha = 1$, $\beta = -2$, $k_1 = -5$ and $k_3 = 7.4074$ (right).

3   A Conservation Law

Until now, we do not know an exact conservation law for the general problem (1), (2). But we managed to prove the existence of a conservation law for the particular set of parameters: $f(u) = 0$, $k_1 \neq 0$ and $k_3 \neq 0$. Let us define the functional

$$E_1(u(\cdot, t)) := \frac{1}{2} \int_{\mathbb{R}} \left((u_t(x,t))^2 + \beta(u_x(x,t))^2 + (u_{xt}(x,t))^2 + (u_{xx}(x,t))^2 \right.$$

$$\left. + k_1(u(x,t))^2 + \frac{k_3}{2}(u(x,t))^4 \right) dx, \tag{7}$$

further referred to as the energy functional. Then the following proposition follows by straightforward calculations.

Proposition 3. *For the problem* (1), (2) *with* $f(u) = 0$, $k_1 \neq 0$, $k_3 \neq 0$ *and initial data* $u(x,0) = u_0(x)$, $u_t(x,0) = u_1(x)$ *we have*

$$E_1(u(\cdot, t)) = E_1(u(\cdot, 0))$$

for every $t \geq 0$, *i.e. the energy* $E_1(u(\cdot, t))$ *is preserved in time.*

Note that the initial energy $E_1(u(\cdot, 0))$ can be evaluated using initial data $u_0(x)$ and $u_1(x)$ only.

In a similar way, the energy for the problem (1), (2) with $f(u) \neq 0, k_1 = 0$, $k_3 = 0$ and initial data $u(x,0) = u_0(x)$, $u_t(x,0) = u_1(x)$, i.e. for Boussinesq paradigm equation, is defined as

$$E_2(u(\cdot, t)) = \frac{1}{2} \int_{\mathbb{R}} \left(((-\partial_x^2)^{-\frac{1}{2}} u_t(x,t))^2 + (u_t(x,t))^2 + \beta(u(x,t))^2 \right.$$

$$\left. + (u_x(x,t))^2 + \frac{2}{3}\alpha(u(x,t))^3 \right) dx \tag{8}$$

and its conservation in time can be found e.g. in [5].

In the next section we define discrete expressions, approximating (7) and (8), and prove their preservation in time. These expressions are used to test the global properties of the numerical solution.

4   Finite Difference Schemes

We consider the discrete problem in the finite space interval $[-L_1, L_2]$. The numbers L_i, $i = 1, 2$ are chosen to be sufficiently large so that the solution and its derivatives are negligible outside this interval. We introduce in $[-L_1, L_2]$ an uniform grid x_i, $i = 0, 1, ..., N$ with step $h = (L_1 + L_2)/N$. Let $T = K\tau$ be the final time and τ be the time step in $[0, T]$. We denote by y_i^k the approximation to the solution $u(x_i, t^k)$ at the grid node x_i on the kth time level. We apply the notations $y_{\bar{x},i}^k = \dfrac{y_i^k - y_{i-1}^k}{h}$, $y_{t,i}^k = \dfrac{y_i^{k+1} - y_i^k}{\tau}$ for the forward and backward finite difference respectively. The following finite differences

$$\Delta_h y_i^k = y_{\bar{x}x,i}^k = \frac{y_{i-1}^k - 2y_i^k + y_{i+1}^k}{h^2}, \quad y_{\bar{t}t,i}^k = \frac{y_i^{k+1} - 2y_i^k + y_i^{k-1}}{\tau^2}$$

approximate the second order space and time derivatives with approximation error $O(h^2)$ and $O(\tau^2)$, respectively. Replacing the derivatives in Eq. (1) with the above finite differences on the main time level k, we obtain at interior grid points $i = 1, ..., N - 1$ for every $k = 1, ..., K - 1$ the following

Explicit Finite Difference Scheme:

$$y_{\bar{t}t,i}^k - \beta \Delta_h y_i^k - \Delta_h y_{\bar{t}t,i}^k + \Delta_h^2 y_i^k - \Delta_h \alpha (y_i^k)^2 + k_1 y_i^k + k_3 (y_i^k)^3 = 0. \quad (9)$$

At every internal node i of the mesh, the Eq. (9) is equivalent to the following system

$$-\frac{1}{\tau^2 h^2} y_{i-1}^{k+1} + \left(\frac{2}{h^2 \tau^2} + \frac{1}{\tau^2}\right) y_i^{k+1} - \frac{1}{\tau^2 h^2} y_{i+1}^{k+1} = \frac{2}{\tau^2} y_i^k - \frac{1}{\tau^2} y_i^{k-1} - \frac{2}{\tau^2} y_{\bar{x}x,i}^k +$$

$$+\frac{1}{\tau^2} y_{\bar{x}x,i}^{k-1} + \beta \Delta_h y_i^k - \Delta_h^2 y_i^k + \Delta_h \alpha (y_i^k)^2 - k_1 y_i^k - k_3 (y_i^k)^3. \quad (10)$$

The system (10) has a three diagonal matrix which is diagonally dominant. Thus the solution of the system can be found, for example, after application of the Thomas algorithm.

We change the approximation to the quadratic and the cubic term in (1) using the expression $u(\cdot, k\tau)^p \approx \frac{1}{(p+1)} \frac{(y^{k+1})^{p+1} - (y^{k-1})^{p+1}}{y^{k+1} - y^{k-1}}$ and keep the other terms in (9) unchanged. This leads to the following

Implicit Finite Difference Scheme:

$$y_{\bar{t}t,i}^k - \beta \Delta_h y_i^k - \Delta_h y_{\bar{t}t,i}^k + \Delta_h^2 y_i^k - \alpha \Delta_h \frac{(y_i^{k+1})^2 + y_i^{k+1} y_i^{k-1} + (y_i^{k-1})^2}{3} +$$

$$+ k_1 y_i^k + k_3 \frac{(y_i^{k+1})^3 + (y_i^{k+1})^2 y_i^{k-1} + y_i^{k+1}(y_i^{k-1})^2 + (y_i^{k-1})^3}{4} = 0 \,(11)$$

Note that the scheme (11) uses the values of the nonlinearities on the upper time level, hence an iterative procedure for evaluation of y^{k+1} is required.

Proposition 4. *Let ϵ be a sufficiently small number. Then the linearized scheme corresponding to (9) or (11) (i.e. (9) or (11) with $\alpha = 0$ and $k_3 = 0$) is stable if the steps h and τ satisfy the following inequality*

$$\tau \leq \frac{2h^2}{\sqrt{(1 + \epsilon)(4\beta h^2 + 16 + k_1 h^4)}}.$$

To obtain the discrete energy law similar to (8) for the case $\alpha = 0$, we multiply both sides of (11) by $y_i^{k+1} - y_i^{k-1}$ and rearrange the terms using summation formulas. In this way we prove the following

Proposition 5. *Let $\alpha = 0$ and y^k be the solution to the discrete implicit finite difference scheme (11). Define the functional $E_{1,h}^k$ as follows*

$$E^k_{1,h} = \left(\frac{1}{2} - \frac{k_1 \tau^2}{8}\right)(y^k_t, y^k_t) + \left(\frac{1}{2} - \frac{\beta \tau^2}{8}\right)(y^k_{t\bar{x}}, y^k_{t\bar{x}}] - \frac{\tau^2}{8}\left(\Delta_h y^k_t, \Delta_h y^k_t\right) +$$

$$+ \frac{\beta}{8}\left(y^{k+1}_{\bar{x}} + y^k_{\bar{x}}, y^{k+1}_{\bar{x}} + y^k_{\bar{x}}\right] + \frac{k_1}{8}\left(y^{k+1} + y^k, y^{k+1} + y^k\right) +$$

$$+ \frac{1}{8}\left(\Delta_h(y^{k+1} + y^k), \Delta_h(y^{k+1} + y^k)\right) + \frac{k_3}{8}\left((y^{k+1})^4 + (y^k)^4, 1\right). \tag{12}$$

Then the discrete energy $E^k_{1,h}$ is conserved in time, i.e.

$$E^k_{1,h} = E^0_{1,h}, \quad k = 1, \cdots, K - 1.$$

Note that the discrete energy $E^k_{2,h}$,

$$E^k_{2,h} = \frac{1}{2}\left((-\Delta_h)^{-1} y^k_t, y^k_t\right) + \left(\frac{1}{2} - \frac{\tau^2 \beta}{8}\right)(y^k_t, y^k_t) - \frac{\tau^2}{8}(y^k_{t\bar{x}}, y^k_{t\bar{x}}] +$$

$$+ \frac{\beta}{8}\left(y^{k+1} + y^k, y^{k+1} + y^k\right) + \frac{1}{8}\left(y^{k+1}_{\bar{x}} + y^k_{\bar{x}}, y^{k+1}_{\bar{x}} + y^k_{\bar{x}}\right] +$$

$$+ \frac{\alpha}{6}\left((y^{k+1})^3 + (y^k)^3, 1\right),$$

for the Boussinesq paradigm equation ($k_1 = k_3 = 0$, $\alpha \neq 0$ in (1)) is also conserved in time exactly, see e.g. [5].

5   Numerical Results

In this section we present some numerical results concerning the accuracy and the convergence of the explicit and implicit finite difference schemes. In the numerical simulations the analytical solution $\tilde{u}$, given in Proposition 1 and Proposition 2 by (4) and (5), is used as an initial condition to (1). The convergence and the accuracy are analysed on embedded grids by Runge's rule.

The first problem we consider is the propagation of a single solitary wave. Let us denote by y_h and $y_{h/2}$ the numerical solutions evaluated on grid with steps h, τ and with steps $h/2$, $\tau/2$, respectively. The maximal difference ψ_u between the exact solution and computed solution, and the order of convergence κ_u are obtained by the expressions

$$\psi_u = \max_{0 \leq k \leq N} \|\tilde{u}^k_h - y^k_h\|, \quad \kappa_u = \log_2 \frac{\|\tilde{u}^k_h - y^k_h\|}{\|\tilde{u}^k_{h/2} - y^k_{h/2}\|}.$$

Problem 1: Propagation of a solitary wave with $f(u) = 0$.

Let the coefficients of (1) be $\alpha = 0$, $\beta = 1$, $k_1 = 0.0816$, $k_3 = -1$ and $c = 0.7$.

Table 1 contains numerical results about the error ψ_u and the order of convergence κ_u of the discrete solution to the exact one. In column *sec* there is a

Table 1. Error and order of convergence of the solution to Problem 1 with $f(u) = 0$, $T = 20, c = 0.7$

h	τ	Explicit FDS			Implicit FDS			
		ψ_u	κ_u	sec	ψ_u	κ_u	sec	MaxIter
0.4	0.0016	0.185554		1	0.185521		109	3
0.2	0.0008	0.044350	2.06	3	0.037733	2.30	270	3
0.1	0.0004	0.008843	2.32	10	0.008841	2.09	1238	3
0.05	0.0002	0.000451	4.29	34	0.000451	4.29	4624	3

computational time and the values in column *MaxIter* are the maximal number of iterations, which are needed to satisfy the stop criteria.

The numerical results show that the computed solution converges to the exact one with error $O(h^2 + \tau^2)$.

The main goal of this example is to analyse the preservation of the discrete energy, which is shown in Table 2. The second table contains the values of the discrete energy E_h^K on the last time layer K, which is given by the expression (12). There is also the absolute difference between this energy and the initial one E_h^0.

Table 2. Numerical results for the discrete energy to Problem 1 with $f(u) = 0$, $T = 20$, $c = 0.7$.

h	τ	Explicit FDS		Implicit FDS	
		E_h^K	$|E_h^0 - E_h^K|$	E_h^K	$|E_h^0 - E_h^K|$
0.4	0.0016	0.04430578	$4.1 * 10^{-10}$	0.04434987	$4.3 * 10^{-10}$
0.1	0.0004	0.04436092	$2.7 * 10^{-9}$	0.04436092	$2.9 * 10^{-9}$
0.05	0.0002	0.04436368	$8.6 * 10^{-9}$	0.04436368	$8.5 * 10^{-9}$

The provided numerical experiments confirm that the discrete energy is preserved in time.

Table 3. Numerical results for the convergence of the discrete energy to Problem 1 with $f(u) = 0$, $T = 20$, $c = 0.7$.

h	τ	Explicit FDS		Implicit FDS	
		ψ_E	κ_E	ψ_E	κ_E
0.4	0.0016				
0.2	0.0008				
0.1	0.0004	$5.9 * 10^{-5}$	1.9967	$5.9 * 10^{-5}$	1.9964
0.05	0.0002	$1.5 * 10^{-5}$	2.0017	$1.5 * 10^{-5}$	2.0017

Table 3 contains results about the error ψ_E and the order of convergence κ_E of the discrete energy on the final time layer. They are obtained by Runge's method:

$$\psi_E = \frac{|E_h^K - E_{h/2}^K|^2}{|E_h^K - E_{h/2}^K| - |E_{h/2}^K - E_{h/4}^K|}, \quad \kappa_E = \log_2 \frac{|E_h^K - E_{h/2}^K|}{|E_{h/2}^K - E_{h/4}^K|}.$$

The discrete energy converges to the exact one with second order of convergence.

Problem 2: Propagation of a solitary wave with $\alpha \neq 0$

Now let us consider the general problem (1) with parameters $\alpha \neq 0$, $k_1 \neq 0$ and $k_3 \neq 0$. The coefficients are $\alpha = 2$, $\beta = 1$, $k_1 = 0.032606620314995$, $k_3 = -0.5$, $c = 0.7$.

The numerical results, shown in Table 4, are similar to the results of the previous problem. The error ψ_u is of order 10^{-4} for the smallest steps and we obtain second order of convergence κ_u of the discrete solution to the exact one.

Table 4. Numerical results for the convergence of the solution to Problem 2 with Implicit FDS at time $T = 20$ and $c = 0.7$.

h	τ	ψ_u	κ_u	sec	*MaxIter*
0.8	0.0032	0.08893882		1	3
0.4	0.0016	0.01928352	2.2054	2	3
0.2	0.0008	0.00478670	2.0103	5	3
0.1	0.0004	0.00120857	1.9857	16	3
0.05	0.0002	0.00027755	2.1225	83	3

Problem 3: Propagation of solitary wave with velocity $c = \pm 1$.

Now we consider the case when the wave velocity c is ± 1. The coefficients $\alpha = -3$, $\beta = 5$, $k_1 = 0.08$, $k_3 = -0.6$, $c = 1$ in problem (1) satisfy the requirements of Proposition 2 for existence of solitary-wave solutions. The numerical results for convergence of the solution to Problem 3 are included in Table 5.

Table 5. Numerical results for the convergence of the solution to Problem 3 by Implicit FDS at time $T = 20$.

h	τ	ψ_u	κ_u	sec	*MaxIter*
0.8	0.0032	0.591005123965154		2	4
0.4	0.0016	0.097438874935317	2.6006	4	3
0.2	0.0008	0.022433906762699	2.1188	13	3
0.1	0.0004	0.004753028515013	2.2388	45	3

Problem 4. Interaction between two waves with $c_1 \neq \pm 1$ and $c_2 \neq \pm 1$.

Here we consider the problem of two waves, traveling toward each other with velocities $c_1 = -c_2 = 0.97$. Let the other coefficients to (1) be $\alpha = 1$, $\beta = 1$, $k_3 = -0.5$. The space interval is $[-130, 130]$. The first initial condition of the problem is given by the expression

$$u(x, 0) = \tilde{u}(x + 30, 0; c_1) + \tilde{u}(x - 40, 0; c_2),$$

where $\tilde{u}$ is the solution of (1), given by (5). For these values of the parameters the waves meet each other at the moment $T = 36$. The results in Table 6 are calculated for $T = 65$, i.e. after the waves have been separated. The position of the two waves and their interaction is shown on Figs. 4, 5 and 6.

Table 6. Numerical results for the solution of Problem 4 by the Implicit FDS with $T = 65$, $c_1 = -0.97$ and $c_2 = 0.97$.

h	τ	$\psi_{h/4}^K$	κ_u	sec	$MaxIter$
0.4	0.0016			198	3
0.2	0.0008			1190	3
0.1	0.0004	0.017565606587319	1.7876	4703	3
0.05	0.0002	0.004820589743132	1.9985	13650	3

This problem does not posses an exact solution, so the error $\psi_{h/4}^K$ and the order of convergence κ_u are obtained by Runge's method.

$$\psi_{h/4}^K = \frac{\|y_{[h]}^K - y_{[h/2]}^K\|_\infty^2}{\|y_{[h]}^K - y_{[h/2]}^K\|_\infty - \|y_{[h/2]}^K - y_{[h/4]}^K\|_\infty}, \quad \kappa_u = \log_2 \left(\frac{\|y_{[h]}^K - y_{[h/2]}^K\|\|_\infty}{\|y_{[h/2]}^K - y_{[h/4]}^K\|_\infty} \right).$$

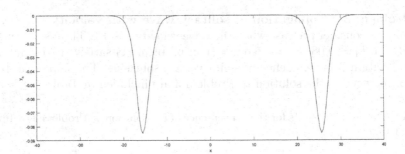

Fig. 4. Position of two waves at $T = 15$

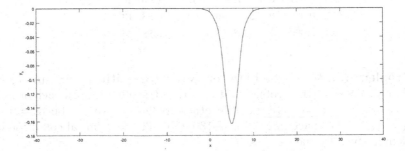

Fig. 5. Interaction of two waves at $T = 36$

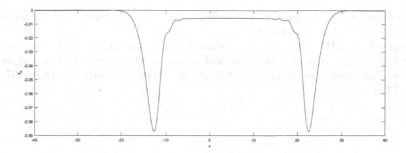

Fig. 6. Position of the waves at $T = 55$

Conclusion. We consider the Boussinesq equation with nonlinear restoring force, which models transverse and longitudinal vibrations of elastic rods laying on a foundation of Winkler-Pasternak type. Some exact solitary wave-like solutions of the regarded equation are given in explicit analytic form. Two FDS with second order of approximation are constructed and studied. The preservation of the discrete energy is investigated for the specific values of the parameters. The presented numerical results show second order of convergence of the discrete solution to the exact solution. The discrete energy also converges to the exact one with second order of convergence.

Acknowledgments. The authors would like to acknowledge the support from the Bulgarian Science Fund under grant KΠ-06-H22/2. The work of the second and the third author has been accomplished with the financial support by Grant No BG05M2OP001-1.002-0011-C02 and Grant No BG05M2OP001-1.001-0003, respectively, financed by the Science and Education for Smart Growth Operational Program (2014–2020) and co-financed by the European Union through the European structural and Investment funds.

References

1. Birman, V.: On the effects of nonlinear elastic foundation on free vibration of beams. J. Appl. Mech. **53**(2), 471–473 (1986)
2. Christou, M.: Soliton interactions on a Boussinesq type equation with a linear restoring force. AIP Conf. Proc. **1404**, 41–48 (2011)
3. Christov, C.I.: An energy-consistent dispersive shallow-water model. Wave Motion **34**, 161–174 (2001)
4. Christov, C.I., Marinov, T.T., Marinova, R.S.: Identification of solitary-wave solutions as an inverse problem: application to shapes with oscillatory tails. Math. Comp. Simulation **80**, 56–65 (2009)
5. Kolkovska, N., Dimova, M.: A new conservative finite difference scheme for Boussinesq paradigm equation. Central Eur. J. Math. **10**, 1159–1171 (2012)
6. Kolkovska, N., Vassilev, V.M.: Solitary waves to Boussinesq equation with linear restoring force. AIP Conf. Proc. **2164**, 110005 (2019)
7. Porubov, A.: Amplification of nonlinear strain waves in solids. World Scientific, Singapore (2003)

8. Samsonov, A.M.: Strain solitons in solids and how to construct them. Chapman and Hall/CRC, Boca Raton (2001)
9. Younesian, D., Hosseinkhani, A., Askari, H., Esmailzadeh, E.: Elastic and viscoelastic foundations: a review on linear and nonlinear vibration modeling and applications. Nonlinear Dyn. **97**(1), 853–895 (2019). https://doi.org/10.1007/s11071-019-04977-9

Simulation of the Mechanical Wave Propagation in a Viscoelastic Media With and Without Stiff Inclusions

Todor Zhelyazov[1]($\boxtimes$) and Sergey Pshenichnov[2]

[1] Technical University of Sofia, 8, Kliment Ohridski boulevard, Sofia, Bulgaria
elovar@yahoo.com
[2] Institute of Mechanics, Lomonosov Moscow State University,
Michurinsky prospect 1, Moscow 119192, Russia

Abstract. The contribution focuses on the numerical simulation of wave propagation in an array of solid inclusions regularly distributed in a viscoelastic matrix. Waves are provoked by a transient load. The study aims to compare wave propagation in the viscoelastic continuum with and without the presence of solid inclusions. To this end, stress and displacement evolutions (in the time domain) are monitored at specified locations at the boundaries of the defined continuum. The case study contributes to a better understanding of the phenomena related to the reflection and diffraction of the mechanical waves by the solid inclusions. The modeled set-up often referred to in the literature as a phononic crystal, will possibly shed light on numerous practical applications. Among others, these are high sound absorption and strategies for the detection of defect locations.

Keywords: Transient wave process · Finite element analysis · Composite · Viscoelastic matrix

1 Introduction

The contribution presents some results of the numerical simulations of the transient wave process in a composite material comprised of a viscoelastic matrix and stiff inclusions. Such materials are the basis of emerging technologies related to the modified wave process attributed to local resonance and wave scattering. For example, they might form scattering resonances at frequencies corresponding to acoustic wavelengths much larger than the dimension and the spacing between inclusions. Theoretically, this could lead to a conversion of longitudinal waves into shear waves, subsequently attenuated due to the material properties of the rubber-like material of the matrix [1–4]. The potential applications of this principle in maritime [1, 2, 5–7] and seismic [8] engineering have been the subject of

Supported by the Russian Foundation for Basic Research (RFBR), project number 20-58-18002, and the Bulgarian National Science Fund, project number KP-06- Russia/5 from 11.12.2020.

I. Georgiev et al. (Eds.): NMA 2022, LNCS 13858, pp. 339–348, 2023.
https://doi.org/10.1007/978-3-031-32412-3_30

active discussion for some time. In the reported study, the viscoelastic response is modeled using the linear Boltzmann-Volterra model. A transient analysis is performed. Technically, viscous effects are accounted for via a modification of the material properties in the function of the (model) time. This strategy is analogous to the action of a hereditary kernel in the analytical description of the problem.

As shown in previous works, the simulation of wave propagation in viscoelastic solids can be based on extending the Volterra principle to the dynamic response of linear viscoelastic materials [9,10]. Some authors (see, for example, [11]) suggest a convolution of the dynamic elastic problem solution with an auxiliary one-dimensional viscoelastic problem, the latter involving a hereditary kernel. Other approaches employ the modal decomposition method [12] or a spectral decomposition of the nonstationary problem solution by biorthogonal systems of eigenfunctions of mutually conjugate bundles of differential operators [13]. Recent trends include fractional derivatives or fractional order operators [14,15]. Problems involving the consideration of linear viscoelasticity usually imply integral Laplace transform in time with subsequent projection in the space of the originals [16–20].

The contribution reports a study of the transient wave process in a composite material containing stiff inclusions, the latter forming a regular array in a viscoelastic matrix. The modification of the transient wave process in the composite is outlined through a comparison with the response of a domain of the same geometry and made of the same material but not containing inclusions.

2   Geometry

Figure 1 shows the geometry of the considered prismatic specimens subjected to a time-dependent load. The boundary conditions depicted in a relatively simplistic manner in Fig. 1 are described in more detail further in the text. One of the specimens (shown in Fig. 1) comprises stiff cylindrical inclusions of linear elastic behavior, aligned along the Z-axis, submerged in a softer material (i.e., matrix) exhibiting viscoelastic properties, whereas the other consists of the matrix material only.

3   Mathematical Formulation of the Problem

3.1   Dynamics of the Viscoelastic Continuum

The equation of motion (1), the constitutive laws (2), and the relationships characterizing the strain tensor within the small displacements assumption (3) define the dynamics of an isotropic, homogeneous, and viscoelastic material

$$\nabla \cdot \bar{\bar{\sigma}} = \frac{\partial^2 \mathbf{u}}{\partial t^2}, \tag{1}$$

$$\bar{\bar{\sigma}} = \hat{\lambda} tr\left(\bar{\bar{\varepsilon}}\right) \bar{\bar{\mathbf{I}}} + 2\hat{\mu}\bar{\bar{\varepsilon}}, \tag{2}$$

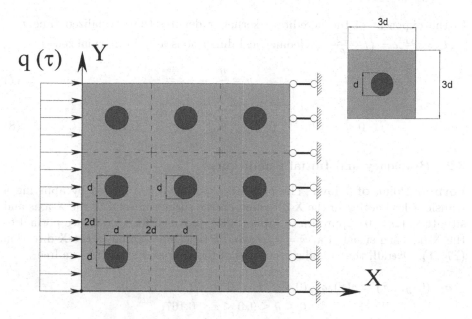

Fig. 1. Geometry and boundary conditions of the specimen with stiff inclusions.

$$\bar{\bar{\varepsilon}} = \frac{1}{2}[\nabla \mathbf{u} + (\nabla \mathbf{u})^T]. \tag{3}$$

In Eqs. (1)–(3) $\bar{\bar{\sigma}}$ denotes the stress tensor, ρ - the material density, $\mathbf{u}$ - the displacement vector, $\bar{\bar{\varepsilon}}$ - the strain tensor, $tr(\bar{\bar{\varepsilon}})$ - the trace of the strain tensor, and $\bar{\bar{\mathbf{I}}}$ - the unit tensor. To model viscoelasticity, in the general case, the initial (or instantaneous) Lamé's coefficients are modified throughout the time-history analysis by introducing operators $\hat{T}_v$ and $\hat{T}_s$

$$\hat{\lambda} = K_0(1 - \hat{T}_v) - \frac{2}{3}G_0(1 - \hat{T}_s), \qquad \hat{\mu} = G_0(1 - \hat{T}_s),$$

$$\hat{T}_j \psi(t) = \int_0^t T_j(t - \chi)\psi(\chi)d\chi, \qquad j = v, s.$$

$$\tag{4}$$

In the sequel, $\lambda_0 = \frac{E_0 \nu_0}{(1+\nu_0)(1-2\nu_0)}$ and $\mu_0 = \frac{E_0}{2(1+\nu_0)}$ ($\mu_0 = G_0$) are the initial values of Lamé's coefficients, E_0 and ν_0 are the initial values of Young's modulus and Poisson's ratio. The numerical analysis uses the following assumptions

$$\hat{T}_v = 0 \qquad \hat{T}_s \xi(t) = \int_0^t T_s(t - \chi)\xi(\chi)d\chi \tag{5}$$

which imply a variable Poisson's ratio since the bulk modulus remains constant. The presented results are obtained by using the following hereditary kernel

$$t_0 T_s = ae^{-b\tau}, \qquad a = 0.2, \qquad b = 0.9. \tag{6}$$

In the exponent of the hereditary kernel, τ denotes the normalized time $\tau = \frac{t}{t_0}, t_0 = \frac{3d}{c}, c = \sqrt{\frac{\lambda_0 + 2\mu_0}{\rho}}$. Geometrical dimensions are also normalized

$$\tilde{x} = \frac{x}{d}, \qquad \tilde{y} = \frac{y}{d}, \qquad \tilde{z} = \frac{z}{d}, \tag{7}$$

thus

$$0 \le \tilde{x} \le 9, \qquad 0 \le \tilde{y} \le 9, \qquad 0 \le \tilde{z} \le 0.167. \tag{8}$$

3.2   Boundary and Initial Conditions

Formal Point of View. The prismatic specimens are loaded by applying a transient load acting in the X-direction on the surface normal to the X-axis and situated at $\tilde{x} = 0$. Appropriate supports are provided for the surface normal to the X-axis and situated at $\tilde{x} = 9$ to avoid a rigid-boy motion in the X-direction (Fig. 1). Overall, the boundary surfaces of the viscoelastic matrix are free

$$\sigma_{xx}(0, \tilde{y}, \tilde{z}, \tau) = q(\tau), \sigma_{xy}(0, \tilde{y}, \tilde{z}, \tau) = 0, \sigma_{xz}(0, \tilde{y}, \tilde{z}, \tau) = 0,$$
$$0 \le \tilde{y} \le 9, 0 \le \tilde{z} \le 0.167,$$
$$u_x(9, \tilde{y}, \tilde{z}, \tau) = 0, \sigma_{xy}(9, \tilde{y}, \tilde{z}, \tau) = 0, \sigma_{xz}(9, \tilde{y}, \tilde{z}, \tau) = 0,$$
$$0 \le \tilde{y} \le 9, 0 \le \tilde{z} \le 0.167,$$
$$\sigma_{yy}(\tilde{x}, 0, \tilde{z}, \tau) = 0, \sigma_{yx}(\tilde{x}, 0, \tilde{z}, \tau) = 0, \sigma_{yz}(\tilde{x}, 0, \tilde{z}, \tau) = 0,$$
$$0 \le \tilde{x} \le 9, 0 \le \tilde{z} \le 0.167,$$
$$\sigma_{yy}(\tilde{x}, 9, \tilde{z}, \tau) = 0, \sigma_{yx}(\tilde{x}, 9, \tilde{z}, \tau) = 0, \sigma_{yz}(\tilde{x}, 9, \tilde{z}, \tau) = 0,$$
$$0 \le \tilde{x} \le 9, 0 \le \tilde{z} \le 0.167,$$
$$\sigma_{zz}(\tilde{x}, \tilde{y}, 0, \tau) = 0, \sigma_{zx}(\tilde{x}, \tilde{y}, 0, \tau) = 0, \sigma_{zy}(\tilde{x}, \tilde{y}, 0, \tau) = 0,$$
$$0 \le \tilde{x} \le 9, 0 \le \tilde{y} \le 9,$$
$$\sigma_{zz}(\tilde{x}, \tilde{y}, 0.167, \tau) = 0, \sigma_{zx}(\tilde{x}, \tilde{y}, 0.167, \tau) = 0, \sigma_{zy}(\tilde{x}, \tilde{y}, 0.167, \tau) = 0,$$
$$0 \le \tilde{x} \le 9, 0 \le \tilde{y} \le 9,$$
$$\tag{9}$$

whereas, at the bottom and top surfaces of the stiff inclusions, the vertical displacements are restrained, and no shear stresses act

$$u_z(\tilde{x}, \tilde{y}, 0, \tau) = 0, \qquad u_z(\tilde{x}, \tilde{y}, 0.167, \tau) = 0,$$
$$\sigma_{zx}(\tilde{x}, \tilde{y}, 0, \tau) = 0, \qquad \sigma_{zy}(\tilde{x}, \tilde{y}, 0, \tau) = 0,$$
$$\sigma_{zx}(\tilde{x}, \tilde{y}, 0.167, \tau) = 0, \qquad \sigma_{zy}(\tilde{x}, \tilde{y}, 0.167, \tau) = 0,$$
$$(\tilde{x} - \tilde{x}_c)^2 + (\tilde{y} - \tilde{y}_c)^2 \le 0.5^2.$$
$$\tag{10}$$

In boundary conditions (10), $\tilde{x}_c$ and $\tilde{y}_c$ take the values 1.5, 4.5, and 7.5. A perfect interface between inclusions and matrix is assumed, i.e., the stress and the displacement vectors are continuous.

The response of the composite containing rigid inclusions is compared with the behavior of a specimen made of viscoelastic material only, with the same geometry, and subjected to comparable boundary conditions. For the configuration without stiff inclusions, boundary conditions (9) apply.

The time-dependent load is modeled using the smoothed Heaviside function factored by a dimensionless constant F_0

$$q(\tau) = F_0\varphi(\tau), \qquad \varphi(\tau) = 1 - e^{-50\tau}, \qquad \tau > 0. \tag{11}$$

For both specimens, the solution is obtained based on the following initial conditions

$$u_i(\tilde{x}, \tilde{y}, \tilde{z}, 0) = 0, \qquad \frac{\partial}{\partial \tau} u_i(\tilde{x}, \tilde{y}, \tilde{z}, 0) = 0, \qquad i = x, y, z,$$

$$0 \leq \tilde{x} \leq 9, \qquad 0 \leq \tilde{y} \leq 9, \qquad 0 \leq \tilde{z} \leq 0.167.$$

$$\tag{12}$$

Computation-related Nuances. In the context of the numerical simulations, to guarantee that no rigid-body motion is possible, displacements in the Y-direction of nodes at specified locations are restrained for both specimens (with and without inclusions, see Fig. 2)

$$u_y(\tilde{x}_s, 0, 0, \tau) = 0, \qquad u_y(\tilde{x}_s, 0, 0.167, \tau) = 0,$$

$$u_y(\tilde{x}_s, 9, 0, \tau) = 0, \qquad u_y(\tilde{x}_s, 9, 0.167, \tau) = 0,$$

$$\tag{13}$$

where $\tilde{x}_s$ takes the values 0, 3, 6 and 9.

Additional discrete supports are defined for the specimen made of the matrix material only (i.e., the one without stiff inclusions) to provide comparable boundary conditions with the specimen in composite material

$$u_z(\tilde{x}_c, \tilde{y}_c, 0, \tau) = 0, \qquad u_z(\tilde{x}_c, \tilde{y}_c, 0.167, \tau) = 0, \tag{14}$$

where $\tilde{x}_c$ and $\tilde{y}_c$ take the values 1.5, 4.5, and 7.5.

4   Transient Analysis

The finite element full transient analysis is carried out in the ANSYS Mechanical APDL environment. As mentioned above, by assumption, only the matrix exhibits viscoelastic behavior, as defined in Eqs. (2), (4)–(6), whereas the response of the inclusions remains purely elastic. $E_i = 210000\,\text{MPa}, \nu_i = 0.3, \rho_i = 7850\,\text{kg/m}^3$ and $E_{m,0} = 2000\,\text{MPa}, \nu_{m,0} = 0.45, \rho_m = 1000\,\text{kg/m}^3$ are the material properties of the stiff inclusions and matrix, respectively. $E_{m,0}$ and $\nu_{m,0}$ denote the initial values of Young's modulus and Poisson's ratio, which are subsequently modified throughout the solution. Several locations are selected to monitor the evolution of the stress components: location

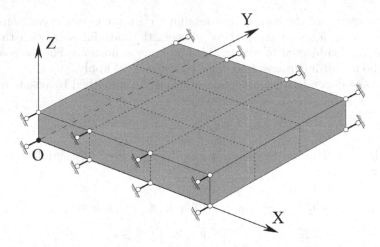

Fig. 2. Restrained nodal displacement in the global Y direction for both specimens; the other boundary conditions are not displayed

1 ($\tilde{x} = 3, \tilde{y} = 3, \tilde{z} = 0.167$) and location 2 ($\tilde{x} = 6, \tilde{y} = 3, \tilde{z} = 0.167$). Additionally, the algorithm keeps track of the shear stress evolution at location M ($\tilde{x} = 4.5, \tilde{y} = 3, \tilde{z} = 0.167$). It should be noted that the results are obtained for the centroid of a finite element having a node coinciding with the specified location. Figure 3 displays the generated finite element mesh and provides visual information on locations where the stress evolution is monitored. Solid 186 (a 3-D, 20-node finite element having three degrees of freedom, translations along the x-, y, and z- nodal directions) is employed.

5   Numerical Results

The figures below present the stress evolution as a function of the normalized time. As previously mentioned, the normalized time is obtained based on estimating the arrival time of the longitudinal wave (or the P-wave). The needed P-wave velocity is calculated using the material properties of the matrix: $\lambda_{m,0}, \mu_{m,0}$, and ρ_m. Figures 4 and 5 depict the evolution of the stress x- and y- components at location 1. The decreasing mechanical waves' amplitudes are related to the viscosity component in the material response. In Fig. 6, the stress x-component evolutions in both specimens, the specimen with stiff inclusions and the other without inclusions, in a broader time range, are compared. Effects due to viscosity for the model without stiff inclusions eventually appear for an even broader time range. Figure 7 provides an insight into the transient wave process in the composite containing inclusions and the specimen without inclusions in terms of shear stresses.

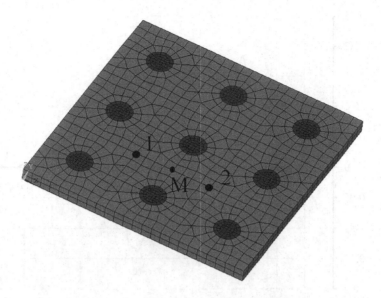

Fig. 3. The generated finite element mesh. The matrix is displayed in light blue, and stiff inclusions are in dark blue. (Color figure online)

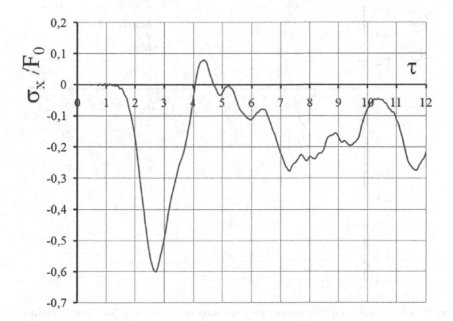

Fig. 4. The evolution of the normalized x-component stress at location 1.

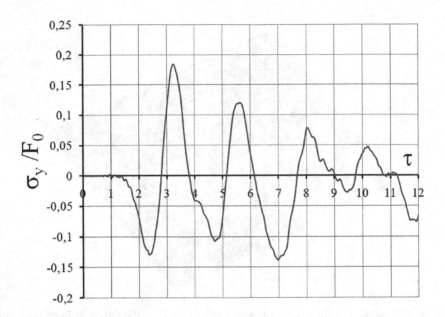

Fig. 5. The evolution of the normalized y-component stress at location 1.

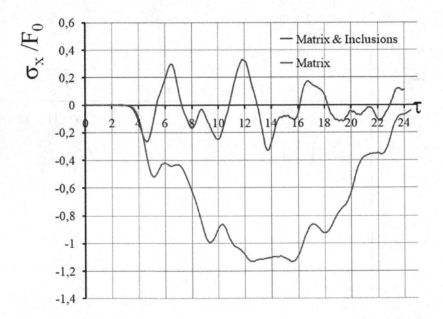

Fig. 6. The evolution of the normalized x-component stress at location 2: comparison between specimens with (blue line) and without (red line) inclusions. (Color figure online)

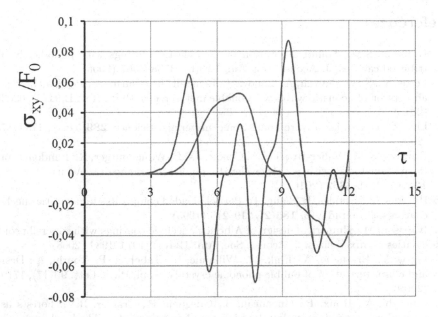

Fig. 7. Comparison between the shear stress, at location M, for the composite (blue line) and the specimens made only of the matrix material (red line). (Color figure online)

6 Conclusion

The contribution has presented some results of the numerical simulation of the transient wave process in a viscoelastic continuum. The propagation of mechanical waves in composite material containing stiff elastic fibers has been compared with the wave process provoked by an identical load in a specimen with the same geometry but containing no inclusions. Two finite element models have been built: one – for the composite material containing stiff inclusions (referred to above as "fibers") of cylindrical form distributed in a viscoelastic matrix and a second one – reproducing a specimen made solely of the viscoelastic material of the matrix. The modification in the wave characteristics (e.g., amplitudes, wavelengths/frequencies) occurring with the waves' interaction with the array of inclusions is apparent. Further specification of the modification rules as a function of the distribution of inclusions is forthcoming in this ongoing study. For possible comparison with results obtained using analytical models, the solution employs non-dimensional quantities where appropriate.

Acknowledgments. This study was performed within the bilateral project funded by the Russian Foundation for Basic Research (RFBR), project number 20-58-18002, and by the Bulgarian National Science Fund, project number KP-06-Russia/5 from 11.XII.2020.

References

1. Ivansson, S.M.: Sound absorption by viscoelastic coatings with periodically distributed cavities. J. Acoust. Soc. Am. **119**(6), 3558–3567 (2006)
2. Leroy, V., Strybulevych, A., Lanoy, M., Lemoult, F., Tourin, A., Page, J.H.: Super-absorption of acoustic waves with bubble metascreens. Phys. Rev. B **91**(2), 020301 (2015)
3. Liu, Z., et al.: Locally resonant sonic materials. Science **289**(5485), 1734–1736 (2000)
4. Duranteau, M., Valier-Brasier, T., Conoir, J.-M., Wunenburger, R.: Random acoustic metamaterial with a subwavelength dipolar resonance. J. Acoust. Soc. Am. **139**(6), 3341–3352 (2016)
5. Hinders, M., Rhodes, B., Fang, T.: Particle-loaded composites for acoustic anechoic coatings. J. Sound Vib. **185**(2), 219–246 (1995)
6. Ivansson, M.: Numerical design of Alberich anechoic coatings with superellipsoidal cavities of mixed sizes. J. Acoust. Soc. Am. **124**(4), 1974–1984 (2008)
7. Leroy, V., Bretagne, A., Fink, M., Willaime, H., Tabeling, P., Tourin, A.: Design and characterization of bubble phononic crystals. Appl. Phys. Lett. **95**(17), 171904 (2009)
8. Colombi, A., Roux, P., Guenneau, S., Gueguen, P., Craster, R.V.: Forests as a natural seismic metamaterial: Rayleigh wave bandgaps induced by local resonances. Sci. Rep. **6**(1), 1–7 (2016)
9. Rozovskii, M.I. : The integral operator method in the hereditary theory of creep. Dokl. Akad. Nauk SSSR **160**(4), 792–795 (1965)
10. Rabotnov, Y.N.: Elements of Hereditary Solid Mechanics. Naka Publishers, Moscow (1977). [in Russian]
11. Ilyasov, M.H.: Dynamical torsion of viscoelastic cone. TWMS J. Pure Appl. Math **2**(2), 203–220 (2011)
12. Jeltkov, V.I., Tolokonnikov, L.A., Khromova, N.G. : Transfer functions in viscoelastic body dynamics. Dokl. Akad. Nauk **329**(6), 718–719 (1993)
13. Lychev, S.A.: Coupled dynamic thermoviscoelasticity problem. Mech. Solids **43**(5), 769–784 (2008)
14. Rossikhin, Y.A., Shitikova, M.V., Trung, P.T. : Analysis of the viscoelastic sphere impact against a viscoelastic Uflyand-Mindlin plate considering the extension of its middle surface. Shock Vibr. **2017**, 5652023 (2017)
15. Mainardi, F.: Fractional calculus and waves in linear viscoelasticity: an introduction to mathematical models. Imperial College Press, London (2010)
16. Achenbach, J.D.: Vibrations of a viscoelastic body. AIAA J. **5**(6), 1213–1214 (1967)
17. Colombaro, I., Giusti, A., Mainardi, F.: On the propagation of transient waves in a viscoelastic Bessel medium. Z. Angew. Math. Phys. **68**(3), 1–13 (2017). https://doi.org/10.1007/s00033-017-0808-6
18. Christensen, R.M.: Theory of Viscoelasticity: An Introduction. Academic Press, New York (1982)
19. Zheng, Y., Zhou, F.: Using Laplace transform to solve the viscoelastic wave problems in the dynamic material property tests. In EPJ Web Conf. **94**, 04021 (2017)
20. Igumnov, L.A., Korovaytseva, E.A., Pshenichnov, S.G.: Dynamics, strength of materials and durability in multiscale mechanics. In: dell'Isola, F., Igumnov, L. (eds.) Advanced Structured Materials, ASM, vol. 137, pp. 89–96. Springer, Cham (2021). https://doi.org/10.1007/978-3-030-53755-5

Correction to: Numerical Methods and Applications

Ivan Georgiev⬤, Maria Datcheva⬤, Krassimir Georgiev,
and Geno Nikolov⬤

Correction to:
I. Georgiev et al. (Eds.): *Numerical Methods and Applications*,
LNCS 13858, https://doi.org/10.1007/978-3-031-32412-3

In an older version of this paper, the presentation of of Maria Datcheva and Krassimir Georgiev was misleading. This has been corrected.

The updated original version of the book can be found at
https://doi.org/10.1007/978-3-031-32412-3

Correction to: Random Sequences in Vehicle Routing Problem

Mehmet Emin Gülşen and Oğuz Yayla⊙

Correction to:
Chapter "Random Sequences in Vehicle Routing Problem"
in: I. Georgiev et al. (Eds.): *Numerical Methods*
and Applications, **LNCS 13858,**
https://doi.org/10.1007/978-3-031-32412-3_14

In the original version of this paper, the last name of Mehmet Emin Gülşen was misspelled. This error was corrected.

The updated original version of this chapter can be found at
https://doi.org/10.1007/978-3-031-32412-3_14

Author Index

I. Georgiev et al. (Eds.): NMA 2022, LNCS 13858, pp. 349–350, 2023.
https://doi.org/10.1007/978-3-031-32412-3

Printed in the United States
by Baker & Taylor Publisher Services